Memristive Devices and Systems: Modelling, Properties & Applications

Memristive Devices and Systems: Modelling, Properties & Applications

Editors

Chun Sing Lai
Zhekang Dong
Donglian Qi

MDPI • Basel • Beijing • Wuhan • Barcelona • Belgrade • Manchester • Tokyo • Cluj • Tianjin

Editors
Chun Sing Lai
Brunel University London
UK

Zhekang Dong
Hangzhou Dianzi University
China

Donglian Qi
Zhejiang University
China

Editorial Office
MDPI
St. Alban-Anlage 66
4052 Basel, Switzerland

This is a reprint of articles from the Special Issue published online in the open access journal *Electronics* (ISSN 2079-9292) (available at: http://www.mdpi.com).

For citation purposes, cite each article independently as indicated on the article page online and as indicated below:

LastName, A.A.; LastName, B.B.; LastName, C.C. Article Title. *Journal Name* **Year**, *Volume Number*, Page Range.

ISBN 978-3-0365-6688-7 (Hbk)
ISBN 978-3-0365-6689-4 (PDF)

Contents

About the Editors

Chun Sing Lai

Chun Sing Lai received his B.Eng. (First Class Hons.) in electrical and electronic engineering from Brunel University London, London, UK, in 2013, and his D.Phil. degree in engineering science from the University of Oxford, Oxford, UK, in 2019. He is currently a Lecturer at the Department of Electronic and Electrical Engineering, Brunel University London. His current research interests are in power system optimization and electric vehicle systems. Dr. Lai was the Publications Co-Chair for both 2020 and 2021 IEEE International Smart Cities Conferences. He is the Vice-Chair of the IEEE Smart Cities Publications Committee and Associate Editor for IET Energy Conversion and Economics. He is the Working Group Chair for IEEE P2814 and P3166 Standards, an Associate Vice President, Systems Science and Engineering of the IEEE Systems, Man, and Cybernetics Society (IEEE/SMCS) and the Chair of the IEEE SMC Intelligent Power and Energy Systems Technical Committee. He is a recipient of the 2022 Meritorious Service Award from the IEEE SMC Society for "meritorious and significant service to IEEE SMC Society technical activities and standards development". He is an IEEE Senior Member, IET Member, Chartered Engineer, and Fellow of the Higher Education Academy.

Zhekang Dong

Zhekang Dong received his B.E. and M.E. degrees in electronics and information engineering in 2012 and 2015, respectively, from South-West University, Chongqing, China and his Ph.D. degree from the School of Electrical Engineering, Zhejiang University, Hangzhou, China, in 2019. He is currently an Associate Professor with Hangzhou Dianzi University, Hangzhou, China. He is also a Research Assistant (Joint-Supervision) with the Hong Kong Polytechnic University, Hong Kong. His research interests cover memristor and memristive systems, artificial neural networks, the design and analysis of nonlinear systems based on memristor and computer simulation. He is an IEEE Member.

Donglian Qi

Donglian Qi received her Ph.D. degree in control theory and control engineering from the School of Electrical Engineering, Zhejiang University, China, in 2002. She is currently a Full Professor and a Ph.D. Advisor with Zhejiang University. Her recent research interests cover intelligent information processing, chaos system, and nonlinear theory and application. She is an IEEE Senior Member.

electronics

Editorial

Memristive Devices and Systems: Modeling, Properties and Applications

Chun Sing Lai [1,*], Zhekang Dong [2,*] and Donglian Qi [3]

[1] Department of Electronic and Electrical Engineering, Brunel University London, Kingston Lane, London UB8 3PH, UK
[2] School of Electronics and Information, Hangzhou Dianzi University, Hangzhou 310018, China
[3] College of Electrical Engineering, Zhejiang University, Hangzhou 310027, China
* Correspondence: chunsing.lai@brunel.ac.uk (C.S.L.); englishp@hdu.edu.cn (Z.D.)

1. Introduction

The memristor is considered to be a promising candidate for next-generation computing systems due to its nonvolatility, high density, low power, nanoscale geometry, nonlinearity, binary/multiple memory capacity, and negative differential resistance. Novel computing architectures/systems based on memristors have shown great potential to replace the traditional von Neumann computing architecture, which faces data movement challenges. As the field of materials science continues to develop, novel preparation and modeling methods for different memristive devices have been recently put forward, which opens up a new path for realizing different computing systems/architectures with practical memristor properties. The purpose of this Special Issue on "Memristive Devices and Systems: Modeling, Properties and Applications" is to provide a comprehensive overview of key computational primitives enabled by these memory devices, as well as their applications in spanning edge computing, signal processing, optimization, machine learning, deep learning, stochastic computing, and so on. More specifically, we invited researchers and practitioners to contribute original research articles that examine challenges that are related, but not limited to, the following topics:

- Memristive device preparation;
- Memristive device modeling and analysis;
- Novel electronic devices that show memristive properties;
- Novel memristive circuit design solutions for neuromorphic systems;
- Memristive circuit fault diagnosis and analysis;
- Memristive systems for different applications (e.g., edge computing, signal processing, optimization, machine learning, deep learning, and stochastic computing);
- Nonvolatile memory solutions with computing capabilities;
- Memory devices and systems for in-memory computing.

2. Short Presentation of the Papers

Ji et al. [1] analyze the mathematical models of memristors and discuss their applications in conventional image processing based on memristive systems, as well as in image processing based on memristive neural networks, to investigate the potential of memristive systems in image processing. In addition, they present recent advances and implications of memristive system-based image processing comprehensively, and explore development opportunities and challenges in different major areas as well. By establishing a complete spectrum of image processing technologies based on memristive systems, this review attempts to provide a reference for future studies in the field, and it is hoped that scholars can promote development in this area through interdisciplinary academic exchanges and cooperation.

Citation: Lai, C.S.; Dong, Z.; Qi, D. Memristive Devices and Systems: Modeling, Properties and Applications. *Electronics* **2023**, *12*, 765. https://doi.org/10.3390/electronics 12030765

Received: 30 January 2023
Accepted: 1 February 2023
Published: 2 February 2023

Romero et al. [2] gathered together the current main alternatives presented in the literature for the emulation of both memcapacitors and meminductors. Different circuit emulators have been thoroughly analyzed and compared in detail, providing a wide range of approaches that could be considered for the implementation of these devices in future designs.

Wang et al. [3] develop an optoelectronic memristor model (containing a mathematical model and circuit model). Moreover, they discuss the composite memristor circuit (series- and parallel-connected configuration) with a rotation mechanism. Further, a multi-valued logic circuit is designed, which is capable of performing multiple logic functions from 0–1, verifying the validity and effectiveness of the established memristor model, as well as opening up a new path for the circuit implementation of fuzzy logic.

Qiu et al. [4] present a novel two-neuron-based memristive Hopfield neural network with a hyperbolic memristor that emulates synaptic crosstalk. The dynamics of the neural networks with varying memristive parameters and crosstalk weights are analyzed via the phase portraits, time-domain waveforms, bifurcation diagrams, and basin of attraction. Complex phenomena, especially coexisting dynamics, chaos, and transient chaos emerge in the neural network. Finally, the circuit simulation results verify the effectiveness of theoretical analyses and mathematical simulation and further illustrate the feasibility of the two-neuron-based memristive Hopfield neural network hardware.

Li et al. [5] propose a globally passive but locally active memristor, which has three stable equilibrium points and two unstable equilibrium points, exhibiting two stable locally active regions and four unstable locally active regions. They found that when the memristor operates in a stable local active region, the memristor-based second-order circuit with a parallel capacitor or a series inductor can produce periodic oscillation. Moreover, the memristor-based third-order circuit with two energy storage elements, a capacitor and an inductor, can produce complex chaotic oscillation, forming the simplest chaotic circuit.

Ying et al. [6] propose a modified Chua corsage memristor endowed with two symmetrical locally active domains. Under the DC bias voltage in the locally active domains, the memristor with an inductor can construct a second-order circuit to generate periodic oscillation. Based on the theories of the edge of chaos and local activity, the oscillation mechanism of the symmetrical periodic oscillations of the circuit is revealed. The third-order memristor circuit is constructed by adding a passive capacitor in parallel with the memristor in the second-order circuit, where symmetrical periodic oscillations and symmetrical chaos emerge either on or near the edge-of-chaos domains. The oscillation mechanisms of the memristor-based circuits are analyzed via domain distribution maps, which include the division of locally passive domains, locally active domains, and the edge-of-chaos domains. Finally, the symmetrical dynamic characteristics are investigated via theory and simulations, including Lyapunov exponents, bifurcation diagrams, and dynamic maps.

Qin et al. [7] investigate a fractional-order memristive model with infinite coexisting attractors. The numerical solution of the system is derived based on the Adomian decomposition method (ADM), and its dynamic behaviors are analyzed by means of phase diagrams, bifurcation diagrams, the Lyapunov exponent spectrum (LES), and dynamic maps based on SE complexity and the maximum Lyapunov exponent (MLE). Simulation results show that it has rich dynamic characteristics, including asymmetric coexisting attractors with different structures and offset boosting. Finally, the digital signal processor (DSP) implementation verifies the correctness of the solution algorithm and the physical feasibility of the system.

Shen and Wang [8] propose a cellular neural network (CNN) based on a VO_2 carbon nanotube memristor. The device is first modeled by SPICE, and then the cell dynamic characteristics based on the device are analyzed. It is pointed out that only when the cell is at the sharp edge of chaos can the cell be successfully awakened after the CNN is formed. In this paper, they provide the example of a 5×5 CNN, set specific initial conditions, and observe the formed pattern. Because the generated patterns are affected by the initial conditions, the cell power supply can be preprogrammed to obtain specific patterns, which can be applied to the future information processing system based on complex space–time patterns, especially in the field of computer vision.

Shen and Wang [9] study the history erase effect of a Hewlett-Packard (HP) TiO_2 memristor and the Self-Directed Channel (SDC) memristor of the Knowm Company. The DC and AC responses of the HP TiO_2 memristor are given, and it is pointed out that there is no AC history erase effect. However, considering the parasitic memcapacitance effect, it is found that it has the effect. Based on the theoretical model of the SDC memristor, its history erase properties are studied by considering and not considering parasitic effects. It should be noted that this study method can be useful for other materials such as Al_2O_3 and MoS_2.

Quesada et al. [10] analyze and evaluate three different RRAM compact models that are implemented in Verilog-A to reproduce the multilevel approach based on the switching capability of experimental devices. These models are integrated in 1T-1R cells to control their analog behavior by means of the compliance current imposed by the NMOS select transistor. Four different resistance levels are simulated and assessed with experimental verification to account for their multilevel capability. Further, an artificial neural network study is carried out to evaluate in a real scenario the viability of the multilevel approach under study.

Author Contributions: Writing—original draft preparation, C.S.L. and Z.D.; writing—review and editing, D.Q. All authors have read and agreed to the published version of the manuscript.

Conflicts of Interest: The authors declare no conflict of interest.

References

1. Ji, X.; Dong, Z.; Zhou, G.; Lai, C.S.; Yan, Y.; Qi, D. Memristive System Based Image Processing Technology: A Review and perspective. *Electronics* **2021**, *10*, 3176. [CrossRef]
2. Romero, F.J.; Ohata, A.; Toral-Lopez, A.; Godoy, A.; Morales, D.P.; Rodriguez, N. Memcapacitor and Meminductor Circuit Emulators: A review. *Electronics* **2021**, *10*, 1225. [CrossRef]
3. Wang, J.; Lin, Y.; Hu, C.; Zhou, S.; Gu, S.; Yang, M.; Ma, G.; Yan, Y. A Kind of Optoelectronic Memristor Model and Its Applications in Multi-Valued Logic. *Electronics* **2023**, *12*, 646. [CrossRef]
4. Qiu, R.; Dong, Y.; Jiang, X.; Wang, G. Two-Neuron Based Memristive Hopfield Neural Network with Synaptic Crosstalk. *Electronics* **2022**, *11*, 3034. [CrossRef]
5. Li, F.; Liu, J.; Zhou, W.; Dong, Y.; Jin, P.; Ying, J.; Wang, G. A Passive but Local Active Memristor and Its Complex Dynamics. *Electronics* **2022**, *11*, 1843. [CrossRef]
6. Ying, J.; Liang, Y.; Li, F.; Wang, G.; Shen, Y. Complex Oscillations of Chua Corsage Memristor with Two Symmetrical Locally Active Domains. *Electronics* **2022**, *11*, 665. [CrossRef]
7. Qin, C.; Sun, K.; He, S. Characteristic Analysis of Fractional-Order Memristor-Based Hypogenetic Jerk System and Its DSP Implementation. *Electronics* **2021**, *10*, 841. [CrossRef]
8. Shen, Y.; Wang, G. VO2 Carbon Nanotube Composite Memristor-Based Cellular Neural Network Pattern Formation. *Electronics* **2021**, *10*, 1198. [CrossRef]
9. Shen, Y.; Wang, G. History Erase Effect of Real Memristors. *Electronics* **2021**, *10*, 303. [CrossRef]
10. Pérez-Bosch, E.; Romero-Zaliz, R.; Perez, E.; Kalishettyhalli Mahadevaiah, M.; Reuben, J.; Schubert, M.A.; Jiménez-Molinos, F.; Roldán, J.B.; Wenger, C. Toward Reliable Compact Modeling of Multilevel 1T-1R RRAM Devices for Neuromorphic Systems. *Electronics* **2021**, *10*, 645. [CrossRef]

Review

Memristive System Based Image Processing Technology: A Review and Perspective

Xiaoyue Ji [1], Zhekang Dong [1,2,3], Guangdong Zhou [4], Chun Sing Lai [5,6], Yunfeng Yan [1,*] and Donglian Qi [1,*]

[1]   College of Electrical Engineering, Zhejiang University, Hangzhou 310027, China; ji.xiaoyue@zju.edu.cn (X.J.); englishp@126.com (Z.D.)
[2]   College of Electronic Information, Hangzhou Dianzi University, Hangzhou 310018, China
[3]   Zhejiang Provincial Key Lab of Equipment Electronics, Hangzhou 310018, China
[4]   College of Artificial Intelligence, Southwest University, Chongqing 400715, China; zhougd@swu.edu.cn
[5]   Department of Electronic and Electrical Engineering, Brunel University London, London UB8 3PH, UK; chunsing.lai@brunel.ac.uk
[6]   Department of Electrical Engineering, School of Automation, Guangdong University of Technology, Guangzhou 510006, China
*   Correspondence: 21210004@zju.edu.cn (Y.Y.); qidl@zju.edu.cn (D.Q.); Tel.: +86-150-5716-6556 (Y.Y.); +86-139-5800-8868 (D.Q.)

Abstract: As the acquisition, transmission, storage and conversion of images become more efficient, image data are increasing explosively. At the same time, the limitations of conventional computational processing systems based on the Von Neumann architecture continue to emerge, and thus, improving the efficiency of image processing has become a key issue that has bothered scholars working on images for a long time. Memristors with non-volatile, synapse-like, as well as integrated storage-and-computation properties can be used to build intelligent processing systems that are closer to the structure and function of biological brains. They are also of great significance when constructing new intelligent image processing systems with non-Von Neumann architecture and for achieving the integrated storage and computation of image data. Based on this, this paper analyses the mathematical models of memristors and discusses their applications in conventional image processing based on memristive systems as well as image processing based on memristive neural networks, to investigate the potential of memristive systems in image processing. In addition, recent advances and implications of memristive system-based image processing are presented comprehensively, and its development opportunities and challenges in different major areas are explored as well. By establishing a complete spectrum of image processing technologies based on memristive systems, this review attempts to provide a reference for future studies in the field, and it is hoped that scholars can promote its development through interdisciplinary academic exchanges and cooperation.

Keywords: memristors; memristive systems; integrated storage and computation; image processing

Citation: Ji, X.; Dong, Z.; Zhou, G.; Lai, C.S.; Yan, Y.; Qi, D. Memristive System Based Image Processing Technology: A Review and Perspective. *Electronics* **2021**, *10*, 3176. https://doi.org/10.3390/electronics10243176

Academic Editor: Chiman Kwan

Received: 6 November 2021
Accepted: 15 December 2021
Published: 20 December 2021

Publisher's Note: MDPI stays neutral with regard to jurisdictional claims in published maps and institutional affiliations.

1. Introduction

With the advent of the Internet of Things, cloud computing, and the big data era, there has been explosive growth in the scale of information. However, the physical separation among perception, computation, and storage in the conventional computing architecture requires frequent data shuttling among the units, thereby causing significant system consumption and speed loss and making it difficult to meet the requirements of information analysis and processing [1–3]. Therefore, developing new electronic components for intelligent processing systems that are closer to the structure and function of biological brains has become a hot research topic in the fields of modern electronic circuits and image processing [4–6].

Image processing technology, which aims to automatically acquire high-level and abstract information from images, after which it simulates how human eyes work with such information, has become increasingly useful in human life and social production.

Exploring the basic structure of human brains, simulating their working mechanisms, and establishing neural network models that integrate perception, storage, and computation as a whole have gradually become research hotspots in the fields of image processing and cognitive computing [7]. The current mainstream neural network models can simulate the reasoning and learning functions of human brains to a certain extent, and they have shown some potential in image processing [8]. However, they are confined to certain types and structures with limited processing capabilities. Additionally, the existing ones lack the process of information perception, transmission, and storage prior to the processing stage. Furthermore, the hardware for neural networks is essential to truly realize the conversion from theoretical studies of brain cognition to new technologies of brain-computer intelligence. Nevertheless, most of the current research focuses on the theoretical analysis of the network structures and algorithms, and the research on implementation schemes for neural network hardware is still in its infancy [9–11]. Influenced by factors, such as device size, energy consumption, and integrability, conventional implementation schemes for image processing cannot well trade-off the relationship between speed, accuracy, and system consumption [12–14]. We schematically compare the traditional image processing systems and memristor-based image processing system as shown in Figure 1. From the perspective of the device, leakage currents become a problem when the channel length and the gate dielectric thickness of a transistor get closer to the scaling limit [3]. With respect to the architecture, the data transfer between processors and memory units significantly reduces both speed and energy efficiency (referred to as the 'von Neumann bottleneck'). Furthermore, the performance mismatch between the memory and processing units leads to great latency (also called the 'memory wall').

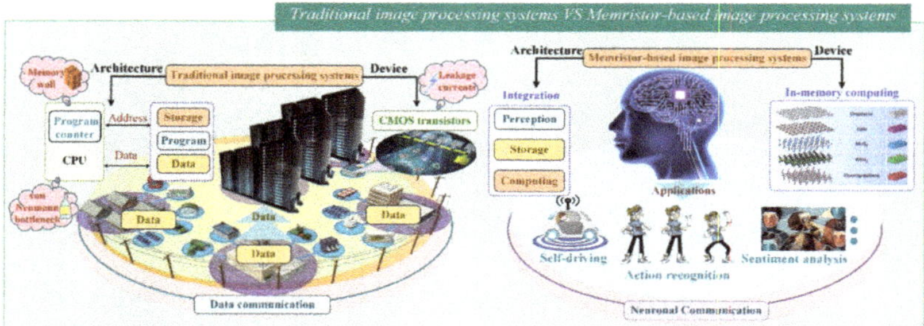

Figure 1. The comparison between the traditional image processing systems and memristor-based image processing system.

The successful preparation of memristors provides a fresh perspective on the hardware implementation of artificial neural networks. It was proposed by Leon Chua, a scientist at the University of California, Berkeley [15] and discovered by Hewlett-Packard (HP) laboratories in 2008 as the fourth fundamental electronic component after the resistor, capacitor, and inductor [16]. Experiments have shown that the memristor has properties, such as non-volatility, variable resistance, nanoscale size, threshold characteristics, low power consumption, and synapse-like structure [17–19]. In particular, by taking full advantage of being synapse-like, the memristor can be used as an "electronic synapse" or an "artificial synapse" in the hardware design of neural networks [20]. Further, after choosing a proper memristor model to simulate the weight of the neural network, a more integrated architecture for hardware implementation can be constructed and applied to different image processing tasks [21,22]. Compared with conventional artificial neural networks, the memristive ones incorporate powerful perception ability, massive storage capacity, and intelligent processing mechanisms to enable deeper analysis and exploration, which are expected to solve slow training speed and insufficient online processing capability in image processing [23–25].

By collating relevant research on memristive system-based image processing technology (including relevant mathematical models and their applications), this paper comprehensively elaborates on the fusion mechanism of memristive systems and image processing from three aspects, namely the mathematical models of memristive systems, the conventional image processing based on memristive systems, and the image processing based on memristive neural networks. Furthermore, the study summarizes the main directions, progresses, and problems in this field, analyses its development law, and strives to establish a complete spectrum for the reference of researchers in various fields.

2. Mathematical Models

The memristor is a two-terminal non-linear passive circuit element in nanometre and with memory characteristics, whose resistance is variable and controlled by the intensity, polarity, and duration of power supply. By applying an external voltage to the memristor, the conductive properties of its internal functional layer can be changed from a high resistance state (HRS) to a low resistance state (LRS). In particular, three types of theory, i.e., ionic migration, quantum tunnelling, and charge trapping/de-trapping, dominate the study of memristors' internal physical mechanisms and dynamic characteristics, and they explain most of the observed memristive phenomena [26].

(1)   Ionic migration: This memristor type usually has the active metal (e.g., Ag) as the top electrode and the inert metal (e.g., Pt) as the bottom electrode. By applying a positive voltage to the top electrode, the active metal will be electrolyzed into metal cations. They will move toward the bottom electrode under the external electric field and then return to metal atoms, the accumulation of which form a metal filament conductive channel for the memristor to transit from the HRS to the LRS. Conversely, by applying a positive voltage to the bottom electrode, the formed conductive channel will gradually break, and the memristor will switch from the LRS to the HRS.

(2)   Quantum tunnelling: The internal functional layer of this type of memristor is mainly a metal oxide (e.g., TiO_x). The Schottky barrier between the metal electrode and the functional layer is adjusted by applying an external voltage to switch the resistive state of the memristor. It disappears when the memristor is in the LRS, whereas it reappears when the memristor is in the HRS.

(3)   Charge trapping/de-trapping: For a memristor whose functional layer is the metal oxide film, there exists an empty state in the film. When a positive voltage is applied to the top electrode, the empty state traps the injected electrons and stores them, and when the empty state is filled, a conductive channel is formed, after which the memristor switches from the HRS to the LRS. By contrast, when a positive voltage is applied to the bottom electrode, the electrons in the empty state are released, the formed conductive channel is broken, and the memristor changes from the LRS to the HRS.

During the memristor fabrication process, a small parameter variation may lead to huge differences between devices, and even significantly affect circuit performance. Meanwhile the unstable performance between memristor cells and the cells themselves makes the integration of the device challenging.

As a result, most applied research on memristors always using their mathematical models. As the fourth circuit element, the memristor represents the interrelationship between the magnetic flux φ and the charge q, as shown in Figure 2.

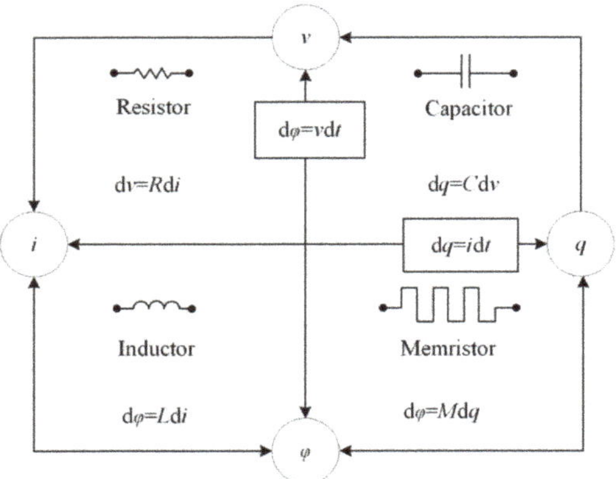

Figure 2. The four fundamental two-terminal circuit elements.

The memristors can be divided into two categories, i.e., being charge-controlled and being flux-controlled. For the charge-controlled ones, their flux φ is a single-valued function of the charge q, which is expressed as follows:

$$\varphi = f(q) \tag{1}$$

Taking the time t derivative of both sides of Equation (1) gives us the following:

$$\frac{d\varphi}{dt} = \frac{d\varphi(q)}{dq} \cdot \frac{dq}{dt} \tag{2}$$

Further, based on the voltage $v = d\varphi/dt$ and the current $i = dq/dt$, the relation between volt and ampere for the memristors can be obtained as follows:

$$v = M(q) \cdot i \tag{3}$$

where the function $M(q)$, which represents the memristance, satisfies the following mathematical expression:

$$M(q) \equiv \frac{d\varphi(q)}{dq} \tag{4}$$

For the flux-controlled memristors, their charge q is a single-valued function of the flux φ, which is expressed as follows:

$$q = f(\varphi) \tag{5}$$

Taking the time t derivative of both sides of Equation (5) gives us the following:

$$\frac{dq}{dt} = \frac{dq(\varphi)}{d\varphi} \cdot \frac{d\varphi}{dt} \tag{6}$$

Based on the voltage $v = d\varphi/dt$ and the current $i = dq/dt$, the relation between voltage v and current i for the two sides of the memristors can be derived as follows:

$$i = G(\varphi) \cdot v \tag{7}$$

where the function $G(q)$, which represents the memristance, satisfies the following mathematical expression:

$$G(\varphi) \equiv \frac{dq(\varphi)}{d\varphi} \tag{8}$$

In 2008, a simple linear memristor model based on the ionic migration theory was proposed by D. Strukov's research team [16], and its structure is shown in Figure 3.

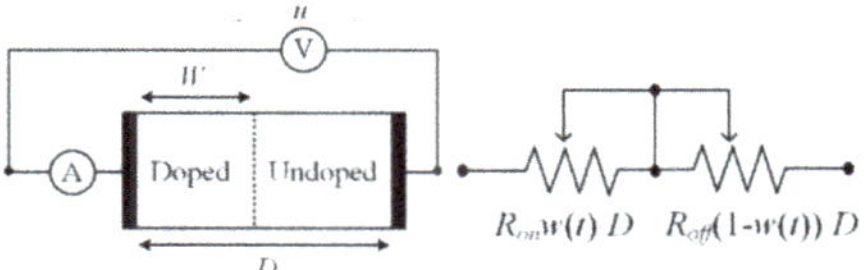

Figure 3. Schematic diagram of HP memristor.

Let us assume that the total thickness of the titanium dioxide functional layer is D, and the one of the doped layer is W. R_{on} denotes the minimal resistance of the memristor, while R_{off} represents the maximum. The resistance $M(t)$ of the HP memristor is expressed as:

$$M(t) = R_{\text{L}} \cdot x(t) + R_{\text{H}} \cdot [1 - x(t)] \tag{9}$$

$$\frac{\mathrm{d}x}{\mathrm{d}t} = ki(t), k = \frac{\mu_v R_{\text{L}}}{D^2} \tag{10}$$

where x represents the internal state variable of the memristor, μ_v represents the average ionic mobility, i represents the current passing through the memristor, and the constant k is the ratio of the rate of change to the current.

On this basis, a nonlinear memristive model with window functions was constructed in the literature [27] to better describe the boundary effect and nonlinear drift of memristors. In a study conducted by [28] a Simmons tunnelling barrier model was proposed based on the quantum tunnelling theory. It accurately presented the properties of memristive devices, but its mathematical model was more complex, and showed no direct explicit relationship between voltage and current, thereby being unconducive to the subsequent research and applications. Additionally, in 2013, S. Kvatinsky's research team at the Technion-Israel Institute of Technology [29] put forward a more simplified mathematical version, which was named the ThrEshold Adaptive Memristor (TEAM) model. Two years later, the team [30] further designed the corresponding Voltage ThrEshold Adaptive Memristor (VTEAM) model, with a simple structure as well as certain generality to simulate the threshold characteristics of voltage-controlled memristive devices. In 2017, Fang Liang's team from the National University of Defense Technology, China [31] brought forward a general TiO_x memristive model by combining the nonlinear drift, ionic migration and negative differential resistance (NDR) effect of memristors. In addition, using traditional analogue circuit components, some researchers [32–34] realized the memristive circuit simulation based on Chua's theory as a way to simulate the basic memristive characteristics. In this paper, the abovementioned mathematical models, which are summarized in Table 1 and compared in Table 2, manifest the fundamental features of memristors to some extent. However, their correlation with the physical realization of memristors is not strong enough, and it cannot fully characterize the electrochemical properties of memristive devices.

Table 1. Mathematical models of memristors.

Model Type	Current-Voltage Relationship	Dynamic Equation of State Variable
HP Memristive Model [16]	$v(t) = \left(R_L \frac{x(t)}{D} + R_H\left(1 - \frac{x(t)}{D}\right)\right)i(t)$	$\frac{dx}{dt} = ki(t), k = \frac{\mu_v R_L}{D^2}$
Nonlinear Memristive Model [27]	$i(t) = w^n(t)\beta\sinh(\alpha v(t)) + x[\exp(\gamma v(t)) - 1]$	$\frac{dw(t)}{dt} = \alpha v^m(t) f(w)$
Simmons Memristive Model [28]	$i(t) = \widetilde{A}(x, v_g)\phi_1(v_g, x)\exp\left(-B(v_g, x)\cdot\phi_1(v_g, x)^{1/2}\right) - \widetilde{A}(x, v_g)(\phi_1(v_z, x) + e|v_z|)\times\exp\left(-B(v_g, x)\cdot(\phi_1(v_g, x) + ev_\varepsilon)^{1/2}\right)$ $v_g = v - i(t)R_s$	$\frac{dx(t)}{dx} = \begin{cases} C_{eft}\sinh\left(\frac{i}{i_{off}}\right)\exp\left[-\exp\left(\frac{x - a_{aff}}{w_c} - \frac{|i|}{b}\right) - \frac{x}{w_c}\right], & i > 0 \\ C_{on}\sinh\left(\frac{i}{i_{on}}\right)\exp\left[-\exp\left(-\frac{x - a_{on}}{w_c} - \frac{|i|}{b}\right) - \frac{x}{w_c}\right], & i < 0 \end{cases}$
TEAM Memristive Model [29]	$v(t) = \left[R_L + \frac{R_H - R_L}{x_{off} - x_{on}}(x - x_{on})\right]\cdot i(t)$ $v(t) = R_L\exp\left(\frac{\lambda}{x_{ofi} - x_{on}}(x - x_{on})\right)\cdot i(t)$	$\frac{dx(t)}{dt} = \begin{cases} k_{off}\left(\frac{i(t)}{i_{off}} - 1\right)^{\alpha_{df}}\cdot f_{off}(x) & 0 < i_{off} < i \\ 0 & i_{an} < i < i_{of} \\ k_{on}\left(\frac{i(t)}{i_{on}} - 1\right)^{\alpha_{on}}\cdot f_{on}(x) & i < i_{on} < 0 \end{cases}$
VTEAM Memristive Model [30]	$i(t) = \left[R_L + \frac{R_H - R_L}{x_H - x_L}\cdot(x - x_L)\right]^{-1}\cdot v(t)$ $i(t) = \frac{e^{-\frac{\lambda}{x_H - x_L}\cdot(x - x_L)}}{R_L}\cdot v(t)$	$\frac{dx}{dt} = \begin{cases} k_{off}\left(\frac{v(t)}{v_{th\,1}} - 1\right)^{\alpha_{of}} & f_{off}(x), 0 < v_{th1} \leq v \\ 0, & v_{th2} < v < v_{th1} \\ k_{on}\left(\frac{v(t)}{v_{th2}} - 1\right)^{\alpha_{on}} f_{on}(x), & v \leq v_{th2} < 0 \end{cases}$
General TiO$_x$ Memristive Model [31]	$i(t) = \begin{cases} x^{n_1}k_{on_1}\sinh\frac{v}{v_{on_1}} + (1 - x^{n_1})k_r\left(e^{v/v_r} - 1\right), v(t) \geq 0 \\ x^{n_2}k_{on_2}\sinh\frac{v}{v_{on_2}} + (1 - x^{n_2})k_{off}\sinh\frac{v}{v_{off}}, v(0) < 0 \end{cases}$	$\frac{dx}{dt} = \begin{cases} \alpha_1\sinh\beta_1 v - \gamma x, v > 0 \\ \alpha_2\sinh\beta_2 v - \gamma x, v < 0 \end{cases}$

Table 2. Comparative information of different memristive models.

	HP Memristive Model [16]	Nonlinear Memristive Model [27]	Simmons Memristive Model [28]	TEAM Memristive Model [29]	VTEAM Memristive Model [30]	General TiO$_x$ Memristive Model [31]
Physical Support	Yes	No	Yes	No	No	No
Physical Mechanism	Ionic Migration	Ionic Migration	Quantum Tunneling	No	No	Ionic Migration
Model Complexity	Simple	Simple	Complex	Moderate	Moderate	Moderate
Applied Range	Wider	Wider	Narrower	Wider	Wider	Wider

3. Traditional Image Processing Based on Memristive Systems

The memristor can perform logic calculations directly on the device, making it possible to achieve a true integration of storage and computing. Therefore, it brings new opportunities for the development of traditional image processing technologies.

3.1. Image Storage Based on Memristive Systems

Image processing is a type of memory access-intensive application, which places high demands on memory, requiring both enough capacities to store large-scale image data and fast access speed to ensure processing performance. Currently, non-volatile memories contain the flash memory (NAND), resistive random-access memory (RRAM), phase-change memory (PCM), spin-transfer torque magnetic random-access memory (STT-RAM), and ferroelectric random-access memory (FeRAM). This paper compares the characteristics of various types of new volatile and non-volatile memory devices in terms of capacity, size, read/write performance, lifetime, power consumption, and current technical bottlenecks, etc., with the specific information summarized in Table 3. It is found that memristor-based RRAM has a series of outstanding advantages, such as small size, non-volatility, low power consumption, high density, fast erasure, and compatibility with CMOS processes, making it one of the most promising memory devices.

Table 3. Comparative information of different memory devices.

Parameter	DRAM	NAND	STT-RAM	RRAM	FeRAM	PCM
Capacity	~16 Gb	~1 Tb	~64 Mb	~1 TB	~64 MB	~8 Gb
Technology level	~20 nm	~16 nm	~32 nm	~11 nm	~65 nm	~5 nm
Feature Size	6–10 F2	4–11 F2	16–60 F2	4–14 F2	15–34 F2	4–8 F2
Read Operation Time	<10 ns	10–50 us	2–20 ns	10–50 ns	20–80 ns	10–100 ns
Write Operation Time	<10 ns	0.1–1 ms	5–35 ns	10–50 ns	10–5 ns	20–120 ns
Lifetime	>1015	104–106	1012–1015	108–1010	1012–1014	108–1012
Data Retention	Refresh	10 Years	>10 Years	10 Years	10 Years	>10 Years
Write Power	0.1 ~0.1 nJ/b	0.1–1 nJ/b	1.6–5 nJ/b	~0.1 nJ/b	<1 nJ/b	<1 nJ/b
Idle Power	High	Low	Low	Low	Low	Low
Non-volatile	Volatile	Non-volatile	Non-volatile	Non-volatile	Non-volatile	Non-volatile
Destructive Read	Destructive	Non-destructive	Non-destructive	Non-destructive	Destructive	Non-destructive
Major Technical Bottlenecks	Memory refresh, volatility, limited memory process	Limited lifetime performance, low storage density	Small capacity, high write power consumption, poor stability	Unclear material storage mechanism	Small capacity, destructive read, low storage density	Small capacity, narrow range of material operable temperature

In 2011, Hu et al. proposed a memristor crossbar array that could be applied to image processing (see Figure 4a) [35]. Together with the peripheral control circuit, the random storage of binary, grey scale and colour images could be successfully realized. When storing binary images, the image information was mapped into pulse sequences of varying amplitudes using a voltage converter as the input of the memristor crossbar array, as shown in Figure 4b. As for the grey scale and colour images, the image information was mapped into pulse sequences of varying widths using the voltage converter, which were then used as the input to the memristor crossbar array. It is worth noting that voltage pulses of different widths were obtained by controlling the timing of write operation, which finally enabled the storage of images, as illustrated in Figure 4c,d.

In the literature [35], a memristor-based resistive random access memory (MRRAM) was mentioned. Through improvement, it stored binary and multi-valued input information with different memristances. The effectiveness of storing ASCII characters and images was verified through simulation experiments, and a new scheme for storing grey scale images was discussed as well. In Tan et al.'s study [36], ITO/CeO$_{2-x}$/AlOy/Al structured memristors were prepared to realize the perception and non-volatile storage of different multispectral images. Wang constructed a storage circuit based on the 2T2M structured memristive synapse to achieve the storage and recovery of binary images [37]. Compared with conventional storage technology, this circuit effectively reduced the storage space and improved the storage efficiency. In 2017, the research team of Prof. Duan [38] at Southwest University, China, successfully prepared a memristor with silver chalcogenide as the functional layer and constructed a memristive synapse with spike rate- and timing-dependent plasticity by analysing its electrochemical properties. Based on this, an improved memristor crossbar array was designed to realize the storage of grey scale images. One year later, Chen et al. designed a vision system on the basis of combining the optical sensor with the memristor, in which the former was used to detect UV light and convert it into voltage pulses of corresponding intensity, and the latter was adopted to store the converted voltage signal, which realized the perception and storage of UV images [32]. In 2020, Wang Xiaoping and her team members from Huazhong University of Science and Technology raised a memristor-CMOS hybrid storage circuit, where the memristor was utilized to store the bit information of images, while CMOS was applied to conduct control, isolation, and logic operations [39]. A series of simulations confirmed that this memory circuit could achieve improved performance in UHF image storage applications. In summary, most of the studies on memristor-based image storage use memristive synapses for crossbar arrays to keep image information, which reduces the storage density to a certain extent. However, the stability of image memory devices is affected by the issue of current leakage in crossbar arrays. Therefore, avoiding or reducing the leakage is one of the problems of memristor-based image storage technology that must be addressed urgently.

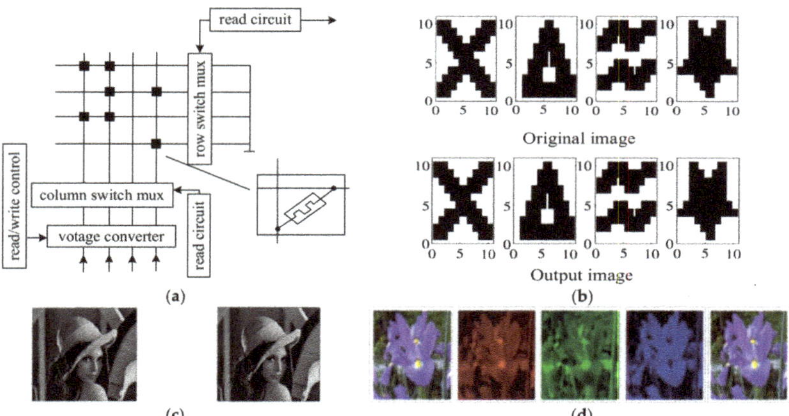

Figure 4. Application of memristor crossbar array in image storage. (**a**) memristor crossbar array; (**b**) memristor crossbar array used to store binary images; (**c**) memristor crossbar array used to store grey scale images; (**d**) memristor crossbar array used to store color images.

3.2. Image Compression Based on Memristive Systems

With the rapid development of sensor technology, the sizes of image data are also expanding rapidly. Meanwhile, higher requirements are put forward on the clarity and transmission rate of images. Applying memristors to image compression can effectively reduce their storage space and improve their transmission speed at the same time. Therefore, the corresponding circuit implementation scheme has been widely studied by scholars in the related fields.

Li et al. constructed a 128×64 memristor crossbar array based on the prepared $Ta/HfO_2/Pd$ memristor, and its circuit structure is presented in Figure 5 [40].

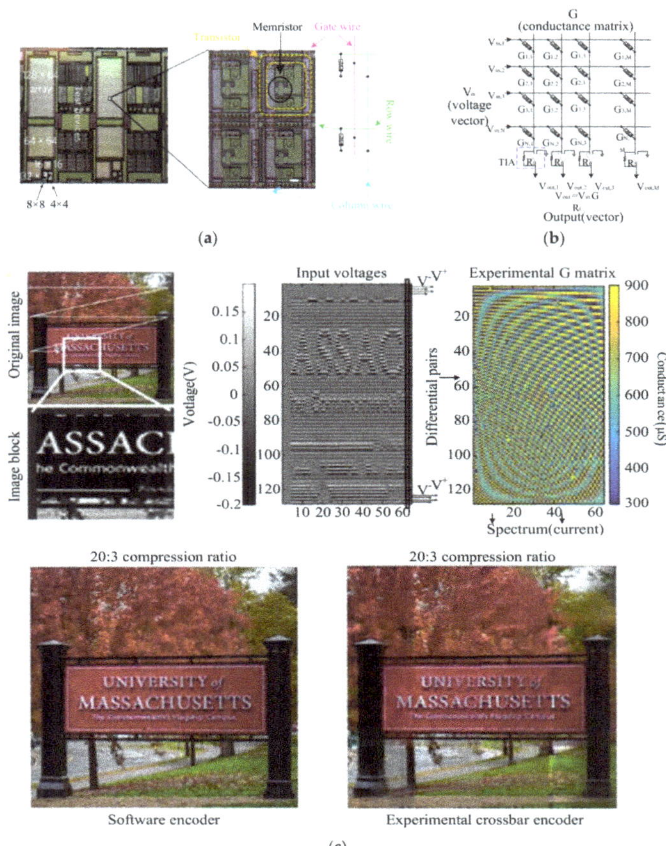

Figure 5. Application of crossbar array based on Ta/HfO$_2$/Pd memristor in image processing. (**a**) memristor hardware structure; (**b**) memristor crossbar array; (**c**) memristor crossbar array to achieve image compression.

Taking advantage of the high parallelism, non-volatility, low power consumption and small size of memristors, this circuit took a single memristor to store image information at the 6-bit precision, which further achieved functions, such as image compression, convolution and filtering. Additionally, a memristor-based image compression framework is presented in Figure 6 [41], considering the loss of two-dimensional discrete wavelet transform. The framework consisted of three memristor crossbar arrays, where the computational one was used to conduct the data multiplication and addition operations, the intermediate array stored the coefficients of row-column transformation, and the final one was used to keep the compressed data of the original image. The image compression could be achieved by taking the generated pulses through a multilayer voltage sensor as input, mapping the image pixels into memristive conductance through the computational array, and then storing them in the other two crossbar arrays. The research conducted by Berco et al. in 2020 proposed a programmable photoelectronic memristor gate circuit, which could perform state switching between optical and electrical signals, to realize in-situ image compression [42]. A research team from Dalian University of Technology [43] designed the simplest fractional-order chaotic memory circuit that identified pseudo-random sequences in image compression through phase diagrams, Lyapunov exponential spectra, and bifurcation diagrams, which achieved image compression for the second time and reduced the storage costs significantly.

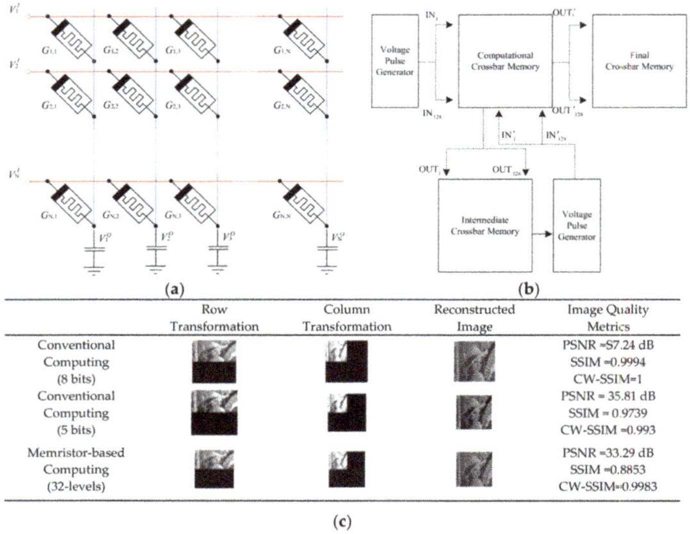
Voltage Pulse Generator
Computational Crossbar Memory
Final Crossbar Memory
Intermediate Crossbar Memory
Voltage Pulse Generator
(a)
(b)
Row Transformation
Column Transformation
Reconstructed Image
Image Quality Metrics
Conventional Computing (8 bits)
PSNR =57.24 dB
SSIM =0.9994
CW-SSIM=1
Conventional Computing (5 bits)
PSNR = 35.81 dB
SSIM = 0.9739
CW-SSIM =0.993
Memristor-based Computing (32-levels)
PSNR =33.29 dB
SSIM =0.8853
CW-SSIM=0.9983
(c)

Figure 6. Application of crossbar array based on memristor in image compression. (**a**) memristor crossbar array; (**b**) image compression framework; (**c**) image compression result.

3.3. *Image Reconstruction Based on Memristive Systems*

High-resolution image information is a prerequisite for the subsequent image processing and analysis. Therefore, effectively and quickly achieving high-resolution image reconstruction has become an urgent problem to be solved in this field. The image reconstruction algorithm based on memristive systems has certain advantages in terms of reconstruction quality and algorithm operation efficiency.

In 2017, Patrick et al. constructed a hardware-implemented sparse coding system using a 32 × 32 memristor crossbar array, as shown in Figure 7. The system input image information as sparsely coded pulses into the array and performed high-resolution reconstruction of the input through online learning. The experimental results demonstrated that the system has the advantage of low power consumption and high speed when performing data-intensive tasks (e.g., real-time video-based reconstruction) [44].

Additionally, a study designed a metal-oxide-based memristive synaptic circuit that enabled "negative (−)", "zero (0)", and "positive (+)" synaptic weights [45]. Based on this, the corresponding neuronal circuit was built to realize the on-chip cyclic learning algorithm, and the super-resolution reconstruction of a single frame was completed, as depicted in Figure 8.

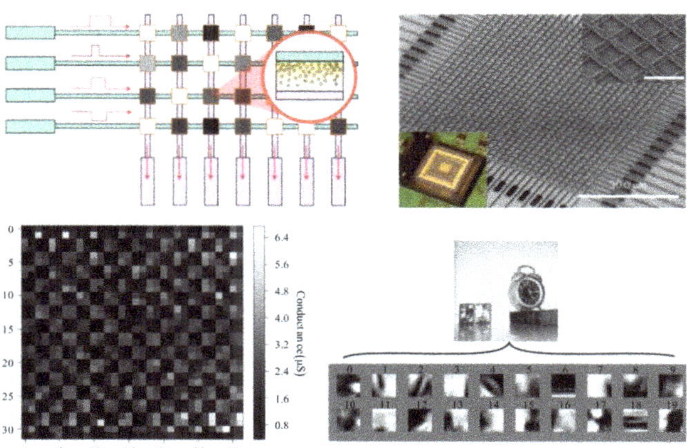

Figure 7. Memristor crossbar array-based computing hardware system.

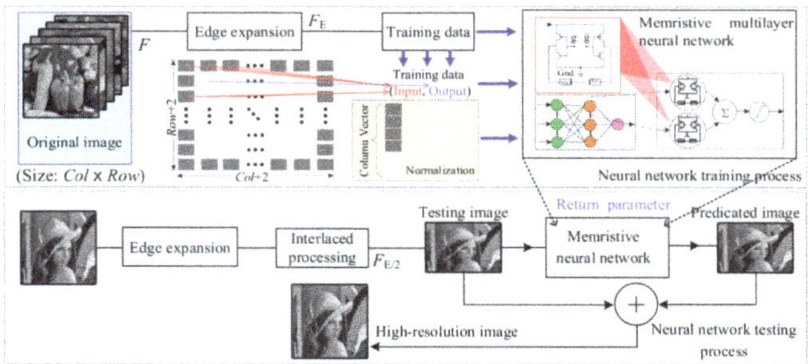

Figure 8. Single frame image super-resolution reconstruction based on memristive synapses.

A memristor-based compressive sampling encoder that could be integrated with an image sensor to achieve super-resolution reconstruction was put forward by Wang et al. [39]. A series of simulations demonstrated the superior performance of the encoder, with low power consumption and low hardware overhead. In addition, Dong et al. [46] designed a multi-channel pulse coupled neural network based on the nanoscale memristor, which effectively solved the problem of parameter estimation in neural networks by simulating the dynamic changes of connection coefficients. The model was further applied to the task of the super-resolution reconstruction of multi-frame images, and its correctness and effectiveness were experimentally demonstrated.

3.4. Others

With the ease of 3D stacking, the memristive system can efficiently complete matrix multiplication and realize the integration of storage and computation. By adjusting the variable parameters and connection methods of the system, and by adding peripheral control circuits, different nonlinear mapping functions are obtained to realize other image processing techniques (e.g., image interpolation, edge detection, image filtering, and image encryption).

Based on the mathematical model of the spintronic memristive device, Dong et al. [47] analysed its electrical characteristics and resistance variation through mathematical derivation and circuit simulation. Additionally, a memristor crossbar array was made by integrating functions, such as image storage and interpolation (as shown in Figure 9).

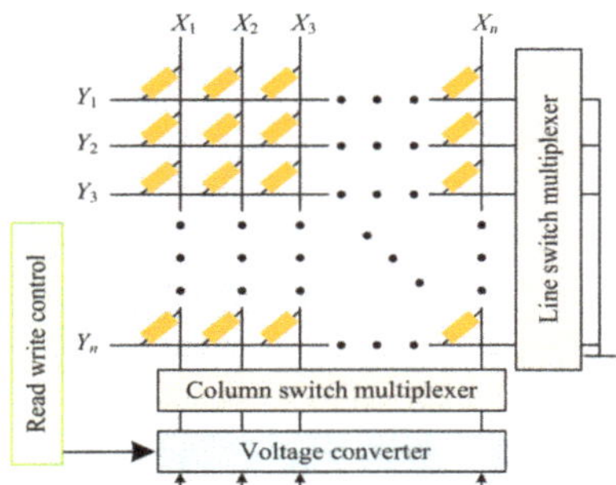

Figure 9. Image interpolation based on memristive system.

A study conducted by Yang et al. [38] showed an improved memristive cell neural network as well as an adaptive thresholding algorithm based on spatial distribution, and they achieved the edge extraction of colour images. The paper [48] discussed a memristive mask circuit based on the computation-in-memory (CIM) architecture, which is shown in Figure 10. The core of this mask circuit was a multi-bit analogue adder based on the memristor crossbar array, which selects the memristive cells to be accessed through the row-column switches. Each of the cells stored 8 bits of data according to the change of memristance, which were defined as pixel values in image processing. By controlling the multi-bit adder, integrator, and input module, the circuit could update the memristance with little dependence on the higher-level computing unit. Additionally, operations, such as image denoising, edge detection, and feature extraction were achieved by constructing different mask operators. The research [49] on the memristor-based 2D convolutional circuit implemented the image colour transformation and compression, while the reference [50] to the structure of the human retina achieved functions, such as image smoothing and edge detection.

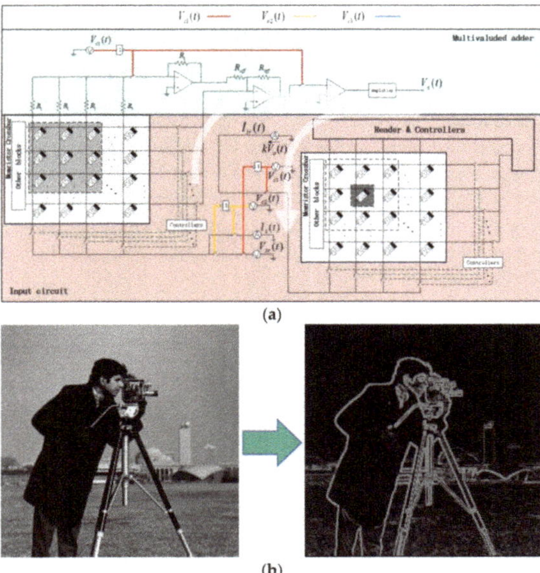

(a)

(b)

Figure 10. The application of mask circuit based on memristor crossbar array in edge extraction. (**a**) memristor crossbar array; (**b**) image edge extraction results.

A memristor-CMOS-based general logic circuit was studied by Yang et al. [51], and furthermore, a new memristor-based full adder circuit and a binary image encryption circuit were designed. Moreover, two available encryption methods were proposed to improve the reliability of encryption results. For Wang et al. [52], they studied a new memristive chaotic circuit to implement image encryption. Through a series of computer simulations, it was proven that the image encryption algorithm based on the new circuit has higher security and better decoding capability compared to the conventional one.

3.5. Summary

Currently, memristive system-based conventional image processing is in a rapid development stage, and some progress has been made in the same field. However, there still exist many problems that must be solved:

(1) The instability and variability of memristive devices have an impact on the accuracy of image processing. Therefore, it will be a significant study to explore the internal physical mechanism of memristive devices and to study their electrochemical properties under the influence of different external factors, to build a mathematical model that can accurately describe their behaviour.

(2) Conventional image processing circuits do not consider the possible faults of memristive circuits in practice. Nevertheless, research on fault diagnosis can effectively help reduce the circuit overhead as well as improve algorithm operation efficiency and image processing accuracy while increasing the robustness and anti-interference capability of the circuit.

(3) On the one hand, the design of the peripheral circuits in some image processing applications is too complex, which increases the power consumption of the system operation. On the other hand, the one with a simple structure and high compatibility can result in enhanced efficiency for complex conventional image processing tasks.

4. Image Processing Based on Memristive Neural Networks

The successful preparation of memristors brings new ideas for simulating the cognitive functions of artificial synapses. By applying memristive synapses to the hardware implementation of neural networks, a new type of neural network with high integration can be built. It possesses powerful image processing capabilities and plays an important role in fields with high computational complexity, such as image recognition, classification, and segmentation.

4.1. Image Recognition Based on Memristive Neural Networks

In the literature [53], an impulsive neural network based on memristors was constructed in which the memristive synapses used STDP rules to update the weights, and the memristive neurons adopted the "winner-take-all" strategy to complete the task of handwriting recognition. It was found that its recognition accuracy could reach 83%. A study conducted by Yakopcic et al. [54] presented a memristor-based convolutional neural network to perform convolutional operations using memristor crossbar arrays, and the accuracy of its handwritten digit recognition reached 94%. Furthermore, a transformation method for neural network models was brought forward [40], as shown in Figure 11.

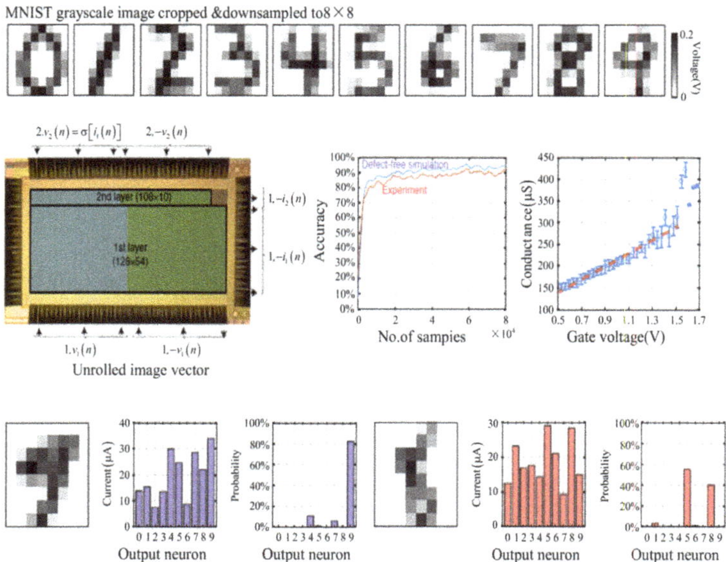

Figure 11. Handwriting recognition based on the ReRAM array.

Specifically, the method of sparsity was taken to divide the original neural network into appropriately sized sub-networks. The limited hardware accuracy was solved by quantizing the input data and somewhat improved to approximately 99.8% from the software side. In the study [55], a 1M structured memristive synapse was introduced to the memristor-based multilayer neural network, and an adaptive backpropagation algorithm was applied to train the neural networks, thereby achieving character recognition. Kang Jinfeng's team at Peking University [56] reported a memristor-based binary neural network. It was trained online, its weight update was achieved using the 2T2R structure of the memristive synapse, and its correctness and effectiveness were verified on the MNIST dataset with a recognition accuracy of 97.4%. In addition, Hu et al. [22] used 2 phase-change memories to construct artificial synapses, based on which a 3-layer perceptron network was built, and they proved its correctness on the MNIST dataset with a recognition accuracy of 82.2%. For Wang et al. [57], they constructed a memristor-based convolutional neural network, which was significantly improved in terms of array area and energy efficiency compared with previous ones for the handwriting recognition task. In 2020, a research team from Tsinghua University [58] designed a memristor-based convolutional neural network (see Figure 12). Meanwhile, a hybrid training method was suggested to enhance the robustness of the network, and the handwriting recognition task realized an accuracy of over 96%.

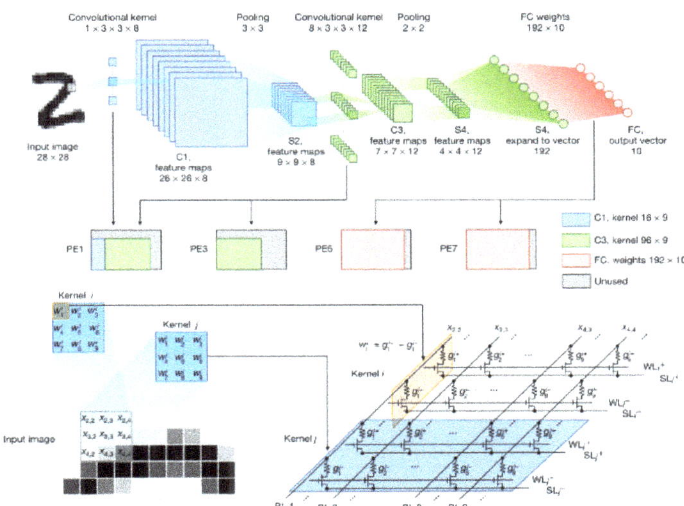

Figure 12. Five-layer mCNN with memristor convolver.

In addition, memristor-based neural networks have been applied to other image recognition tasks. For instance, Professor Wu Huaqiang and his team members from Tsinghua University constructed a multilayer perceptron neural network based on 1T1R memristive synapses [59], as shown in Figure 13. The network achieved grey scale face image recognition from the Yale Face Database through online learning, and the recognition rate could reach 88.08% for 9000 test images with noise added. Other researchers [60] investigated a hierarchical temporal memory (HTM) network based on memristors, which applied sparse distributed representations to obtain spatial information of input signals, after which they used parallel learning to adjust the network weights and finally verified the correctness and effectiveness of the network through face recognition tasks. As for the memristor-based probabilistic neural network [61], it carried out product multiplication using memristor crossbar arrays as well as normalization operations on weights to reduce the complexity of the circuit. The network was validated on the Iris Flower dataset with a recognition accuracy of 98%. Furthermore, the multilayer perceptron neural network studied by Yu et al. [62] showed increased adaptive capability by introducing nonlinear features in the learning process and superior performance on general datasets, such as MNIST, Iris, and Car Evaluation.

4.2. Image Classification Based on Memristive Neural Networks

In 2013, Alibart et al. [63] successfully prepared a TiO_{2-x}-based memristor, after which they developed a single-layer perceptron (SLP) neural network based on the TiO_{2-x} memristor crossbar array to achieve image classification. Its circuit structure is displayed in Figure 14.

Another (SLP) neural network was made based on 2M memristive synapses [10], and its circuit structure is presented in Figure 15. The network, which was trained using delta rules, achieved the classification of 3×3-pixel black-and-white images. Professor Strukov's team at the University of California, Santa Barbara [64] prepared a 20×20 memristor crossbar array, as depicted in Figure 16. The array adopted TiO_{2-x} and Al_2O_3 as the functional and stacked layers, respectively, after which it was interconnected with traditional CMOS peripheral circuits, thereby constructing an SLP neural network to achieve the image classification with an accuracy of more than 97%.

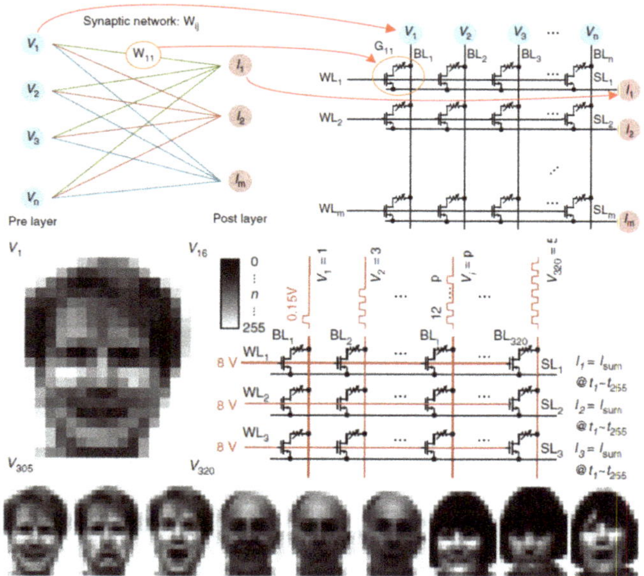

Figure 13. Face recognition task is realized in 1T1R array.

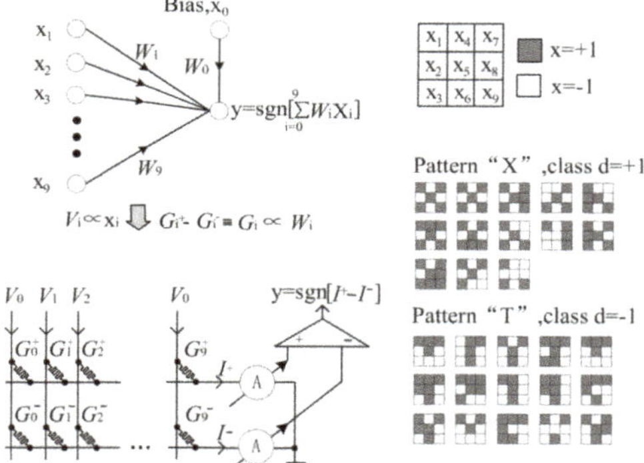

Figure 14. Single-layer perceptron network memristor circuit.

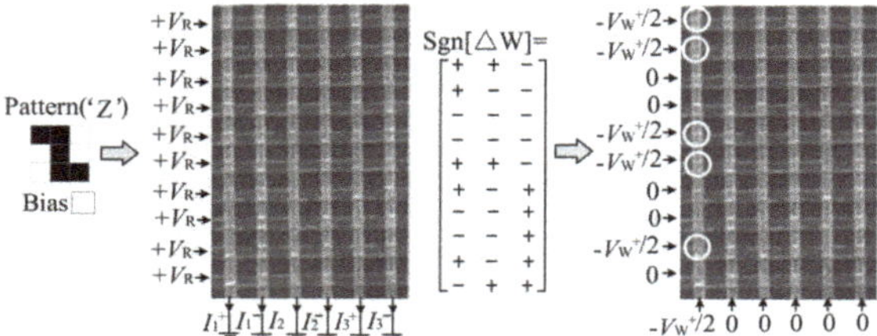

Figure 15. Single layer perceptron implemented using 10×6 memristor crossbar array.

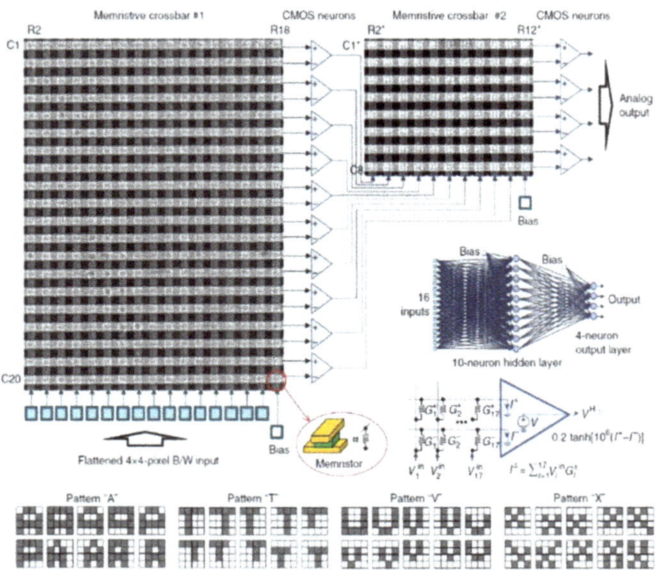

Figure 16. Three-layer fully connected perceptron network realized though $Pt/Al_2O_3/TiO_{2-x}/Ti/Pt$ memristor arrays.

Wang et al. [65] prepared a three-dimensional structured memristor and applied it to image classification, which improved the operational efficiency of the algorithm and opened a new path for the in-depth integration of computer vision and novel nanodevices. Additionally, a memory computing framework based on memristors was proposed by Zhang et al. in 2021 [66], which used a greedy search algorithm to improve the robustness and anti-interference capability of the system, and its accuracy reached 92.3% on the classification tasks involving the CIFAR-10 dataset.

4.3. Image Segmentation Based on Memristive Neural Networks

As early as 2014, Myonglae et al. [67] proposed a memristor-based visual recognition system, where the system used a programmable gate array to convert image signals into pulse signals and performed weight updates based on STDP learning rules. As a result, the foreground and background segmentation of figure images from "0" to "9" were achieved. One year later, Chiu and his team members [68] constructed a differential 2R crossbar array, which applied RRAM as a cache to reduce system energy consumption, and they verified its correctness and effectiveness using image segmentation tasks. In the literature [69], a fully convolutional neural network based on memristors was introduced. It utilized voltage selectors and memristor arrays to construct its max-pooling layers as well as a sliding window approach to enhance operation efficiency. Moreover, the weight updates of memristor arrays were implemented through the ex-situ training method, and the effectiveness of the proposed network was finally verified through image segmentation. The study [70] designed a memristor-based cell neural network based on the fractional-order calculus theory, as illustrated in Figure 17.

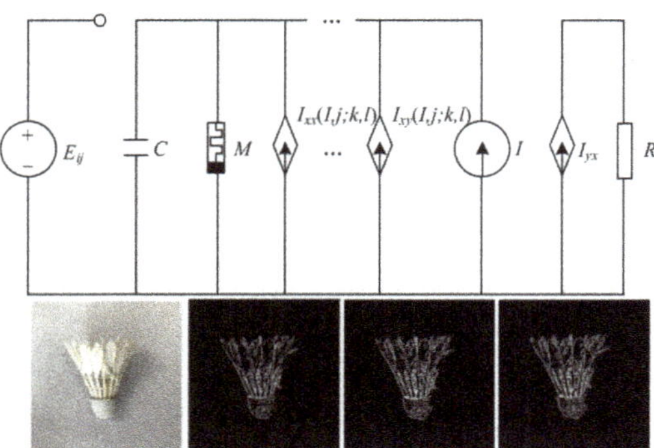

Figure 17. Memristor-based cell neural network.

In the process of image edge extraction, it took the fractional-order control method to increase the high-frequency information and retain more low-texture information. The simulation results proved that the edge images extracted by this network had more complete and clear contour information and richer texture detail information. Another example is the prepared memristor with NbO_x as its functional layer [71]. An artificial sensory neuron was constructed, then in combination with an $InGaZnO_4$ optical sensor (see Figure 18), which encoded optical information into impulses, image segmentation in complex backgrounds was achieved by such a pulse-coupled neural network. It is believed that this study has paved the way for the integration of neuromorphology and bioelectronics. In 2021, Chen et al. [72] proposed an efficient memristor-based fully convolutional neural network, which adopted a convolutional kernel-first (CKF) algorithm to achieve effective parameter pruning, thereby significantly reducing circuit power consumption and demonstrating high accuracy and adaptiveness for medical image segmentation tasks.

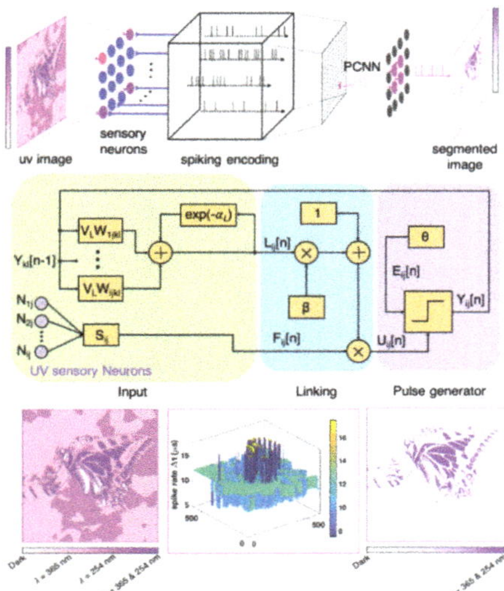

Figure 18. Impulse coupled neural network based on memristor.

4.4. Others

Tsai et al. [73] reported a long short-term memory network, which mapped and programmed the network weights into the phase-change memory devices, as demonstrated in Figure 19. Compared to other methods, this network realized the software-equivalent text prediction as well as a larger improvement in the accuracy of weight mapping and text prediction.

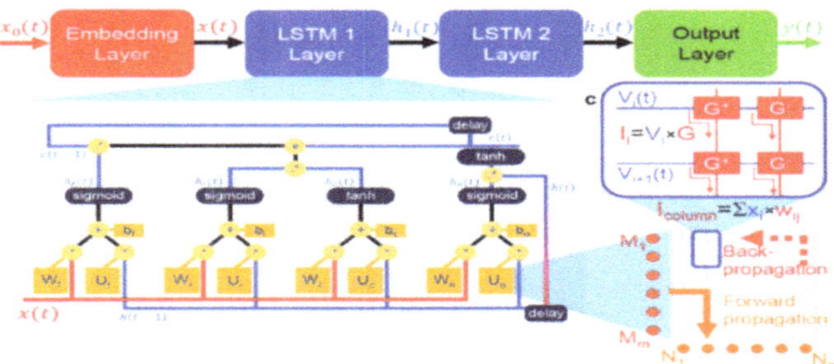

Figure 19. Realization method of long short-term memory network based on phase-change memory device unit.

Another long short-term memory network was built on a 128×64 1T1R memristor array [74], as shown in Figure 20. Through utilizing the memristor arrays to store synaptic weights for different time steps, the network performed the prediction task of the number of global airline passengers and the recognition task of human gait, and it verified the feasibility of the memristor-based long short-term memory neural network in performing tasks, such as linear regression and pattern recognition.

Moreover, Farkhani et al. [75] designed a neuromorphic computing system based on spintronic memristors, where the read circuit was replaced with a proposed real-time sensing circuit, and the input signals were turned into the switching of magnetic moments, thereby substantially reducing circuit energy consumption, providing system operational efficiency, and achieving the real-time tracking of targets. As for the study [76], the chaotic trajectories of memristive circuits were included, which combined the homotopy analysis method (HAM) and multi-objective optimization (MO) to tackle the high computational complexity and low computational efficiency of traditional analysis methods.

In this paper, the architectural characteristics of several image processing algorithms based on memristive neural networks are comprehensively summarized, including their input coding patterns, weight representations and the data types of interlayer communication. The specific comparative information is summarized in Table 4.

The above key research questions will provide references for building the next generation of novel memristive neural networks with integrated perception-storage-computation architectures.

Table 4. Comparative information of memristive neural network-based image processing.

Reference	Architectural Characteristics of Image Processing Algorithms Based on Memristive Neural Networks		
	Input Coding	**Weight Representation**	**Neural Network Communication**
[67]	Amplitude Encoding/Time Encoding Analogue Signal	Differential Amplifier	Multi-precision
[69]	Amplitude Encoding Analogue Signal	Multi-precision	MSB
[38]	Amplitude Encoding Analogue Signal	Differential Amplifier	Multi-precision
[40]	Amplitude Encoding Analogue Signal	Differential Amplifier	Multi-precision
[62]	Amplitude Encoding Analogue Signal	Peripheral Circuit Processing	MSB

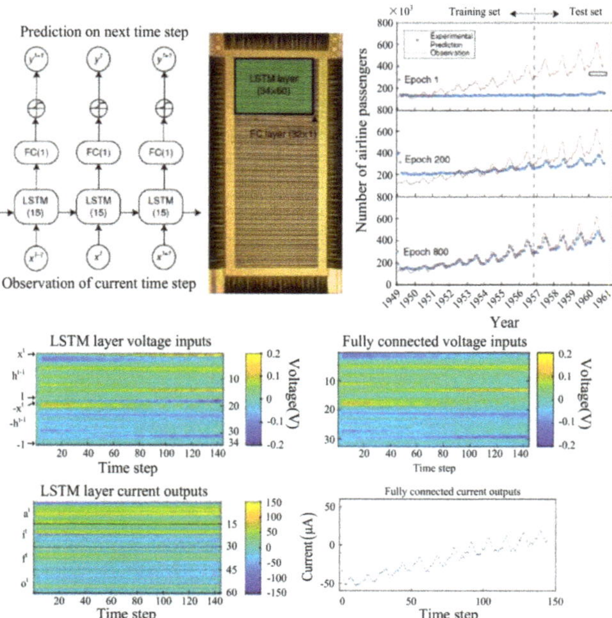

Figure 20. LSTM network based on memristor synaptic array.

4.5. Summary

With the expanded research in nanomaterials science and image processing technology, image processing based on memristive neural networks has become one of the hot issues in the study on neural network hardware implementation schemes. Currently, there are the following problems that must be solved timely.

(1)   The existing memristive synaptic circuits can only simulate the basic functions and behavioural characteristics of biological synapses, and they lack enough theoretical support from computational neuroscience. Therefore, it is crucial to design a fully functional and simple structured memristive synaptic circuit, which can address the problems of insufficient portray, unclear mechanism, and single plasticity of the conventional ones.

(2)   There is the accumulation of computational errors in memristor-based neural network circuits, which is mainly owing to the discrete nature of memristors, and it is difficult to avoid at the device level. Therefore, designing a newly structured memristor crossbar array that offsets the accumulated errors can provide a new perspective for the hardware implementation of neural networks.

(3)   The current research on image processing based on memristive neural networks is still stuck in the simulation of existing artificial neural networks. Therefore, the next research hotspot will involve taking both the advantages of neurocomputing science and image processing studies, exploring brain-inspired neural network training algorithms, and building memristive neural networks with brain-like memory.

5. Prospects

As the fourth passive circuit component, the memristor has certain memory properties, with its resistance value changing dynamically with the flowing charge and its high similarity with the synapse in the human brain. Using memristors to construct artificial synapses for neuromorphic computation is of great significance to the new intelligent information processing systems and integrated image storage and computation. Memristive system-based image processing technology is an interdisciplinary field of research, covering materials, devices, circuits, architectures, algorithms and integration technologies. We list major challenges and potential solutions for memristive system-based image processing technology, as shown in Table 5.

(1) At the device level, the device stability is critical to the computing accuracy, as the drift of conductance states with time or environmental changes will result in undesired synaptic weight changes. On the one hand, more reliable and eco-friendly memory devices and memristive arrays are required. On the other hand, the construction of scalable and highly stable memristive mathematical models, following the physical mechanisms of memristor devices and the special properties of memristors, is one of the future directions to further promote image processing research based on memristive systems.

(2) At the hardware level, in the short term, memristors will be specially utilized to accelerate the construction of artificial neural networks. Compared with conventional computer processors, their analogue signals are processed in a massively parallel manner, which increases the computational speed and fault tolerance simultaneously and significantly reduces the system power consumption. This parallel computing and low power consumption feature is well suited for image processing tasks with large data volumes and high computational complexity. In the long term, artificial synapses built on memristors will be one of the new approaches for facilitating the hardware implementation of brain-like neural networks. Nevertheless, the current memristive synaptic circuits can merely simulate the basic functions and behavioural characteristics of biological synapses, and they receive insufficient theoretical support from computational neuroscience. Therefore, the design of the memristive synaptic circuits with multiple biological synaptic properties can provide a new idea and platform for exploring a general memristive system-based image processing architecture to address the problems of insufficient portray, unclear mechanism, single plasticity, etc. Meanwhile, peripheral circuits control the read/write process in the memristor-based image processing systems. memristor-based image processing systems are expected to further improve the performance of online learning and reduce the complexity of peripheral programming circuits in the future.

(3) At the algorithm level, the learning algorithms of memristor based image processing systems are still under development. The conventional computing system has the problems of high cost and difficult training when simulating impulsive neural networks, whereas the unique dynamic memory and reconfigurable characteristics of memristors can realize not only the diverse biological synaptic plasticity for artificial synapses but also the natural compatibility of artificial neural networks and impulsive neural networks. The image processing algorithm based on memristive systems can learn from deep learning and computational neuroscience to solve the problems of slow training speed and the insufficient online processing capability of conventional artificial neural networks in image processing applications. With better understanding of neuronal communications and functionalities, general learning algorithms should be designed to promote hardware development as well.

Table 5. Key challenges and possible strategies of memristive system-based image processing technology on the device, hardware, and algorithm levels.

		Key Challenges	Possible Strategies
Device level	Materials	Fabricate standard-process and compatible new materials and interconnect materials with high conductance	Use alternative organic materials, 2D, and functional materials, and develop new processes for new materials
	Models	Less computational complexity and high physics fidelity for large-scale system simulation	Build mathematical models of memristors, combined physical and empirical behavior of devices
Hardware level	Peripheral circuits	Efficient read/write scheme for digital/analog mode	Use analog circuits, field programmable gate array (FPGA), and look-up-table (LUT) connected to the chips and approximate circuits
	Synaptic circuits	The operating mechanism is still obscure, the cognition function modeling is not good, and the fault diagnosis system is still in progress.	Develop the novel memristive synapse circuit will possess biological synaptic features
Algorithm level	Operations	Develop a general computing system for data mapping, dot product, and STDP	Experimentally build applications with a memristive crossbar
	Training and testing accuracies	Develop practical network topology and learning algorithm	Develop hybrid algorithms, and brain-inspired systems consist of both ANNs and SNNs

6. Conclusions

Memristors have been widely studied in image possessing for their synapse-like properties, low power consumption, high efficiency, integrability, etc. Two of their major applications are memristive system-based traditional image processing, including image compression, reconstruction, and edge extraction, and memristive neutral network-based image processing, including image recognition, classification, and segmentation. In neural networks, memristors are mainly adopted as synaptic devices to realize the hardware mapping of synaptic weights under pulse stimulation and to store the synaptic weights in real time for in-situ computation. The parallel computing capability of the memristor array improves the operational efficiency of the neural network and reduces the energy consumption of the system. Additionally, it is believed that the image processing technology based on memristive systems has very promising prospects in terms of its computational speed, computational energy efficiency, and processing accuracy, etc. Therefore, to develop a new type of energy-efficient memristor-based image processing system, collaborative innovations are needed in areas, such as mathematical modelling, architecture, and algorithm implementation.

Author Contributions: X.J.: writing—original draft, writing—review & editing, conceptualization, supervision. Z.D.: methodology, writing—review & editing, implementation, funding acquisition. G.Z.: investigation, visualization. C.S.L.: writing—original draft, writing—review & editing. D.Q.: writing—review & editing, formal analysis, supervision, funding acquisition. Y.Y.: writing—review & editing, implementation. All authors have read and agreed to the published version of the manuscript.

Funding: This work was supported in part by the National Natural Science Foundation of China under Grant U1909201, Grant 62001149, and Natural Science Foundation of Zhejiang Province under Grant LQ21F010009.

Conflicts of Interest: The authors declare no conflict of interest.

References

1. Zidan, M.A.; Strachan, J.P.; Lu, W.D. The future of electronics based on memristive systems. *Nat. Electron.* **2018**, *1*, 22–29. [CrossRef]
2. Wang, Z.; Joshi, S.; Savel'ev, S.E.; Jiang, H.; Midya, R.; Lin, P.; Hu, M.; Ge, N.; Strachan, J.P.; Li, Z.; et al. Memristors with diffusive dynamics as synaptic emulators for neuromorphic computing. *Nat. Mater.* **2017**, *16*, 101–108. [CrossRef]
3. Dong, Z.; Sing Lai, C.; Zhang, Z.; Qi, D.; Gao, M.; Duan, S. Neuromorphic extreme learning machines with bimodal memristive synapses. *Neurocomputing* **2021**, *453*, 38–49. [CrossRef]
4. Schuman, C.D.; Potok, T.E.; Patton, R.M.; Birdwell, J.D.; Dean, M.E.; Rose, G.S.; Plank, J.S. A survey of neuromorphic computing and neural networks in hardware. *arXiv* **2017**, arXiv:1705.06963.

5. Davies, M.; Srinivasa, N.; Lin, T.H.; Chinya, G.; Cao, Y.; Choday, S.H.; Dimou, G.; Joshi, P.; Imam, N.; Jain, S.; et al. Loihi: A neuromorphic manycore processor with on-chip learning. *IEEE Micro* **2018**, *38*, 82–99. [CrossRef]

6. Jo, S.H.; Chang, T.; Ebong, I.; Bhadviya, B.B.; Mazumder, P.; Lu, W. Nanoscale memristor device as synapse in neuromorphic systems. *Nano Lett.* **2010**, *10*, 1297–1301. [CrossRef] [PubMed]

7. Schmidhuber, J. Deep Learning in neural networks: An overview. *Neural Netw.* **2015**, *61*, 85–117. [CrossRef]

8. Ji, X.; Qi, D.; Dong, Z.; Lai, C.S.; Zhou, G.; Hu, X. TSSM: Three-state switchable memristor model based on Ag/TiOx nanobelt/Ti configuration. *Int. J. Bifurc. Chaos* **2021**, *31*, 2130020. [CrossRef]

9. Yang, J.J.; Strukov, D.B.; Stewart, D.R. Memristive devices for computing. *Nat. Nanotechnol.* **2013**, *8*, 13–24. [CrossRef]

10. Prezioso, M.; Merrikh-Bayat, F.; Hoskins, B.D.; Adam, G.C.; Likharev, K.K.; Strukov, D.B. Training and operation of an integrated neuromorphic network based on metal-oxide memristors. *Nature* **2015**, *521*, 61–64. [CrossRef]

11. Pi, S.; Li, C.; Jiang, H.; Xia, W.; Xin, H.; Yang, J.J.; Xia, Q. Memristor crossbar arrays with 6-nm half-pitch and 2-nm critical dimension. *Nat. Nanotechnol.* **2019**, *14*, 35–39. [CrossRef] [PubMed]

12. Gokmen, T.; Onen, M.; Haensch, W. Training deep convolutional neural networks with resistive cross-point devices. *Front. Neurosci.* **2017**, *11*, 538. [CrossRef] [PubMed]

13. Esser, S.K.; Merolla, P.A.; Arthur, J.V.; Cassidy, A.S.; Appuswamy, R.; Andreopoulos, A.; Berg, D.J.; McKinstry, J.L.; Melano, T.; Barch, D.R.; et al. Convolutional networks for fast, energy-efficient neuromorphic computing. *Proc. Natl. Acad. Sci. USA* **2016**, *113*, 11441–11446. [CrossRef]

14. Choi, S.; Shin, J.H.; Lee, J.; Sheridan, P.; Lu, W.D. Experimental demonstration of feature extraction and dimensionality reduction using memristor networks. *Nano Lett.* **2017**, *17*, 3113–3118. [CrossRef] [PubMed]

15. Chua, L.O. Memristor—The missing circuit element. *IEEE Trans. Circuit Theory* **1971**, *18*, 507–519. [CrossRef]

16. Strukov, D.B.; Snider, G.S.; Stewart, D.R.; Williams, R.S. The missing memristor found. *Nature* **2008**, *453*, 80–83. [CrossRef]

17. Zhang, C.; Ye, W.B.; Zhou, K.; Chen, H.Y.; Yang, J.Q.; Ding, G.; Chen, X.; Zhou, Y.; Zhou, L.; Li, F.; et al. Bioinspired artificial sensory nerve based on nafion memristor. *Adv. Funct. Mater.* **2019**, *29*, 1970133. [CrossRef]

18. Hu, L.; Fu, S.; Chen, Y.; Cao, H.; Liang, L.; Zhang, H.; Gao, J.; Wang, J.; Zhuge, F. Ultrasensitive memristive synapses based on lightly oxidized sulfide films. *Adv. Mater.* **2017**, *29*, 6927. [CrossRef]

19. Xu, W.; Lee, Y.; Min, S.Y.; Park, C.; Lee, T.W. Simple, inexpensive, and rapid approach to fabricate cross-shaped memristors using an inorganic-nanowire-digital-alignment technique and a one-step reduction process. *Adv. Mater.* **2016**, *28*, 527–532. [CrossRef]

20. Dong, Z.; Lai, C.S.; He, Y.; Qi, D.; Duan, S. Hybrid dual-complementary metal-oxide-semiconductor/memristor synapse-based neural network with its applications in image super-resolution. *IET Circuits Devices Syst.* **2019**, *13*, 1241–1248. [CrossRef]

21. Sheri, A.M.; Hwang, H.; Jeon, M.; Lee, B.G. Neuromorphic character recognition system with two PCMO memristors as a synapse. *IEEE Trans. Ind. Electron.* **2014**, *61*, 2933–2941. [CrossRef]

22. Hu, M.; Graves, C.E.; Li, C.; Li, Y.; Ge, N.; Montgomery, E.; Davila, N.; Jiang, H.; Williams, R.S.; Yang, J.J.; et al. Memristor-Based analog computation and neural network classification with a dot product engine. *Adv. Mater.* **2018**, *30*, 5914. [CrossRef] [PubMed]

23. Xie, L.; Nguyen, H.A.D.; Yu, J.; Kaichouhi, A.; Taouil, M.; Alfailakawi, M.; Hamdioui, S. Scouting logic: A novel memristor-based logic design for resistive computing. In Proceedings of the IEEE Computer Society Annual Symposium on VLSI, ISVLSI, Bochum, Germany, 3–5 July 2017; IEEE Computer Society: Bochum, Germany, 2017; pp. 176–181.

24. Kvatinsky, S.; Satat, G.; Wald, N.; Friedman, E.G.; Kolodny, A.; Weiser, U.C. Memristor-based material implication (IMPLY) logic: Design principles and methodologies. *IEEE Trans. Very Large Scale Integr. Syst.* **2014**, *22*, 2054–2066. [CrossRef]

25. James, A.P. Memristor threshold logic: An overview to challenges and applications. *arXiv* **2016**, arXiv:1612.01711.

26. Wang, Z.; Wu, H.; Burr, G.W.; Hwang, C.S.; Wang, K.L.; Xia, Q.; Yang, J.J. Resistive switching materials for information processing. *Nat. Rev. Mater.* **2020**, *5*, 173–195. [CrossRef]

27. Biolek, D.; Biolek, Z.; Biolkova, V. SPICE modeling of memristive, memcapacitative and meminductive systems. In Proceedings of the ECCTD 2009—European Conference on Circuit Theory and Design Conference Program, Sofia, Bulgaria, 7–10 September 2020; IEEE: Antalya, Turkey, 2009; pp. 249–252.

28. Wang, X.; Chen, Y.; Xi, H.; Li, H.; Dimitrov, D. Spintronic memristor through spin-thorque-induced magnetization motion. *IEEE Electron Device Lett.* **2009**, *30*, 294–297. [CrossRef]

29. Kvatinsky, S.; Friedman, E.G.; Kolodny, A.; Weiser, U.C. TEAM: Threshold adaptive memristor model. *IEEE Trans. Circuits Syst. I Regul. Pap.* **2013**, *60*, 211–221. [CrossRef]

30. Kvatinsky, S.; Ramadan, M.; Friedman, E.G.; Kolodny, A. VTEAM: A general model for voltage-controlled memristors. *IEEE Trans. Circuits Syst. II Express Briefs* **2015**, *62*, 786–790. [CrossRef]

31. Zhang, J.; Tang, Z.; Xu, N.; Wang, Y.; Sun, H.; Wang, Z.; Fang, L. A generalized model of TiOx-based memristive devices and its application for image processing. *Chin. Phys. B* **2017**, *26*, 502. [CrossRef]

32. Chen, M.; Bao, B.; Jiang, T.; Bao, H.; Xu, Q.; Wu, H.; Wang, J. Flux-charge analysis of initial state-dependent dynamical behaviors of a memristor emulator-based chua's circuit. *Int. J. Bifurc. Chaos* **2018**, *28*, e1850120. [CrossRef]

33. Xie, X.; Zou, L.; Wen, S.; Zeng, Z.; Huang, T. A flux-controlled logarithmic memristor model and emulator. *Circuits Syst. Signal Process.* **2019**, *38*, 1452–1465. [CrossRef]

34. Ginoux, J.M.; Muthuswamy, B.; Meucci, R.; Euzzor, S.; Di Garbo, A.; Ganesan, K. A physical memristor based Muthuswamy–Chua–Ginoux system. *Sci. Rep.* **2020**, *10*, 6108. [CrossRef] [PubMed]

35. Hu, X.; Wang, L.; Duan, S.; Liao, X. Memristor cross array and its application in image processing. *Sci. Sin. Informationis* **2011**, *41*, 500–512. [CrossRef]

36. Tan, H.; Liu, G.; Zhu, X.; Yang, H.; Chen, B.; Chen, X.; Shang, J.; Lu, W.D.; Wu, Y.; Li, R.W. An optoelectronic resistive switching memory with integrated demodulating and arithmetic functions. *Adv. Mater.* **2015**, *27*, 2797–2803. [CrossRef] [PubMed]

37. Wang, Z.Y. Research on Memristor-Based Multilevel Storage Circuit Design and Applications. Master's Thesis, Huazhong University of Science & Technology, Wuhan, China, 2016.

38. Liu, Q.; Wang, L.; Yang, J.; Wang, Y.; Duan, S. Fusion of image storage and operation based on ag-chalcogenide memristor with synaptic plasticity. *J. Circuits Syst. Comput.* **2017**, *26*, 1614. [CrossRef]

39. Wang, T.Y.; Meng, J.L.; Li, Q.X.; Chen, L.; Zhu, H.; Sun, Q.Q.; Ding, S.J.; Zhang, D.W. Forming-free flexible memristor with multilevel storage for neuromorphic computing by full PVD technique. *J. Mater. Sci. Technol.* **2021**, *60*, 21–26. [CrossRef]

40. Li, C.; Hu, M.; Li, Y.; Jiang, H.; Ge, N.; Montgomery, E.; Zhang, J.; Song, W.; Dávila, N.; Graves, C.E.; et al. Analogue signal and image processing with large memristor crossbars. *Nat. Electron.* **2018**, *1*, 52–59. [CrossRef]

41. Halawani, Y.; Mohammad, B.; Al-Qutayri, M.; Al-Sarawi, S.F. Memristor-based hardware accelerator for image compression. *IEEE Trans. Very Large Scale Integr. Syst.* **2018**, *26*, 2749–2758. [CrossRef]

42. Berco, D.; Ang, D.S.; Kalaga, P.S. Programmable photoelectric memristor gates for in situ image compression. *Adv. Intell. Syst.* **2020**, *2*, 2000079. [CrossRef]

43. Hu, H.; Cao, Y.; Xu, J.; Ma, C.; Yan, H. An image compression and encryption algorithm based on the fractional-order simplest chaotic circuit. *IEEE Access* **2021**, *9*, 22141–22155. [CrossRef]

44. Sheridan, P.M.; Cai, F.; Du, C.; Ma, W.; Zhang, Z.; Lu, W.D. Sparse coding with memristor networks. *Nat. Nanotechnol.* **2017**, *12*, 784–789. [CrossRef]

45. Dong, Z.; Lai, C.S.; Xu, Z.; Qi, D. Single image super-resolution via the implementation of the hardware-friendly sparse coding. In Proceedings of the 2018 37th Chinese Control Conference (CCC), Wuhan, China, 25–27 July 2018; IEEE Computer Society: Wuhan, China, 2018; Volume 2018, pp. 8132–8137.

46. Dong, Z.; Du, C.; Lin, H.; Lai, C.S.; Hu, X.; Duan, S. Multi-channel Memristive Pulse Coupled Neural Network Based Multi-frame Images Super-resolution Reconstruction Algorithm. *J. Electron. Inf. Technol.* **2020**, *42*, 835–843. [CrossRef]

47. Dong, Z.K.; Yan, Y.F.; Qi, D.L.; Chen, J.; Diam, S.C. Transmemristive cross array and its application in image processing. In Proceedings of the 36th China Control Conference, Dalian, China, 26–28 July 2017; Dalian University of Technology: Dalian, China, 2017.

48. Shang, L.; Duan, S.; Wang, L.; Huang, T. SRMC: A multibit memristor crossbar for self-renewing image mask. *IEEE Trans. Very Large Scale Integr. Syst.* **2018**, *26*, 2830–2841. [CrossRef]

49. Athreyas, N.; Song, W.; Perot, B.; Xia, Q.; Mathew, A.; Gupta, J.; Gupta, D.; Yang, J.J. Memristor-CMOS analog coprocessor for acceleration of high-performance computing applications. *ACM J. Emerg. Technol. Comput. Syst.* **2018**, *14*, 9985. [CrossRef]

50. Pajouhi, Z.; Roy, K. Image edge detection based on swarm intelligence using memristive networks. *IEEE Trans. Comput. Des. Integr. Circuits Syst.* **2018**, *37*, 1774–1787. [CrossRef]

51. Yang, H.; Duan, S.C.; Dong, Z.K.; Wang, L.D.; Hu, X.F.; Shang, L.T. General logic circuit based on memristor-cmos and its application. *Sci. China Inf. Sci.* **2020**, *50*, 14.

52. Ye, X.; Wang, X.; Gao, S.; Mou, J.; Wang, Z.; Yang, F. A new chaotic circuit with multiple memristors and its application in image encryption. *Nonlinear Dyn.* **2020**, *99*, 1489–1506. [CrossRef]

53. Wu, X.; Saxena, V.; Zhu, K. Homogeneous spiking neuromorphic system for real-world pattern recognition. *IEEE J. Emerg. Sel. Top. Circuits Syst.* **2015**, *5*, 254–266. [CrossRef]

54. Yakopcic, C.; Alom, M.Z.; Taha, T.M. Memristor crossbar deep network implementation based on a Convolutional neural network. In Proceedings of the 2016 International Joint Conference on Neural Networks (IJCNN), Vancouver, BC, Canada, 24–29 July 2016; Institute of Electrical and Electronics Engineers Inc.: Vancouver, BC, Canada, 2016; Volume 2016, pp. 963–970.

55. Zhang, Y.; Wang, X.; Friedman, E.G. Memristor-based circuit design for multilayer neural networks. *IEEE Trans. Circuits Syst. I Regul. Pap.* **2018**, *65*, 677–686. [CrossRef]

56. Zhou, Z.; Huang, P.; Xiang, Y.C.; Shen, W.S.; Zhao, Y.D.; Feng, Y.L.; Gao, B.; Wu, H.Q.; Qian, H.; Liu, L.F.; et al. A new hardware implementation approach of BNNs based on nonlinear 2T2R synaptic cell. *Tech. Dig. Int. Electron Devices Meet. IEDM* **2019**, *18*, 71–74. [CrossRef]

57. Wang, Z.; Li, C.; Lin, P.; Rao, M.; Nie, Y.; Song, W.; Qiu, Q.; Li, Y.; Yan, P.; Strachan, J.P.; et al. In situ training of feed-forward and recurrent convolutional memristor networks. *Nat. Mach. Intell.* **2019**, *1*, 434–442. [CrossRef]

58. Yao, P.; Wu, H.; Gao, B.; Tang, J.; Zhang, Q.; Zhang, W.; Yang, J.J.; Qian, H. Fully hardware-implemented memristor convolutional neural network. *Nature* **2020**, *577*, 641–646. [CrossRef]

59. Yao, P.; Wu, H.; Gao, B.; Eryilmaz, S.B.; Huang, X.; Zhang, W.; Zhang, Q.; Deng, N.; Shi, L.; Wong, H.S.P.; et al. Face classification using electronic synapses. *Nat. Commun.* **2017**, *8*, 5199. [CrossRef]

60. Liu, X.; Huang, Y.; Zeng, Z.; Wunsch, D.C. Memristor-based HTM spatial pooler with on-device learning for pattern recognition. *IEEE Trans. Syst. Man Cybern. Syst.* **2020**, *3*, 5612. [CrossRef]

61. Krestinskaya, O.; James, A.P. Approximate probabilistic neural networks with gated threshold logic. In Proceedings of the 2018 IEEE 18th International Conference on Nanotechnology, Cork, Ireland, 23–26 July 2018; Volume 18, p. 6302. [CrossRef]

62. Yu, Y.; Adu, K.; Tashi, N.; Anokye, P.; Wang, X.; Ayidzoe, M.A. RMAF: Relu-Memristor-Like Activation Function for Deep Learning. *IEEE Access* **2020**, *8*, 72727–72741. [CrossRef]
63. Alibart, F.; Zamanidoost, E.; Strukov, D.B. Pattern classification by memristive crossbar circuits using ex situ and in situ training. *Nat. Commun.* **2013**, *4*, 3072. [CrossRef]
64. Bayat, F.M.; Prezioso, M.; Chakrabarti, B.; Nili, H.; Kataeva, I.; Strukov, D. Implementation of multilayer perceptron network with highly uniform passive memristive crossbar circuits. *Nat. Commun.* **2018**, *9*, 4482. [CrossRef]
65. Lin, P.; Li, C.; Wang, Z.; Li, Y.; Jiang, H.; Song, W.; Rao, M.; Zhuo, Y.; Upadhyay, N.K.; Barnell, M.; et al. Three-dimensional memristor circuits as complex neural networks. *Nat. Electron.* **2020**, *3*, 225–232. [CrossRef]
66. Zhang, W.; Gao, B.; Yao, P.; Tang, J.; Qian, H.; Wu, H. Array-level boosting method with spatial extended allocation to improve the accuracy of memristor based computing-in-memory chips. *Sci. China Inf. Sci.* **2021**, *64*, 3198. [CrossRef]
67. Chu, M.; Kim, B.; Park, S.; Hwang, H.; Jeon, M.; Lee, B.H.; Lee, B.G. Neuromorphic hardware system for visual pattern recognition with memristor array and CMOS neuron. *IEEE Trans. Ind. Electron.* **2015**, *62*, 2410–2419. [CrossRef]
68. Chiu, P.F.; Nikolić, B. A differential 2R crosspoint RRAM array with zero standby current. *IEEE Trans. Circuits Syst. II Express Briefs* **2015**, *62*, 461–465. [CrossRef]
69. Wen, S.; Wei, H.; Zeng, Z.; Huang, T. Memristive fully convolutional network: An accurate hardware image-segmentor in deep learning. *IEEE Trans. Emerg. Top. Comput. Intell.* **2018**, *2*, 324–334. [CrossRef]
70. Xiu, C.; Li, X. Edge extraction based on memristor cell neural network with fractional order template. *IEEE Access* **2019**, *7*, 90750–90759. [CrossRef]
71. Wu, Q.; Dang, B.; Lu, C.; Xu, G.; Yang, G.; Wang, J.; Chuai, X.; Lu, N.; Geng, D.; Wang, H.; et al. Spike encoding with optic sensory neurons enable a pulse coupled neural network for ultraviolet image segmentation. *Nano Lett.* **2020**, *20*, 8015–8023. [CrossRef]
72. Chen, J.; Wu, Y.; Yang, Y.; Wen, S.; Shi, K.; Bermak, A.; Huang, T. An efficient memristor-based circuit implementation of squeeze-and-excitation fully convolutional neural networks. *IEEE Trans. Neural Netw. Learn. Syst.* **2021**, *4*, 4047. [CrossRef]
73. Tsai, H.; Ambrogio, S.; MacKin, C.; Narayanan, P.; Shelby, R.M.; Rocki, K.; Chen, A.; Burr, G.W. Inference of long-short term memory networks at software-equivalent accuracy using 2.5M analog phase change memory devices. *Dig. Tech. Pap.—Symp. VLSI Technol.* **2019**, *2019*, T82–T83. [CrossRef]
74. Li, C.; Wang, Z.; Rao, M.; Belkin, D.; Song, W.; Jiang, H.; Yan, P.; Li, Y.; Lin, P.; Hu, M.; et al. Long short-term memory networks in memristor crossbar arrays. *Nat. Mach. Intell.* **2019**, *1*, 49–57. [CrossRef]
75. Farkhani, H.; Tohidi, M.; Farkhani, S.; Madsen, J.K.; Moradi, F. A low-power high-speed spintronics-based neuromorphic computing system using real-time tracking method. *IEEE J. Emerg. Sel. Top. Circuits Syst.* **2018**, *8*, 627–638. [CrossRef]
76. Hu, W.; Luo, H.; Chen, C.; Wei, R. A multi-interval homotopy analysis method using multi-objective optimization for analytically analyzing chaotic dynamics in memristive circuit. *IEEE Access* **2019**, *7*, 116328–116341. [CrossRef]

electronics

MDPI

Review

Memcapacitor and Meminductor Circuit Emulators: A Review

Francisco J. Romero [1,2,*], Akiko Ohata [3], Alejandro Toral-Lopez [1,2], Andres Godoy [1,2], Diego P. Morales [1,4] and Noel Rodriguez [1,2,*]

[1] Department of Electronics and Computer Technology, Faculty of Sciences, University of Granada, 18071 Granada, Spain; atoral@ugr.es (A.T.-L.); agodoy@ugr.es (A.G.); diegopm@ugr.es (D.P.M.)
[2] Pervasive Electronics Advanced Research Laboratory, University of Granada, 18071 Granada, Spain
[3] Institute of Space and Astronautical Science, Japan Aerospace Exploration Agency, Kanagawa 252-5210, Japan; ohata@post.kek.jp
[4] Biochemistry and Electronics as Sensing Technologies Group, University of Granada, 18071 Granada, Spain
[*] Correspondence: franromero@ugr.es (F.J.R.); noel@ugr.es (N.R.)

Abstract: In 1971, Prof. L. Chua theoretically introduced a new circuit element, which exhibited a different behavior from that displayed by any of the three known passive elements: the resistor, the capacitor or the inductor. This element was called memristor, since its behavior corresponded to a resistor with memory. Four decades later, the concept of mem-elements was extended to the other two circuit elements by the definition of the constitutive equations of both memcapacitors and meminductors. Since then, the non-linear and non-volatile properties of these devices have attracted the interest of many researches trying to develop a wide range of applications. However, the lack of solid-state implementations of memcapacitors and meminductors make it necessary to rely on circuit emulators for the use and investigation of these elements in practical implementations. On this basis, this review gathers the current main alternatives presented in the literature for the emulation of both memcapacitors and meminductors. Different circuit emulators have been thoroughly analyzed and compared in detail, providing a wide range of approaches that could be considered for the implementation of these devices in future designs.

Keywords: emulator; gyrator; memcapacitor; meminductor; memristor

Citation: Romero, F.J.; Ohata, A.; Toral-Lopez, A.; Godoy, A.; Morales, D.P.; Rodriguez, N. Memcapacitor and Meminductor Circuit Emulators: A Review. *Electronics* **2021**, *10*, 1225. https://doi.org/10.3390/electronics10111225

Academic Editors: Chun Sing Lai, Zhekang Dong and Donglian Qi

Received: 15 April 2021
Accepted: 18 May 2021
Published: 21 May 2021

Publisher's Note: MDPI stays neutral with regard to jurisdictional claims in published maps and institutional affiliations.

1. Introduction

Prof. Leon L. Chua presented in 1971 the theoretical definition of the two terminal device which defined the relation between the time-integral of its input voltage (ϕ, flux) and its electric charge (q) [1]. This element was called *memristor* given that its behavior corresponds to a nonlinear resistor in which the current through its terminals at an instant t_1 depends not only on the input voltage at t_1, but also on the input voltage from $t = -\infty$ to $t = t_1$ (i.e., a resistor whose resistance depends on the history of its input). It was also demonstrated that this element was passive and that, contrary to capacitors and inductors, it cannot store energy. Therefore, as a manifestation of these characteristics, the current of the memristor is zero whenever the input voltage is zero and, for a periodic current input, the memristive systems show a "closed pinched hysteretic loop" in their *i-v* characteristic [2].

However, until 2008 the investigation into the memristor concept was very limited due to the lack of a solid-state implementation of this device [3–6]. However, it was in 2008 that a group of researchers of Hewlett Packard Labs announced the first solid-state device fulfilling the theoretical definition of the memristor [7], which constituted a turning point in the research of memristors and its applications. Since then, thanks to its non-volatility and non-linear behavior, the memristor is expected to play a disruptive role in diverse fields, such as neuromorphic circuits and neural networks [8–12], analog programmable circuits and arithmetic circuits [13–16], logic gates [17], crossbar classifiers [18–20], adaptive filters [21], chaotic circuits [22,23] and non-volatile memories [24–26]. This had led to intensive studies of the memristive behavior in a wide range of materials, such as transition metal oxides (e.g., NiO and TaO$_x$) [27,28], polymers [29], 2D materials [30] or graphene oxide [31–33], among others. The success of the memristor led Di Ventra, Pershin and Chua to extended the concept of the memory circuit elements to capacitive and inductive systems, thus defining the memcapacitor and the meminductor, respectively [34]. In this way, together with the memristor, they established the electrical relations between the time-integral of the charge (σ) and the flux (ϕ) with the memcapacitor; and between the time-integral of the flux (ρ) and the charge (q) with the meminductor (see Figure 1).

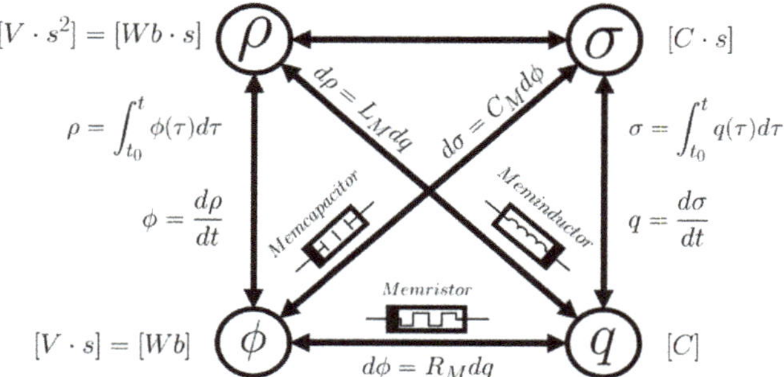

Figure 1. Mem-elements definition on the basis of their fundamental physical magnitudes (units are presented in brackets). Memristor: relation between the charge (q) and the time-integral of the voltage (ϕ); memcapacitor: relation between time-integral of the charge (σ) and the time-integral of the voltage (ϕ); meminductor: relation between time-integral of the flux (ρ) and the charge (q) [35].

As in the case of memristors, memcapacitors and meminductors also present a memory ability manifested through a closed pinched hysteresis loop in the characteristic of their two constitutive variables; with the additional advantage of being capable of storing energy in capacitive and inductive forms, respectively [36]. These devices are expected to be the key for the emergence of a new form of computation called neuromorphic computing, since their essential properties are envisaged to allow them to mimic biological computing. Thanks to their ability to both store and process information simultaneously, computers based on these mem-elements would offer capabilities and power consumption comparable to those of the human brain [37–39]. However, the lack of solid-state implementations of memcapacitors and meminductors hinders the exploitation of the prominent features of these devices in practical implementations. Due to this, in recent years there has been an emerging line of research dedicated to the development of emulators of these devices, i.e., circuits that satisfy the constitutive equations of the emulated mem-element.

In this context, this work reviews the different models and practical memcapacitor and meminductor emulators presented in the literature. Thus, the different approaches followed for the emulation of the memory effect and nonlinear behavior of these devices have been analyzed in detail and in a comparative way. The manuscript is structured as follows: after this introduction, Section 2 presents the concept of memcapacitance as well as the different approaches proposed for the emulation of memcapacitors. Similarly, Section 3 introduces the concept of meminductive system and the different alternatives adopted for the emulation of meminductors. Moreover, those circuits that based on the same design are able to emulate either a memcapacitor or a meminductor with minimal changes in their design have been grouped in Section 4. Finally, the main conclusions of the different emulation approaches are drawn in Section 5.

2. Memcapacitor Emulators

The general memcapacitance (C_M) is defined as the nth-order system that establishes a nonlinear relation between the charge of the device (q) and its input voltage (v) [34]. It can be either voltage-controlled or charge-controlled depending on its constitutive input variable. Therefore, an *n*th-order voltage-controlled memcapacitive system can be defined by Equation (1):

$$q(t) = C_M\left(\vec{x_N},\, v, t\right)\cdot v(t) \tag{1}$$

whereas the *n*th-order charge-controlled memcapacitance systems are defined by Equation (2):

$$v(t) = C_M^{-1}\left(\vec{x_N},\, q, t\right)\cdot q(t) \tag{2}$$

being $\vec{x_N}$ a vector that represents the *n* internal state variables of the system.

The memcapacitor is a particular case of memcapacitive system with one single state variable; the voltage in the case of voltage-controlled memcapacitors, Equation (3), or the charge in the case of charge-controlled memcapacitors, Equation (4).

$$q(t) = C_M\left[\int_{t_0}^{t} v(\tau)d\tau\right]\cdot v(t) \tag{3}$$

$$v(t) = C_M^{-1}\left[\int_{t_0}^{t} q(\tau)d\tau\right]\cdot q(t) \tag{4}$$

In the previous equations, the initial instant of time, t_0, may be selected to ensure that $\int_{-\infty}^{t_0} v(\tau)d\tau = 0$ and $\int_{-\infty}^{t_0} q(\tau)d\tau = 0$, respectively.

Therefore, the memcapacitors are nothing but capacitors whose capacitance depends on the history of the constitutive variable that acts as input (either charge or voltage) and whose *q-v* characteristic presents a closed-pinched hysteresis loop in which *v = 0* whenever *q = 0* (and vice versa) for bipolar sine wave-like excitations. In this way, the memcapacitor emulators must be able to monitor the control variable (*q* or *v*) and then change its input capacitance according to the history of this variable. Therefore, the memcapacitor emulators can also be either voltage- or charge-controlled.

An example of charge-controlled memcapacitor emulator is the one proposed by Fouda and Radwan in Ref. [40], and shown in Figure 2. This circuit is based on the mathematical model of charge-controlled memcapacitance introduced by Biolek et al. [41], which is given by Equation (5):

$$\frac{1}{C_M(t)} = \frac{1}{C_0} + k'\int_{0}^{t} q(\tau)\tau \tag{5}$$

where C_0 corresponds to the initial capacitance and k' is the mobility factor.

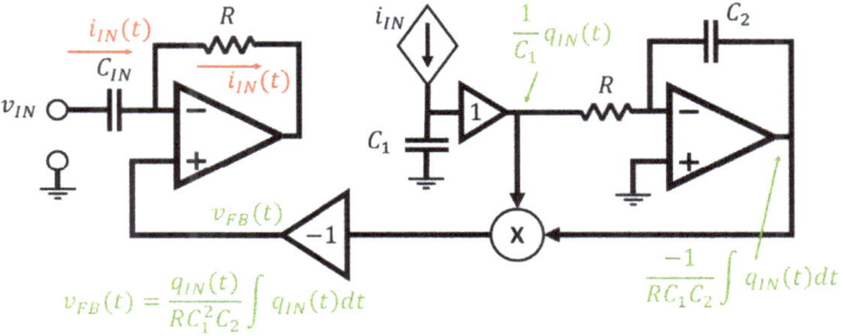

Figure 2. Memcapacitor emulator circuit proposed by Fouda and Radwan [40].

This emulator is designed to achieve the behavior indicated in Equation (5) from the input current of the circuit, then:

$$v_{IN}(t) = \frac{1}{C_{IN}} \int i_{IN}(t)dt + v_{FB}(t) = \frac{q(t)}{C_{IN}} + v_{FB}(t) = \frac{q(t)}{C_{IN}} + k'q(t) \int_0^t q(\tau)d\tau$$
$$= \frac{q(t)}{C_{IN}} + \frac{q(t)}{RC_2C_1^2} \int_0^t q(\tau)d\tau \tag{6}$$

Note that this circuit requires implementing a copy of the injected current in order to obtain the input charge and its integration; besides, it is limited for the emulation of grounded memcapacitors. The circuit of Figure 2 was simulated using SPICE, demonstrating that it certainly behaves as a charge-controlled memcapacitor for a frequency of 10 Hz resulting in a good agreement with the mathematical derivation. However, there is a lack of physical implementation of this design demonstrating its actual performance.

A similar approach, but without the drawback of requiring a copy of the input current, was proposed by Sah et al. in Ref. [42] and it is presented in the circuit of Figure 3 which, following the same principle than the previous design, can be modelled as follows:

$$v_{IN}(t) = \frac{1}{C_1} \int i_{IN}(t)dt - v_{FB}(t) = \frac{q(t)}{C_1} - v_{FB}(t) = \frac{q(t)}{C_1} - k'q(t) \int_0^t q(\tau)d\tau$$
$$= \frac{q(t)}{C_1} + \frac{q(t)}{RC_2C_1^2} \int_0^t q(\tau)d\tau \tag{7}$$

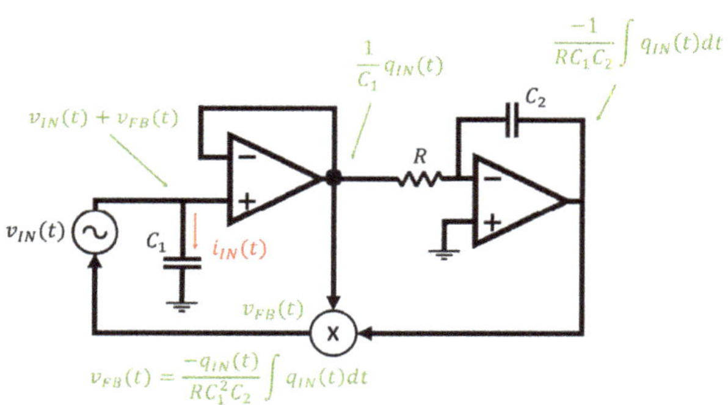

Figure 3. Memcapacitor emulator circuit proposed by Sah et al. [42].

Their authors validated this proposal through both SPICE simulations and experimental results, demonstrating that this model is able to emulate a charge-controlled memcapacitor at input frequencies ranging from 0.1 Hz to 25 Hz. Hence, this circuit was able to provide a similar behavior to the previous one with a simplified design.

Another alternative to emulate grounded memcapacitors was proposed by Romero et al. in Ref. [43], although in this case for voltage-controlled memcapacitors. This emulator was implemented by relating the memcapacitance concept with the Miller effect, which accounts for the amplification of the feedback capacitance in inverting voltage amplifiers Equation (8).

$$Z_{IN} = \frac{V_{IN}}{I_{IN}} = \frac{V_{IN}}{j\omega C_1(V_{IN} - V_{OUT})} = \frac{1}{j\omega C_1(1 + A)} = \frac{1}{j\omega C_{IN}} \tag{8}$$

On the basis of Equation (8), the authors proposed a gain, A, which depends on the time-integral of the input voltage (i.e., the flux). To do so, they used a voltage-controlled resistor, as shown in Figure 4a, to change the amplifier's voltage gain according to the flux, hence satisfying the definition of the voltage-controlled memcapacitor, as derived in Equation (9).

$$\frac{d\sigma_{IN}}{dt} = q_{IN}(t) = \int i_{IN}(t)dt$$
$$= \int C_1 \frac{dv_{C_1}(t)}{dt}dt = C_1(v_{IN}(t) - v_{out}(t)) = C_1(1 + A(\phi))v_{in}(t) \qquad (9)$$
$$= C_M(\phi)v_{IN}(t)$$

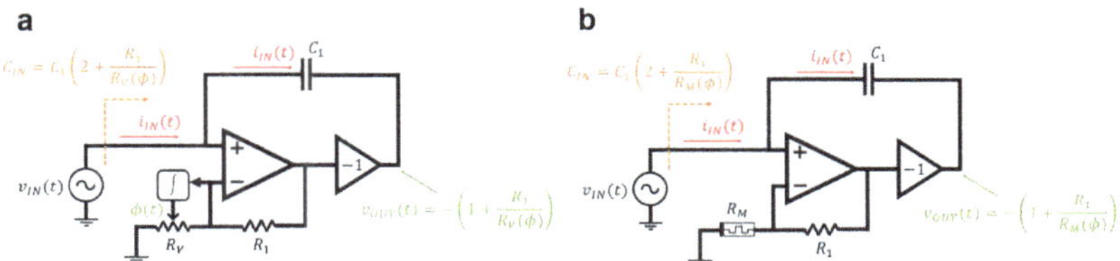

Figure 4. Memcapacitor emulator circuit proposed by Romero et al. [43] implemented with a voltage-controlled resistor (**a**) and with a voltage-controlled memristor (**b**).

In the case of the implementation shown in Figure 4a, the resulting memcapacitance is given by:

$$C_M(\phi) = C_1\left(2 + \frac{R_1}{R_V(\phi)}\right) \qquad (10)$$

Additionally, having a voltage-controlled resistor whose value changes according to the input flux (which is actually the time-integral of its input) makes also feasible the implementation of this circuit by means of a memristor, as depicted in Figure 4b. In this case, the memcapacitance could be expressed as indicated in Equation (11). Circuits such as this one are considered as electrical mutators since, according to Equation (9) and Equation (11), they transform the constitutive equation of the memristor ($R_M = \frac{d\phi}{dq}$) into a memcapacitor with its own constitutive relation ($C_M = \frac{d\sigma}{d\phi}$).

$$C_M(\phi) = C_1\left(2 + \frac{R_1}{R_M(\phi)}\right) \qquad (11)$$

The feasibility of this implementation was demonstrated by SPICE simulations for different input waveforms at a frequency of 50 Hz, as well as by means of its physical implementation in a field-programmable analog array (FPAA) using a controlled-gain amplifier.

Actually, the use of mutators is a common approach for the implementation of memcapacitor emulators. Another example of this kind is the design proposed by Wang et al. in Ref. [44] to emulate voltage-controlled memcapacitors. In this work, the authors relied on the use of two commercially available second-generation current conveyors (CCII) AD844 in combination with a memristor, as shown in Figure 5a.

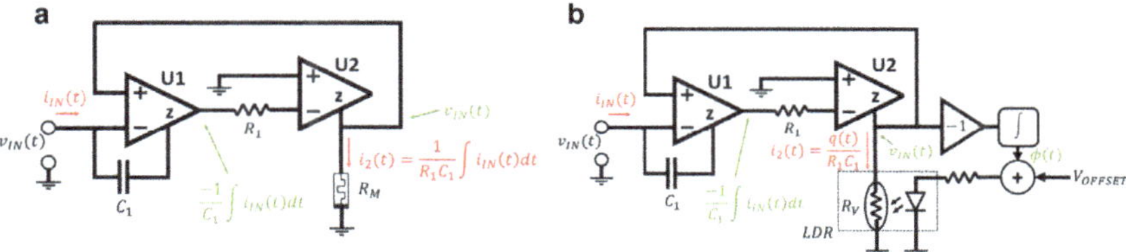

Figure 5. (**a**) Memcapacitor emulator circuit proposed by Wang et al. [44] implemented with a voltage-controlled memristor. (**b**) Memcapacitor emulator schematic using a memristor emulator based on a LED optically coupled to a LDR (light-dependent resistor).

In this circuit, the capacitor C_1 and the first CCII are used to obtain a voltage proportional to the integration of the input current (i.e., proportional to the charge). After that, the second CCII allows to convert that voltage to current, given that $I_Z = -I_-$ and $V_- = V_+$:

$$i_2(t) = I_{Z_2} = \frac{-v_{out_1}}{R_1} = \frac{1}{R_1 C_1} \int_{t_0}^{t} i_{IN}(\tau)d\tau = \frac{q(t)}{R_1 C_1} \tag{12}$$

Therefore, the relation between the current and the voltage across the memristor can be expressed as follows:

$$R_M(\phi) = \frac{v_{IN}(t)}{i_2} = \frac{v_{IN}(t)R_1 C_1}{q(t)} \tag{13}$$

As seen, from the constitutive equation of the memristor we can get the equivalent memcapacitance of this circuit, which is given by Equation (14).

$$C_M(\phi) = \frac{R_1 C_1}{R_M(\phi)} \tag{14}$$

Moreover, the authors presented in this work a novel approach for dealing with both voltage-dependent resistors and/or voltage-controlled memristors (see Figure 5b). This approach is based on a LED optically coupled with a LDR (light-dependent resistor) and, as it will be shown later, it has been adopted for other authors for the implementation of their emulators. However, it is important to highlight that this approach limits the upper frequency of the emulator, since the LDRs usually suffer from a slow time-response.

The use of current conveyors to implement mutators was theoretically introduced by Pershin and Di Ventra in Ref. [45], and since then it has been adopted by many authors in the literature. One of the benefits of using current conveyors relies on the possibility to implement floating memcapacitors, as shown in Figure 6.

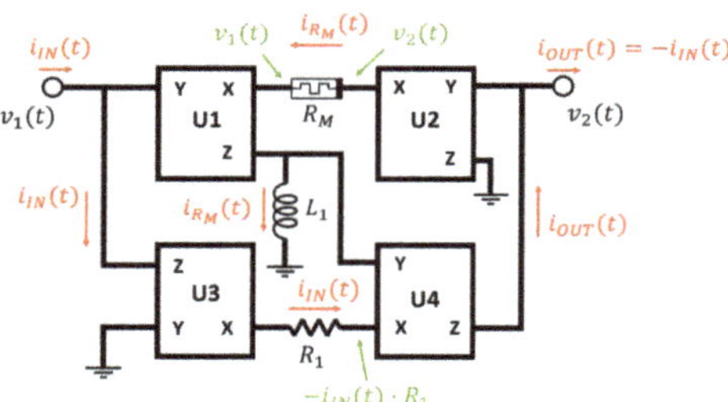

Figure 6. Memcapacitor emulator circuit proposed by Pershin and Di Ventra [45].

In the circuit of Figure 6, the current through the memristor corresponds to the current through the inductor L_1, and therefore:

$$v_{L1}(t) = -i_{IN}(t)\cdot R_1 = L_1 \cdot \frac{d\left(i_{R_M}(t)\right)}{dt} = \frac{L_1}{R_M(\phi)} \cdot \frac{d(v_2(t) - v_1(t))}{dt} = \frac{-L_1}{R_M(\phi)} \cdot \frac{d(v_{IN}(t))}{dt} \tag{15}$$

which indicates that this circuit emulates a voltage-controlled memcapacitor whose memcapacitance is given by Equation (16).

$$C_M(\phi) = \frac{L_1}{R_1 R_M(\phi)} \tag{16}$$

A similar approach to the one proposed in this work was followed by Yu et al. for the implementation of a practical emulator based on this model [46]. However, their proposal presents the drawback of requiring the use of a custom implementation of memristor emulator, which does not guarantee the equality between the input and output current of its two terminals.

There are additional works that also make use of current conveyors for the practical implementation of emulators without the requirement of including any memristor or memristor emulator. This is the case of the grounded memcapacitor emulator presented by Yesil and Babacan in Ref. [47] and schematized in Figure 7.

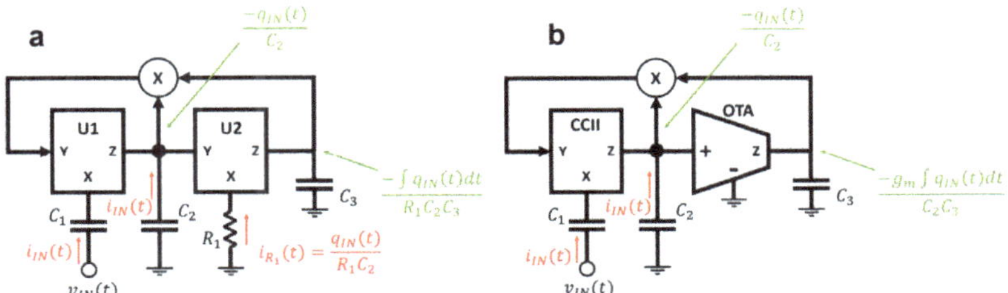

Figure 7. Memcapacitor emulator circuit proposed by Yesil and Babacan [47]. (**a**) Implementation based on only CCII, (**b**) implementation based on a CCII-OTA combination.

In this emulator, the memcapacitance can be derived from the input current, which can be expressed as:

$$i_{IN}(t) = C_1 \frac{d\left(v_{IN}(t) - \frac{q(t)}{R_1 C_2^2 C_3} \int q(t)dt\right)}{dt} \tag{17}$$

and, therefore, the equivalent charge-controlled memcapacitance corresponds to the following expression (see Equation (5)):

$$\frac{1}{C_M(q)} = \frac{1}{C_1} + \frac{1}{R_1 C_2^2 C_3} \int_{t_0}^{t} q(t)dt \tag{18}$$

Moreover, this emulator could also be implemented by replacing the second CCII with an operational transconductance amplifier (OTA), as shown in Figure 7b. In that case, the resulting memcapacitance would be given by:

$$\frac{1}{C_M(q)} = \frac{1}{C_1} + \frac{g_m}{C_2^2 C_3} \int_{t_0}^{t} q(t)dt \tag{19}$$

where g_m is the OTA's transconductance gain.

The experimental results obtained using off-the-shelf components demonstrated that the circuits of Figure 7 was able to emulate a grounded charge-controlled memcapacitor at frequencies up to 48 Hz.

The emulation of mem-elements using OTAs-based circuits is also common in the literature. For instance, in Ref. [48] Vista and Ranjan presented a memcapacitor emulator using a dual X current conveyor differential input transconductance amplifier (DXCCDITA).

Their emulator is based on a DXCCDITA modeled as indicated in Figure 8.

On this basis, the memcapacitance can be derived from the voltage at the three different passive elements, R_1, C_1 and C_2 as:

$$\begin{cases} v_{C_1} = v_{Z_-} = \frac{1}{C_1} \int_{t_0}^{t} i_{C_1}(t)dt = \frac{1}{C_1} \int_{t_0}^{t} i_{Z_-}(t)dt = \frac{\alpha}{C_1} \int_{t_0}^{t} i_{X_-}(t)dt = \frac{\alpha q_{IN}(t)}{C_1} \\ v_{C_2}(t) = V_{O_+}(t) = V_{B_{O_-}}(t) = \frac{1}{C_2} \int_{t_0}^{t} i_{O_+}(t)dt = \frac{g_m}{C_1} \int_{t_0}^{t} V_{Z_-}(t)dt = \frac{\alpha g_m}{C_2 C_1} \int_{t_0}^{t} q_{IN}(t)dt \\ V_Y(t) = \frac{V_{X_+}(t)}{\beta} = \frac{-V_{X_-}(t)}{\beta} = V_{O_-}(t) = I_{O_-}(t)R_1 = -g_m V_{Z_-}(t)R_1 = \frac{-g_m \alpha q_{IN}(t)R_1}{C_1} \end{cases} \tag{20}$$

being α and β the current transfer gain and voltage transfer gain, respectively. On the other hand, the transconductance (g_m) can be expressed as $g_m = K\left(V_{B_{O_-}} + V_{DD} - V_t\right)$, where

V_{DD} is the positive supply voltage and both V_t and K are parameters that depend on the CMOS technology used.

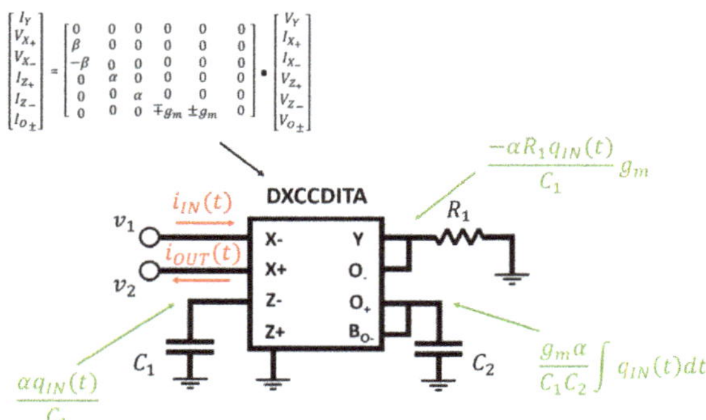

Figure 8. Memcapacitor emulator circuit proposed by Vista and Ranjan [48] based on a DXCCDITA.

Therefore, the constitutive equation of this charge-controlled memcapacitor can be obtained as:

$$v_{in}(t) = V_{X_-}(t) - V_{X_+}(t) = -2\beta V_Y = \frac{2\alpha\beta KR_1}{C_1} \cdot q(t) \cdot \left(V_{DD} - V_T + \frac{\alpha g_m}{C_1 C_2} \int_{t_0}^{t} q_{IN}(t) dt \right) \tag{21}$$

Hence, the charge-controlled memcapacitance is given by:

$$\frac{1}{C_M(q)} = \frac{2\alpha\beta KR_1}{C_1} \cdot \left(V_{DD} - V_T + \frac{\alpha g_m}{C_1 C_2} \int_{t_0}^{t} q_{IN}(t) dt \right) \tag{22}$$

The feasibility of this floating charge-controlled memcapacitor model has been verified by means of SPICE simulation, and additionally, the practicability of this model is examined in an adaptative neuromorphic structure [48].

Finally, a brief comparison of the different memcapacitor emulators presented in this section is summarized in Table 1. The comparison has been carried out in terms of their key components and mode of operation (grounded or floating), among other parameters.

Table 1. Comparison of the different memcapacitor emulators presented in this review.

Reference	Mutator	Configuration	Control Variable	Key Components	Experimental
Fouda and Radwan [40]	No	Grounded	Charge	Op amps Analog multiplier Copy of the input current	No
Sah et al. [42]	No	Grounded	Charge	Op amps Analog multiplier	Yes
Romero et al. [43]	Yes	Grounded	Voltage	Op amps Memristor [1]	Yes
Wang et al. [44]	Yes	Grounded	Voltage	Current conveyors Memristor [1]	Yes
Pershin and Di Ventra [45]	Yes	Floating	Voltage	Current conveyors Inductor Memristor	No
Yesil and Babacan [47]	No	Grounded	Charge	Current conveyor OTA Analog multiplier	Yes
Vista and Ranjan [48]	No	Floating	Charge	Custom DXCCDITA	No

[1] Or memristor emulator (applicable in all cases).

3. Meminductor Emulators

The meminductance (L_M) is defined as the nth-order system that establishes a non-linear relation between the current across the terminal of the device (I) and its input flux (ϕ) [34]. It can be either current-controlled or flux-controlled depending on its constitutive input variable. Therefore, the nth-order current-controlled meminductive systems are defined by Equation (23), whereas the flux-controlled ones are defined by Equation (24).

$$\phi(t) = L_M\left(\overrightarrow{x_N},\, I, t\right) \cdot I(t) \tag{23}$$

$$I(t) = L_M^{-1}\left(\overrightarrow{x_N},\, \phi, t\right) \cdot \phi(t) \tag{24}$$

being $\overrightarrow{x_N}$ a vector which represents the n internal state variables of the system.

The meminductor is a particular case of meminductive system with one single state variable; the current in the case of current-controlled meminductors (Equation (25)) or the flux in the case of flux-controlled meminductors (Equation (26)):

$$\phi(t) = L_M\left[\int_{t_0}^{t} I(\tau)d\tau\right] \cdot I(t) \tag{25}$$

$$I(t) = L_M^{-1}\left[\int_{t_0}^{t} \phi(\tau)d\tau\right] \cdot \phi(t) \tag{26}$$

where the initial instant of time, t_0, may be selected to ensure that $\int_{-\infty}^{t_0} I(\tau)d\tau = 0$ and $\int_{-\infty}^{t_0} \phi(\tau)d\tau = 0$, respectively.

Therefore, the meminductance of meminductors depends on either the current or the flux depending on whether they are current-controlled or flux-controlled, respectively. In addition, their i-ϕ characteristic presents a closed-pinched hysteresis loop in which $i = 0$ whenever $\phi = 0$ (and vice versa) for bipolar sine wave-like excitations. The usual approaches followed to implement meminductors emulators are quite similar to those used to emulate memcapacitors. One of these common approaches employs mutators in order to transform memristors into meminductors in both grounded and floating configurations. This is the case of the grounded meminductor shown in Figure 9, which was proposed by Wang in Ref. [49].

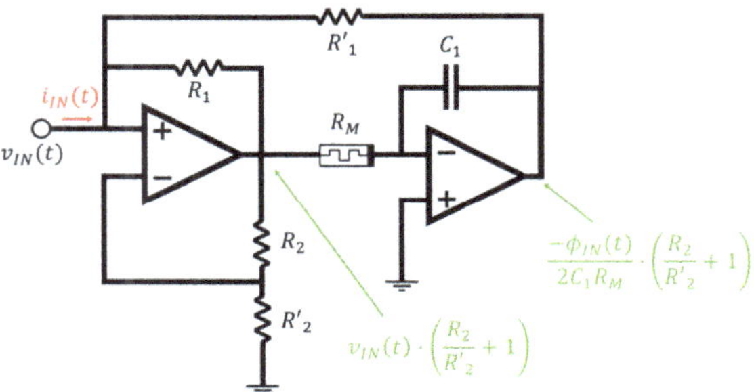

Figure 9. Meminductor emulator circuit proposed by Wang [49].

In this circuit, the input current can be expressed as follows:

$$i_{IN}(t) = i_{R_1} + i_{R'_1} = v_{IN}(t)\cdot\left(\frac{1}{R'_1} - \frac{R_2}{R_1 R'_2}\right) + \frac{\phi_{IN}(t)}{2R_2 R_M(\phi)C_1}\cdot\left(\frac{R_2}{R'_2}+1\right) \tag{27}$$

As seen, Equation (27) can be directly related to the constitutive equation of a flux-controlled meminductor with the condition of cancelling the term associated with the input voltage, i.e., with $R'_1 = R_1$ and $R'_2 = R_2$. In that case, the resulting input current can be expressed as indicated in Equation (28) and, therefore, the circuit would emulate the behavior of a flux-controlled meminductance modelled by Equation (29).

$$i_{IN}(t) = \frac{\phi_{IN}(t)}{R_2 R_M(\phi) C_1} \tag{28}$$

$$L_M(\phi) = R_2 R_M(\phi) C_1 \tag{29}$$

This simple model was verified by means of simulations; however, it was studied neither in the frequency-domain nor with an experimental implementation.

Another example of mutator, based on a gyrator, was presented by Romero et al. upon the design of the Antoniou's circuit, as depicted in Figure 10 [35].

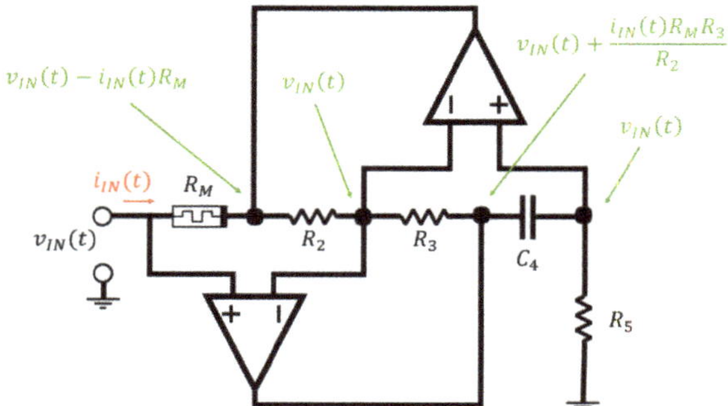

Figure 10. Grounded meminductor emulator circuit proposed by Romero et al. [35].

In this case, the meminductance can be derived from the current through R_5, given that $i_{R_5} = i_{C_4}$. Therefore:

$$\frac{v_{IN(t)}}{R_5} = \frac{C_4 R_M R_3}{R_2} \cdot \frac{d(i_t(t))}{dt} \rightarrow i_{IN}(t) = \phi_{IN}(t) \cdot \frac{R_2}{R_M(\phi) R_3 R_5 C_4} \tag{30}$$

which indicates that the circuit behaves as a flux-controlled meminductor whose value is given by Equation (31).

$$L_M(\phi) = \frac{R_M(\phi) R_3 R_5 C_4}{R_2} \tag{31}$$

This circuit was validated using SPICE simulations for various input signals and frequencies. For the simulations, the memristor was implemented by means of a LDR, as shown in previous implementations. In addition, the practicability of the meminductor model was also exhibited with a long-term potentiation (LTP) and long-term depression (LTD) example [35]. However, this circuit also presents the disadvantage of being restricted to grounded configurations.

Following the same approach, Romero et al. also presented a floating meminductor emulator based on the Riordan gyrator. In this case, the meminductor emulator is based on the schematic shown in Figure 11.

In order to emulate a floating meminductor, the input current at the first terminal must be equal to the output current of terminal two, therefore:

$$\boldsymbol{I}_{IN} = -\boldsymbol{I}_{OUT} = \boldsymbol{V}_{IN} \cdot \frac{Z_2 Z_4}{Z_5 Z_M Z_1} = -\boldsymbol{V}_{IN} \cdot \left(\frac{Z_7}{Z_8 Z_6} + \frac{Z_2 Z_4 Z_7}{Z_M Z_5 Z_6 Z_8} - \frac{1}{Z_5} \right) \tag{32}$$

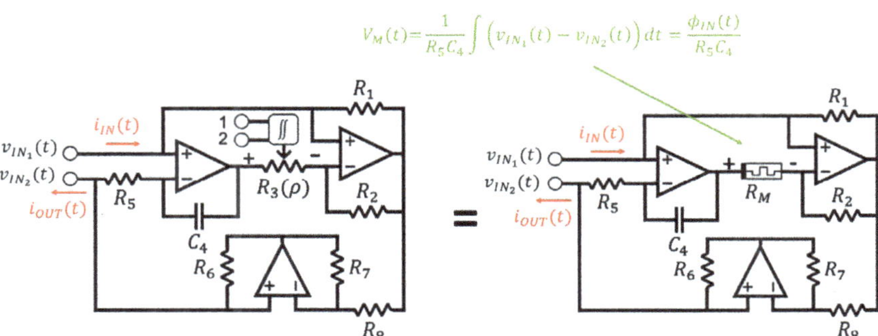

Figure 11. Floating meminductor emulator circuit proposed by Romero et al. [50].

Thus, the circuit of Figure 11 needs to fulfill the following condition:

$$\frac{1}{Z_5} = \frac{1}{Z_1} = \frac{Z_7}{Z_6 Z_8} \tag{33}$$

where Z_i represents the impedance of the passive element i. On this basis, considering $R_1 = R_2 = R_5 = R_7 = R_6 = R_8 = R$, Equation (32) can be expressed as given in Equation (34).

$$I_{IN} = \frac{V_{IN}}{s} \cdot \frac{1}{RC_4 R_M(\phi(s))} \tag{34}$$

Finally, the constitutive equation of this floating meminductor emulator can be obtained by transforming Equation (34) to the time domain:

$$i_{IN}(t) = \phi_{IN}(t) \cdot \frac{1}{RC_4 R_M(\phi)} = \phi_{IN}(t) \cdot \frac{1}{L_M(\phi)} \tag{35}$$

Therefore, with this implementation we can avoid the drawback of being subject to grounded configurations when implementing a meminductor emulator. The feasibility of this circuit was proved by a practical implementation, besides, an example of application in which the emulator is used in an adaptative low-pass filter was also shown.

As in the case of memcapacitor emulators, some authors also rely on the use of current conveyors for the implementation of their emulators. An example of this practice is the model proposed by Sah et al. in Ref. [51], whose schematic is shown in Figure 12.

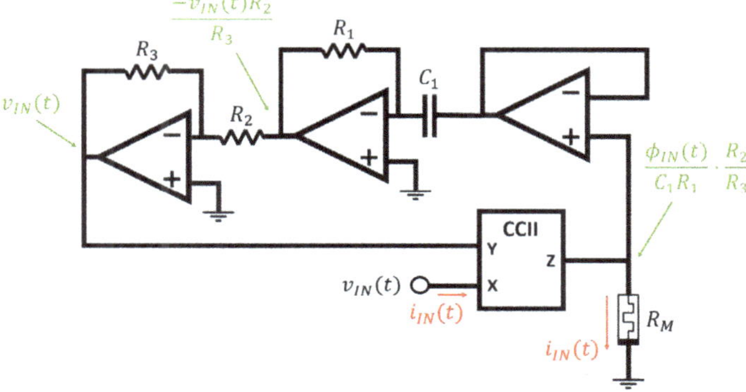

Figure 12. Meminductor emulator circuit proposed by Sah et al. [51].

In this circuit, the current relations $i_{C_1} = i_{R_1}$ and $i_{R_2} = i_{R3}$ allow extracting the constitutive equation of the equivalent flux-controlled meminductor:

$$i_{IN}(t) = \phi_{IN}(t) \cdot \frac{R_2}{C_1 R_1 R_3 R_M(\phi)} \tag{36}$$

which results in the following meminductance:

$$L_M(\phi) = \frac{C_1 R_1 R_3 R_M(\phi)}{R_2} \tag{37}$$

as it was demonstrated by means of both SPICE and experimental results for different input frequencies. Alternatively, in Ref [52] the same authors presented an equivalent circuit based on two current conveyors (see Figure 13).

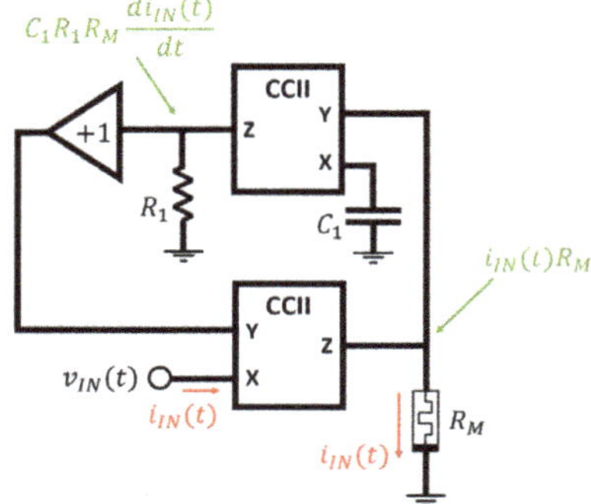

Figure 13. Meminductor emulator circuit based on current conveyors proposed by Sah et al. [52].

In this emulator, the meminductance can be derived from the relation between the current passing through the different passive elements, resistor, capacitor and memristor:

$$i_{C_1} = i_{R_1} = C_1 R_M \frac{di_{IN}(t)}{dt} = \frac{v_{IN}(t)}{R_1} \tag{38}$$

Therefore, the flux-controlled meminductance is given by Equation (39), as demonstrated experimentally by the authors.

$$i_{IN}(t) = \frac{\phi_{IN}(t)}{R_1 R_M(\phi)C_1} = \frac{\phi_{IN}(t)}{L_M(\phi)} \tag{39}$$

Another example of mutator based on current conveyors was the circuit proposed by Liang et al. [36] to emulate floating flux-controlled meminductors (Figure 14).

As it is shown, the equivalent input meminductance of this mutator can be extracted from the current through the memristor:

$$i_{R_M}(t) = i_{R_2}(t) = \frac{\phi_{IN}(t)}{R_1 C_1 R_M} = \frac{i_{IN}(t)R_3}{R_2} \tag{40}$$

Thus, the flux-controlled meminductance can be calculated as:

$$L_M(\phi) = \frac{R_2}{R_1 C_1 R_3 R_M(\phi)} \tag{41}$$

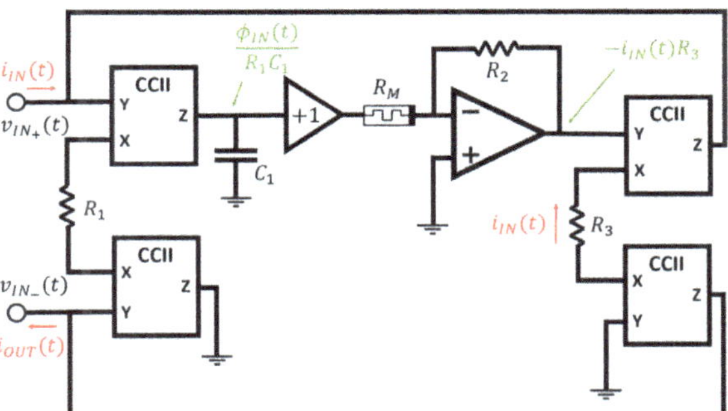

Figure 14. Floating meminductor emulator circuit based on current conveyors proposed by Liang et al. [36].

Contrary to others practical mutators, in this work, the authors opted for the use of an analog multiplier rather than a LDR for the implementation of the memristor with the goal of achieving a better control over its memristance. Their proposed circuit was validated experimentally using a sinusoidal input voltage for two different frequencies, 28.3 Hz and 36.9 Hz. A similar approach to the followed in this latter work was presented in Ref. [53] by the same authors, and by Sozen and Cam in Ref. [54], although in this latter case the authors made use of an OTA instead of a current conveyor to obtain the input flux.

All the meminductor emulators presented so far require the use of either a memristor or a memristor emulator for their practical implementations. An alternative also based on current conveyors, but without the need of implementing a memristor, can be found in Ref. [55], in which Fouda and Radwan proposed the circuit depicted in Figure 15 to emulate grounded current-controlled meminductors.

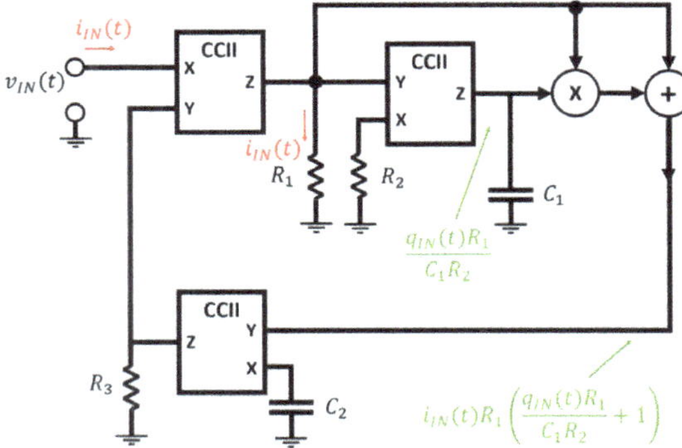

Figure 15. Meminductor emulator circuit based on current conveyors proposed by Fouda and Radwan [55].

This circuit is designed to fulfill the constitutive equation of the current-controlled meminductors as defined in Equation (25) [56]:

$$\phi(t) = (L_0 + kq(t)) \cdot i(t) \tag{42}$$

being L_0 the initial inductance and k the mobility factor.

Therefore, considering that $i_{R_3} = i_{C_2}$, we can obtain:

$$\phi_{IN}(t) = \left(R_1 R_3 C_2 + \frac{R_1^2 R_3 C_2}{C_1 R_2} q_{IN}(t) \right) \cdot i_{IN}(t) \tag{43}$$

By comparing the two previous equations, the current-controlled meminductance can be expressed as indicated in Equation (44), which was demonstrated by SPICE simulations and a circuit implementation at a frequency of 10 Hz.

$$L_M(q) = L_0 + kq_{IN}(t) = R_1 R_3 C_2 + \frac{R_1^2 R_3 C_2}{C_1 R_2} q_{IN}(t) \tag{44}$$

Moreover, as in the case of the memcapacitors emulators, some authors resorted to the use of custom CMOS-based circuits to implement memristor-less meminductor emulators. Some examples of these circuits are the works presented by Konal and Kacar in Ref. [57], where the authors proposed a CMOS realization of multi-output OTAs for the emulation of grounded meminductors; or the work presented by Vistan and Ranjan in Ref. [58], where a voltage difference transconductance amplifier (VDTA) implemented with CMOS technology is revealed to be also used for the emulation of grounded meminductors.

Finally, a brief comparison of the different meminductor emulators presented in this section is given in Table 2. The comparison has been carried out in terms of their key components and mode of operation (grounded or floating), among other parameters.

Table 2. Comparison of the different meminductor emulators presented in this work.

Reference	Mutator	Configuration	Control Variable	Key Components	Experimental
Wang [49]	Yes	Grounded	Flux	Op amps, Memristor [1]	No
Romero et al. [35]	Yes	Grounded	Flux	Op amps, Memristor	No
Romero et al. [50]	Yes	Floating	Flux	Op amps, Memristor	Yes
Sah et al. [51]	Yes	Grounded	Flux	Current conveyor, Op amps, Memristor [1]	Yes
Sha et al. [52]	Yes	Grounded	Flux	Current conveyors, Memristor	No
Liang et al. [36]	Yes	Floating	Flux	Current conveyor, Op amps, Memristor	Yes
Fouda and Radwan [55]	No	Grounded	Current	Current conveyor, Analog multiplier, Adder	No

[1] Or memristor emulator (applicable in all cases).

4. Universal Emulators: Memcapacitors and Meminductor

In this section, we select some of the remarkable circuits available in the literature that are able to emulate either a memcapacitor or a meminductor by minor changes in their structure or by a proper configuration of their passive elements. For instance, the circuits shown in Figure 16a,b were proposed by Babacan for the emulation of memcapacitors and meminductors, respectively [59]. In the first case, the memcapacitance behavior is achieved by the feedback provided by the capacitors connected to the outputs of the OTA:

$$i_{IN}(t) = C_1 \frac{d\left(v_{IN}(t) - \frac{q_{IN}(t) \int q_{IN}(t)dt}{C_2^2} \right)}{dt} \tag{45}$$

and therefore, the equivalent input charge-controlled memcapacitance can be derived as:

$$C_M(q) = \frac{1}{C_1} + \frac{\int_{t_0}^{t} q_{IN}(\tau)d\tau}{C_2^2} \tag{46}$$

Similarly, in the circuit depicted in Figure 16b, the feedback provided in the negative input of the OTA and the combination of the voltage in both R_1 and C_1 allows to express the input voltage as follows:

$$v_{IN}(t) = L_1 \frac{di_{IN}(t)}{dt} + \frac{R_1}{C_1} \cdot \frac{d(i_{IN}(t) \cdot q_{IN}(t))}{dt} \tag{47}$$

and therefore, according to Equation (25), the current-controlled equivalent input meminductance of this circuit corresponds to Equation (48).

$$\phi_{IN}(t) = \left(L_1 + \frac{R_1}{C_1} q_{IN}(t) \right) i_{IN}(t) = L_M(q) i_{IN}(t) \tag{48}$$

The mutation of memristive systems into universal memcapacitive and meminductive emulators have also been considered by some authors, as the case of Taşkiran et al. [60]. In this work, the authors proposed a simple current backward transconductance amplifier (CBTA) to implement a universal mutator based on the scheme exhibited in Figure 17a, whose equivalent input impedance in the Laplace domain can be expressed as:

$$Z_{IN}(s) = \frac{V_{IN}}{I_{IN}} = \frac{Z_W}{Z_Z} \cdot \frac{1}{\mu_W g_m \alpha} \tag{49}$$

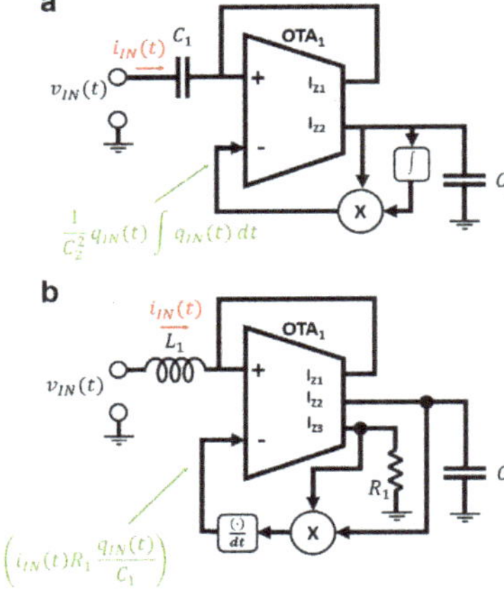

Figure 16. Memcapacitor (**a**) and meminductor emulator (**b**) circuits proposed by Babacan [59].

Thus, by means of the substitutions shown in Figure 17b,c, the circuit of Figure 17a can be used to emulate either a memcapacitor or a meminductor, respectively. In that case, the equivalent input memcapacitance and meminductance can be derived as follows:

$$Z_{IN_{MC}}(s) = \frac{1}{R_M(\phi(s))C_1 s} \cdot \frac{1}{\mu_W g_m \alpha} \rightarrow C_M(\phi) = R_M(\phi)C_1 \mu_W g_m \alpha \tag{50}$$

$$Z_{IN_{MI}}(s) = \frac{R_M(\phi(s))C_1 s}{\mu_W g_m \alpha} \rightarrow L_M(\phi) = \frac{R_M(\phi)C_1}{\mu_W g_m \alpha} \tag{51}$$

where g_m, μ_W, and α are the transconductance gain and both voltage and current gains, respectively.

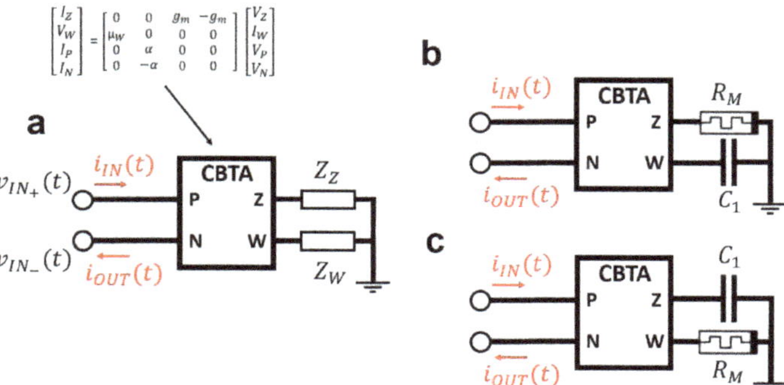

Figure 17. (**a**) CBTA-based circuit for the emulation of floating memcapacitors and meminductors devices proposed by Taşkiran et al. [60]. (**b**) Mutator for the emulation of memcapacitors, (**c**) mutator for the emulation of meminductors.

Similarly, Yu et al. [61] proposed an universal mutator based on commercial current conveyors for emulating grounded mem-elements (Figure 18a). As in the case of the circuit proposed by in Ref. [60], this mutator can be used to achieve a straightforward transformation between a memristor and either a memcapacitor or a meminductor by just modifying the combination of its different impedances. In both cases, either the memcapacitor (Figure 18b) or the meminductor (Figure 18c), the constitutive equation of the emulated device can be derived from the relation between the current and the voltage in Z_1. Thus, for the memcapacitive ciruit:

$$q_{IN}(t) = \frac{C_1}{R_2 R_3 R_4 R_M(\phi)} v_{IN}(t) = C_M(\phi) \cdot v_{IN}(t) \tag{52}$$

whereas for the meminductive circuit:

$$i_{IN}(t) = \frac{R_4}{R_1 R_2 R_M(\phi) C_3} \phi_{IN}(t) = L_M^{-1}(\phi) \cdot \phi_{IN}(t) \tag{53}$$

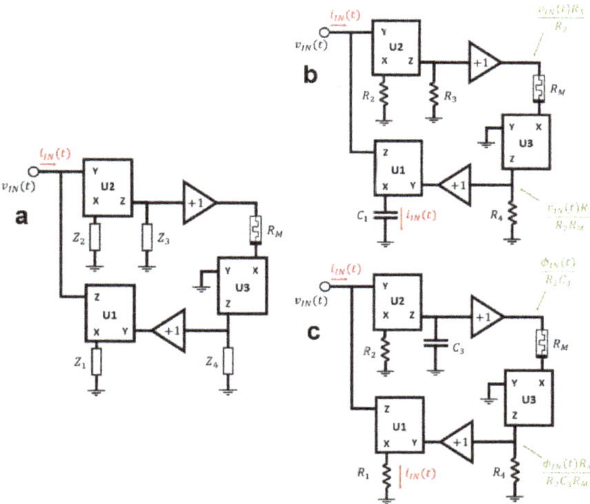

Figure 18. (**a**) Universal circuit for the emulation of grounded mem-elements proposed by Yu et al. [61]. (**b**) Mutator for the emulation of memcapacitors, (**c**) mutator for the emulation of meminductors.

Recently, Yu et al. [62] revisited this circuit aiming to emulate not only grounded mem-elements but also their floating configurations, with the additional advantage of avoiding the inclusion of a memristor (or its emulator) for its implementation. The behavior of these circuits is derived from the relation between the current through the resistor R_2 and the varactor diode C_{VD}, given that $i_{R_2} = i_{C_{VD}}$. On this basis, the memcapacitance of the circuit displayed in Figure 19a can be extracted as:

$$i_{R_2}(t) = \frac{q_{IN}(t)}{C_1 R_2} = C_{VD}(\phi) \frac{d\left(\frac{\phi_{IN}(t)}{R_1 C_2} - V_{OFFSET}\right)}{dt} = C_{VD}(\phi) \frac{v_{IN}(t)}{R_1 C_2}$$
$$\rightarrow \quad C_M(\phi) = \frac{C_1 R_2 C_{VD}(\phi)}{R_1 C_2} \tag{54}$$

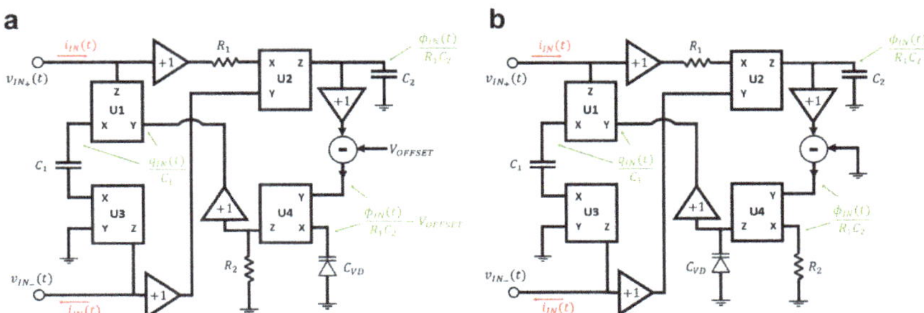

Figure 19. Universal circuit proposed by Yu et al. [62] for the emulation of floating memcapacitors (**a**), and floating meminductor (**b**).

In the same way, the equivalent meminductance of the circuit shown in Figure 19b can be expressed as:

$$i_{VD}(t) = \frac{C_{VD}(q)}{C_1} \cdot \frac{dq_{IN}(t)}{dt} = \frac{C_{VD}(q)}{C_1} i_{IN}(t) = \frac{\phi_{IN}(t)}{R_1 R_2 C_2}$$
$$\rightarrow \quad L_M(q) = \frac{R_1 R_2 C_2 C_{VD}(q)}{C_1} \tag{55}$$

The feasibility of these circuits has been proved by means of experimental results for sinusoidal input signals and in a wide range of frequencies (up to 22 kHz).

A similar circuit was proposed recently by Zhao et al. [63] as an alternative of this latter emulator. The circuit presented by Zhao et al., shown in Figure 20, makes use of an additional current conveyor and an analog multiplier in order to avoid the inclusion of a varactor diode, thus also escaping from the necessity of an external offset voltage. In both cases, memcapacitor emulator (Figure 20a) and meminductor emulator (Figure 20b), the constitutive equations can be extracted relating the voltage at the output terminal Z of both current conveyors, U3 and U4. Therefore, for the memcapacitor emulator we can write:

$$v_{IN}(t) = v_{IN+} - v_{IN-} = q_{IN}(t) \cdot \left(\frac{R_4}{C_0 R_5} - \frac{R_2}{C_0 R_3} + \frac{R_2}{C_0^2 C_1 R_1 R_3} \int_{t_0}^{t} q(\tau) d\tau\right) \tag{56}$$

while for the meminductor emulator:

$$\phi_{IN}(t) = \phi_{IN+} - \phi_{IN-} = i_{IN}(t) \cdot \left(R_0 R_4 C_2 - R_0 R_2 C_0 + \frac{R_0^2 R_2 C_0}{R_1 C_1} q_{IN}(t)\right) \tag{57}$$

Therefore, the equivalent charge-controlled memcapacitance and the current-controlled meminductance of these circuits can be expressed as indicated in Equation (58) and Equation (59), respectively.

$$C_M^{-1}(q) = \frac{R_4}{C_0 R_5} - \frac{R_2}{C_0 R_3} + \frac{R_2}{C_0^2 C_1 R_1 R_3} \int_{t_0}^{t} q(\tau)d\tau \tag{58}$$

$$L_M(q) = R_0 R_4 C_2 - R_0 R_2 C_0 + \frac{R_0^2 R_2 C_0}{R_1 C_1} q_{IN}(t) \tag{59}$$

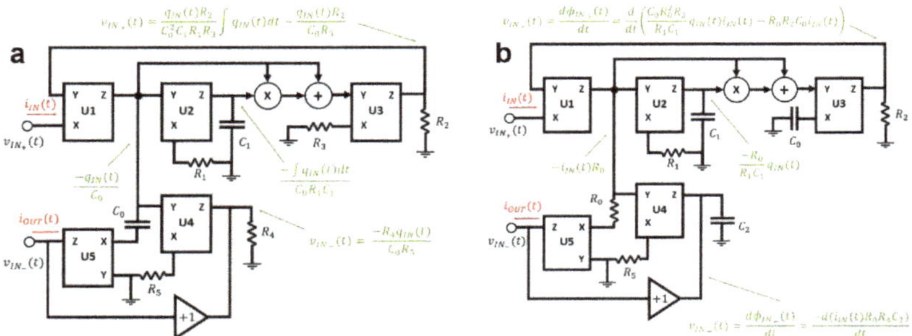

Figure 20. Universal circuit proposed by Zhao et al. [63] for the emulation of floating memcapacitors (**a**), and floating meminductors (**b**).

To sum up, Table 3 presents a brief comparison of the main features of the emulators presented in this section for both memcapacitor and meminductor configurations.

Table 3. Comparison among of the different universal emulators cited in this review.

Reference	Emulator	Mutator	Conf.	Control Variable	Key Components	Experimental
Babacan [59]	Memcap.	No	Grounded	Charge	OTA Integrator Differentiator Multiplier	No
	Memind.			Current		
Taşkiran et al. [60]	Memcap.	Yes	Floating	Voltage	Custom CBTA Memristor [1]	No
	Memind.			Flux		
Yu et al. [61]	Memcap.	Yes	Grounded	Voltage	Current Conveyor Memristor	Yes
	Memind.			Flux		
Yu et al. [62]	Memcap.	No	Floating	Voltage	Current Conveyor Varactor diode Subtractor	Yes
	Memind.			Flux		
Zhao et al. [63]	Memcap.	No	Floating	Charge	Current Conveyor Multiplier Adder	Yes
	Memind.			Current		

[1] Or memristor emulator (applicable in all cases).

5. Conclusions

In this work, different approaches proposed in the literature for the emulation of memcapacitors and meminductors are reviewed in detail. The selected emulator circuits have been theoretically analyzed to infer their constitutive equations and their equivalent memcapacitance or meminductance. It has been reported that most of the emulators presented in the literature are based on mutators, i.e., circuits that transform the constitutive equation of memristors into the corresponding constitutive equation of the emulated device.

Moreover, there are also a set of emulators that does not require the use of a memristor (or its emulator) for their implementation, providing a reliable and simpler alternative to emulate mem-elements. The main features of the analyzed mem-elements emulators have been gathered in three tables to offer a complete overview of the technological options. So that, we firmly consider that this study provides a useful guide for those researchers trying to choose the appropriate emulator restricted by the requirements and constraints of their practical implementations.

Author Contributions: Conceptualization, F.J.R., A.O., A.G., D.P.M. and N.R.; methodology, A.O., D.P.M. and N.R.; formal analysis, F.J.R. and A.O.; investigation, F.J.R. and A.T.-L.; writing—original draft preparation, F.J.R., A.O. and N.R.; writing—review and editing, A.T.-L., A.G. and D.P.M.; funding acquisition, A.O., D.P.M. and N.R. All authors have read and agreed to the published version of the manuscript.

Funding: This research was funded by the Japanese KAKENHI through Grant Number JP18k04275 and Spanish Ministry of Education, Culture, and Sport (MECD), through Project TEC2017-89955-P and Grant Numbers: FPU16/01451 and FPU16/04043.

Conflicts of Interest: The authors declare no conflict of interest.

References

1. Chua, L. Memristor-The Missing Circuit Element. *IEEE Trans. Circuit Theory* **1971**, *18*, 507–519. [CrossRef]
2. Chua, L.O.; Kang, S.M. Memristive Devices and Systems. *Proc. IEEE* **1976**, *64*, 209–223. [CrossRef]
3. Chua, L.O.; Kocarev, L.; Eckert, K.; Itoh, M. Experimental Chaos Synchronization in Chua's Circuit. *Int. J. Bifurc. Chaos* **1992**, *2*, 705–708. [CrossRef]
4. Chua, L.O. State Space Theory of Nonlinear Two- Terminal Higher-Order Elements. *J. Frankl. Inst.* **1983**, *316*, 1–50. [CrossRef]
5. Chua, L.O. Chua's Circuit: An Overview Ten Years Later. *J. Circuits Syst. Comput.* **1994**, *4*, 117–159. [CrossRef]
6. Süsse, R.; Domhardt, A.; Reinhard, M. Calculation of Electrical Circuits with Fractional Characteristics of Construction Elements. *Forsch. Ing.* **2005**, *69*, 230–235. [CrossRef]
7. Strukov, D.B.; Snider, G.S.; Stewart, D.R.; Williams, R.S. The Missing Memristor Found. *Nature* **2008**, *453*, 80–83. [CrossRef]
8. Jo, S.H.; Chang, T.; Ebong, I.; Bhadviya, B.B.; Mazumder, P.; Lu, W. Nanoscale Memristor Device as Synapse in Neuromorphic Systems. *Nano Lett.* **2010**, *10*, 1297–1301. [CrossRef] [PubMed]
9. Pershin, Y.V.; Di Ventra, M. Experimental Demonstration of Associative Memory with Memristive Neural Networks. *Neural Netw.* **2010**, *23*, 881–886. [CrossRef]
10. Azghadi, M.R.; Linares-Barranco, B.; Abbott, D.; Leong, P.H.W. A Hybrid CMOS-Memristor Neuromorphic Synapse. *IEEE Trans. Biomed. Circuits Syst.* **2017**, *11*, 434–445. [CrossRef]
11. Prezioso, M.; Merrikh-Bayat, F.; Hoskins, B.D.; Adam, G.C.; Likharev, K.K.; Strukov, D.B. Training and Operation of an Integrated Neuromorphic Network Based on Metal-Oxide Memristors. *Nature* **2015**, *521*, 61–64. [CrossRef] [PubMed]
12. Kozma, R.; Pino, R.E.; Pazienza, G.E. (Eds.) *Advances in Neuromorphic Memristor Science and Applications*; Springer: Dordrecht, The Netherlands, 2012; ISBN 978-94-007-4490-5.
13. Shin, S.; Kim, K.; Kang, S. Memristor Applications for Programmable Analog ICs. *IEEE Trans. Nanotechnol.* **2011**, *10*, 266–274. [CrossRef]
14. Merrikh-Bayat, F.; Shouraki, S.B. Memristor-Based Circuits for Performing Basic Arithmetic Operations. *Procedia Comput. Sci.* **2011**, *3*, 128–132. [CrossRef]
15. Pershin, Y.V.; Ventra, M.D. Practical Approach to Programmable Analog Circuits With Memristors. *IEEE Trans. Circuits Syst. I Regul. Pap.* **2010**, *57*, 1857–1864. [CrossRef]
16. Pershin, Y.V.; Sazonov, E.; Di Ventra, M. Analogue-to-Digital and Digital-to-Analogue Conversion with Memristive Devices. *Electron. Lett.* **2012**, *48*, 73. [CrossRef]
17. Vourkas, I.; Sirakoulis, G.C. Emerging Memristor-Based Logic Circuit Design Approaches: A Review. *IEEE Circuits Syst. Mag.* **2016**, *16*, 15–30. [CrossRef]
18. Chen, S.; Mahmoodi, M.R.; Shi, Y.; Mahata, C.; Yuan, B.; Liang, X.; Wen, C.; Hui, F.; Akinwande, D.; Strukov, D.B.; et al. Wafer-Scale Integration of Two-Dimensional Materials in High-Density Memristive Crossbar Arrays for Artificial Neural Networks. *Nat. Electron.* **2020**, *3*, 638–645. [CrossRef]
19. Yuan, B.; Liang, X.; Zhong, L.; Shi, Y.; Palumbo, F.; Chen, S.; Hui, F.; Jing, X.; Villena, M.A.; Jiang, L.; et al. 150 Nm × 200 Nm Cross-Point Hexagonal Boron Nitride-Based Memristors. *Adv. Electron. Mater.* **2020**, *6*, 1900115. [CrossRef]
20. Zhu, K.; Liang, X.; Yuan, B.; Villena, M.A.; Wen, C.; Wang, T.; Chen, S.; Hui, F.; Shi, Y.; Lanza, M. Graphene–Boron Nitride–Graphene Cross-Point Memristors with Three Stable Resistive States. *ACS Appl. Mater. Interfaces* **2019**, *11*, 37999–38005. [CrossRef]
21. Driscoll, T.; Quinn, J.; Klein, S.; Kim, H.T.; Kim, B.J.; Pershin, Y.V.; Di Ventra, M.; Basov, D.N. Memristive Adaptive Filters. *Appl. Phys. Lett.* **2010**, *97*, 093502. [CrossRef]

22. Buscarino, A.; Fortuna, L.; Frasca, M.; Valentina Gambuzza, L. A Chaotic Circuit Based on Hewlett-Packard Memristor. *Chaos* **2012**, *22*, 023136. [CrossRef] [PubMed]
23. Muthuswamy, B.; Kokate, P.P. Memristor-Based Chaotic Circuits. *IETE Tech. Rev.* **2009**, *26*, 417–429. [CrossRef]
24. Xu, C.; Dong, X.; Jouppi, N.P.; Xie, Y. Design Implications of Memristor-Based RRAM Cross-Point Structures. In Proceedings of the 2011 Design, Automation Test in Europe, Grenoble, France, 14–18 March 2011; pp. 1–6.
25. Secco, J.; Corinto, F.; Sebastian, A. Flux–Charge Memristor Model for Phase Change Memory. *IEEE Trans. Circuits Syst. II Express Briefs* **2018**, *65*, 111–114. [CrossRef]
26. Almurib, H.A.F.; Kumar, T.N.; Lombardi, F. Design and Evaluation of a Memristor-Based Look-up Table for Non-Volatile Field Programmable Gate Arrays. *IET Circuits Devices Syst.* **2016**, *10*, 292–300. [CrossRef]
27. Ting, Y.-H.; Chen, J.-Y.; Huang, C.-W.; Huang, T.-K.; Hsieh, C.-Y.; Wu, W.-W. Observation of Resistive Switching Behavior in Crossbar Core–Shell Ni/NiO Nanowires Memristor. *Small* **2018**, *14*, 1703153. [CrossRef] [PubMed]
28. Miao, F.; Yi, W.; Goldfarb, I.; Yang, J.J.; Zhang, M.-X.; Pickett, M.D.; Strachan, J.P.; Medeiros-Ribeiro, G.; Williams, R.S. Continuous Electrical Tuning of the Chemical Composition of TaOx-Based Memristors. *ACS Nano* **2012**, *6*, 2312–2318. [CrossRef] [PubMed]
29. Chen, Y.; Liu, G.; Wang, C.; Zhang, W.; Li, R.-W.; Wang, L. Polymer Memristor for Information Storage and Neuromorphic Applications. *Mater. Horiz.* **2014**, *1*, 489–506. [CrossRef]
30. Zhang, L.; Gong, T.; Wang, H.; Guo, Z.; Zhang, H. Memristive Devices Based on Emerging Two-Dimensional Materials beyond Graphene. *Nanoscale* **2019**, *11*, 12413–12435. [CrossRef]
31. Romero, F.J.; Toral, A.; Medina-Rull, A.; Moraila-Martinez, C.L.; Morales, D.P.; Ohata, A.; Godoy, A.; Ruiz, F.G.; Rodriguez, N. Resistive Switching in Graphene Oxide. *Front. Mater.* **2020**, *7*. [CrossRef]
32. Romero, F.J.; Toral-Lopez, A.; Ohata, A.; Morales, D.P.; Ruiz, F.G.; Godoy, A.; Rodriguez, N. Laser-Fabricated Reduced Graphene Oxide Memristors. *Nanomaterials* **2019**, *9*, 897. [CrossRef]
33. Sahu, D.P.; Jetty, P.; Jammalamadaka, S.N. Graphene Oxide Based Synaptic Memristor Device for Neuromorphic Computing. *Nanotechnology* **2021**, *32*, 155701. [CrossRef] [PubMed]
34. Ventra, M.D.; Pershin, Y.V.; Chua, L.O. Circuit Elements With Memory: Memristors, Memcapacitors, and Meminductors. *Proc. IEEE* **2009**, *97*, 1717–1724. [CrossRef]
35. Romero, F.J.; Escudero, M.; Medina-Garcia, A.; Morales, D.P.; Rodriguez, N. Meminductor Emulator Based on a Modified Antoniou's Gyrator Circuit. *Electronics* **2020**, *9*, 1407. [CrossRef]
36. Liang, Y.; Chen, H.; Yu, D.S. A Practical Implementation of a Floating Memristor-Less Meminductor Emulator. *IEEE Trans. Circuits Syst. II Express Briefs* **2014**, *61*, 299–303. [CrossRef]
37. The Computer That Stores and Processes Information at the Same Time. Available online: https://www.technologyreview.com/2012/11/21/181520/the-computer-that-stores-and-processes-information-at-the-same-time/ (accessed on 15 May 2021).
38. Pershin, Y.V.; Di Ventra, M. Memcomputing: A Computing Paradigm to Store and Process Information on the Same Physical Platform. In Proceedings of the 2014 International Workshop on Computational Electronics (IWCE), Paris, France, 3–6 June 2014; pp. 1–2.
39. Di Ventra, M.; Pershin, Y.V. The Parallel Approach. *Nat. Phys.* **2013**, *9*, 200–202. [CrossRef]
40. Fouda, M.E.; Radwan, A.G. Charge Controlled Memristor-Less Memcapacitor Emulator. *Electron. Lett.* **2012**, *48*, 1454–1455. [CrossRef]
41. Biolek, D.; Biolek, Z.; Biolkova, V. SPICE Modelling of Memcapacitor. *Electron. Lett.* **2010**, *46*, 520–522. [CrossRef]
42. Sah, M.P.; Yang, C.; Budhathoki, R.K.; Kim, H.; Yoo, H.J. Implementation of a Memcapacitor Emulator with Off-the-Shelf Devices. *Elektron. Elektrotechnika* **2013**, *19*, 54–58. [CrossRef]
43. Romero, F.J.; Morales, D.P.; Godoy, A.; Ruiz, F.G.; Tienda-Luna, I.M.; Ohata, A.; Rodriguez, N. Memcapacitor Emulator Based on the Miller Effect. *Int. J. Circuit Theory Appl.* **2019**, *47*, 572–579. [CrossRef]
44. Wang, X.Y.; Fitch, A.L.; Iu, H.H.C.; Qi, W.G. Design of a Memcapacitor Emulator Based on a Memristor. *Phys. Lett. A* **2012**, *376*, 394–399. [CrossRef]
45. Pershin, Y.V.; Ventra, M.D. Emulation of Floating Memcapacitors and Meminductors Using Current Conveyors. *Electron. Lett.* **2011**, *47*, 243–244. [CrossRef]
46. Yu, D.S.; Liang, Y.; Chen, H.; Iu, H.H.C. Design of a Practical Memcapacitor Emulator Without Grounded Restriction. *IEEE Trans. Circuits Syst. II Express Briefs* **2013**, *60*, 207–211. [CrossRef]
47. Yesil, A.; Babacan, Y. Electronically Controllable Memcapacitor Circuit with Experimental Results. *IEEE Trans. Circuits Syst. II Express Briefs* **2021**, *68*, 1443–1447. [CrossRef]
48. Vista, J.; Ranjan, A. Simple Charge Controlled Floating Memcapacitor Emulator Using DXCCDITA. *Analog. Integr. Circuits Signal Process.* **2020**, *104*, 37–46. [CrossRef]
49. Wang, S.-F. The Gyrator for Transforming Nano Memristor into Meminductor. *Circuit World* **2016**, *42*, 197–200. [CrossRef]
50. Romero, F.J.; Medina-Garcia, A.; Escudero, M.; Morales, D.P.; Rodriguez, N. Design and Implementation of a Floating Meminductor Emulator upon Riordan Gyrator. *AEU Int. J. Electron. Commun.* **2021**, *133*, 153671. [CrossRef]
51. Sah, M.P.; Budhathoki, R.K.; Yang, C.; Kim, H. Mutator-Based Meminductor Emulator for Circuit Applications. *Circuits Syst. Signal Process.* **2014**, *33*, 2363–2383. [CrossRef]
52. Sah, M.P.; Budhathoki, R.K.; Yang, C.; Kim, H. A Mutator-Based Meminductor Emulator Circuit. In Proceedings of the 2014 IEEE International Symposium on Circuits and Systems (ISCAS), Melbourne, VIC, Australia, 1–5 June 2014; pp. 2249–2252.

53. Liang, Y.; Ying, S.; Bingmeng, H.; Lu, C.; Jing, S. Design and Characteristic Analysis of Floating Flux-Controlled Meminductor Emulator. *J. Syst. Simul.* **2018**, *30*, 1337. [CrossRef]
54. Sozen, H.; Cam, U. A Novel Floating/Grounded Meminductor Emulator. *J. Circuits Syst. Comput.* **2020**, *29*, 2050247. [CrossRef]
55. Fouda, M.E.; Radwan, A.G. Memristor-Less Current- and Voltage-Controlled Meminductor Emulators. In Proceedings of the 2014 21st IEEE International Conference on Electronics, Circuits and Systems (ICECS), Marseille, France, 7–10 December 2014; pp. 279–282.
56. Elwakil, A.S.; Fouda, M.E.; Radwan, A.G. A Simple Model of Double-Loop Hysteresis Behavior in Memristive Elements. *IEEE Trans. Circuits Syst. II Express Briefs* **2013**, *60*, 487–491. [CrossRef]
57. Konal, M.; Kacar, F. Electronically Tunable Meminductor Based on OTA. *AEU Int. J. Electron. Commun.* **2020**, *126*, 153391. [CrossRef]
58. Vista, J.; Ranjan, A. High Frequency Meminductor Emulator Employing VDTA and Its Application. *IEEE Trans. Comput. Aided Des. Integr. Circuits Syst.* **2020**, *39*, 2020–2028. [CrossRef]
59. Babacan, Y. An Operational Transconductance Amplifier-Based Memcapacitor and Meminductor. *Istanb. Univ. J. Electr. Electron. Eng.* **2018**, *18*, 36–38. [CrossRef]
60. Çam Taşkıran, Z.G.; Sağbaş, M.; Ayten, U.E.; Sedef, H. A New Universal Mutator Circuit for Memcapacitor and Meminductor Elements. *AEU Int. J. Electron. Commun.* **2020**, *119*, 153180. [CrossRef]
61. Yu, D.; Liang, Y.; Iu, H.H.C.; Chua, L.O. A Universal Mutator for Transformations Among Memristor, Memcapacitor, and Meminductor. *IEEE Trans. Circuits Syst. II Express Briefs* **2014**, *61*, 758–762. [CrossRef]
62. Yu, D.; Zhao, X.; Sun, T.; Iu, H.H.C.; Fernando, T. A Simple Floating Mutator for Emulating Memristor, Memcapacitor, and Meminductor. *IEEE Trans. Circuits Syst. II Express Briefs* **2020**, *67*, 1334–1338. [CrossRef]
63. Zhao, Q.; Wang, C.; Zhang, X. A Universal Emulator for Memristor, Memcapacitor, and Meminductor and Its Chaotic Circuit. *Chaos* **2019**, *29*, 013141. [CrossRef]

 *electronics*

MDPI

Article

A Kind of Optoelectronic Memristor Model and Its Applications in Multi-Valued Logic

Jiayang Wang [1,2], Yuzhe Lin [1], Chenhao Hu [1,3], Shiqi Zhou [1,3], Shenyu Gu [1,3], Mengjie Yang [1,3], Guojin Ma [3,*] and Yunfeng Yan [2,*]

1. School of Electronics and Information, Hangzhou Dianzi University, Hangzhou 310018, China
2. College of Electrical Engineering, Zhejiang University, Hangzhou 310027, China
3. Zhejiang Provincial Key Lab of Equipment Electronics, Hangzhou 310018, China
* Correspondence: magj@hdu.edu.cn (G.M.); 21210004@zju.edu.cn (Y.Y.); Tel.: +86-13958173588 (G.M.); +86-15057166556 (Y.Y.)

Abstract: Memristors have been proved effective in intelligent computing systems owing to the advantages of non-volatility, nanometer size, low power consumption, compatibility with traditional CMOS technology, and rapid resistance transformation. In recent years, considerable work has been devoted to the question of how to design and optimize memristor models with different structures and physical mechanisms. Despite the fact that the optoelectronic effect inevitably makes the modelling process more complex and challenging, relatively few research works are dedicated to optoelectronic memristor modelling. Based on this, this paper develops an optoelectronic memristor model (containing mathematical model and circuit model). Moreover, the composite memristor circuit (series- and parallel-connected configuration) with a rotation mechanism is discussed. Further, a multi-valued logic circuit is designed, which is capable of performing multiple logic functions from 0–1, verifying the validity and effectiveness of the established memristor model, as well as opening up a new path for the circuit implementation of fuzzy logic.

Keywords: optoelectronic memristor; composite circuit; multi-valued logic; rotation mechanism

Citation: Wang, J.; Lin, Y.; Hu, C.; Zhou, S.; Gu, S.; Yang, M.; Ma, G.; Yan, Y. A Kind of Optoelectronic Memristor Model and Its Applications in Multi-Valued Logic. *Electronics* **2023**, *12*, 646. https://doi.org/10.3390/electronics12030646

Academic Editor: Flavio Canavero

Received: 21 December 2022
Revised: 15 January 2023
Accepted: 25 January 2023
Published: 28 January 2023

1. Introduction

Memristors that are non-volatile and nano-sized, and that have low power consumption, compatibility with conventional CMOS technology, and variable resistance have been widely used in intelligent computing systems [1–5]. This novel circuit component, was first proposed by Chua L in 1971, represents the relationship between charge and flux [6]. The existence of a physical memristor was first verified by Hewlett-Packard Lab in 2008, and it was confirmed to be a nanoscale passive two-terminal circuit component [7]. The successful preparation of memristors has attracted a lot of attention among research scholars around the world, and numerous memristor devices with different structures and physical mechanisms have been prepared [8–10].

Owing to stringent manufacturing processes and high costs, actual memristors are difficult to prepare outside the laboratory [11–13]. Considerable work has been devoted to the study of mathematical and circuit models that can reproduce the complex dynamics of memristors, such as the HP model [14], the spintronic model [15], the threshold adaptive model (TEAM) [16], the voltage threshold adaptive model (VTEAM) [17], etc. However, the present modelling is mostly based on two factors: the voltage applied to the memristor and the current flowing through it, which cannot accommodate the physical memristor properties prepared with evolving materials [18–20]. Indeed, physical memristors are affected by multiple factors such as light, temperature, humidity, and magnetic field, increasing the complexity and difficulty of modelling [21–24], and relatively little and incomplete work has been done in this area. In order to approximate the electrical characteristics possessed by physical amnesic resistors, this paper develops a model for optoelectronic memristors affected by both optical and electrical signals.

With the exploration of the binary characteristics of memristors, the application of memristors in logic circuits has received a lot of attention [25–28]. Existing research in the field of memristor-based logic implementation is mainly aimed at binary logic (e.g., material implication, memristor-aided logic, and memristor-ratioed logic), and ternary logic (e.g., balanced ternary logic and unbalanced ternary logic) [29–33]. However, relatively little research has been conducted on the implementation of multi-valued logic circuits based on memristors. Multi-valued logic with more logic states is an extension of traditional binary (or ternary) logic, which has only two (or three) logical states. Compared with binary (or ternary) logic, multi-valued logic carries more information in the process of processing a large amount of data, and has a faster operational speed, smaller area and size, and lower power consumption [34–36]. This paper presents the design of a multivalued logic circuit based on the established optoelectronic model. The research gaps and the main contributions of this paper are summarized in Table 1.

Table 1. The research gaps and the main contributions.

Research Gaps	Contributions
• Most of the existing memristor models are based on voltage and current factors, which fail to closely approximate the physical memristors that are affected by multiple factors.	• The illumination factor is introduced as a variable for modelling optoelectronic memristor, which provides a new idea for modelling memristors affected by multiple factors such as temperature, humidity, and magnetic field.
• The electrical characteristics of physical memristors are less unstable owing to vulnerability to the external environment.	• The electrical characteristics of the optoelectronic memristor and its composite circuit are analyzed from the perspective of the model.
• Existing research on memristor-based logic implementation mainly focuses on binary logic and ternary logic.	• A multi-valued logic circuit with 10 logic functions from 0–1 is proposed, which demonstrates the validity of the model and offers the possibility to explore fuzzy logic in the future.

The rest of the paper is organized as follows. Section 2 details the mathematical and PSIPCE models of a kind of optoelectronic memristor. Moreover, a series of tests and analysis on the electrical characteristics of the model is carried out in the same section. Section 3 discusses a composite circuit incorporating a rotation mechanism and proposes a multivalued logic circuit based on this circuit. The section concludes with a series of simulation experiments and analysis to verify the correctness of the proposed multi-valued logic and the validity of the model. Section 4 discusses the limitations of the designed multi-valued logic circuit and provides future research directions. Finally, Section 5 summarizes the whole work.

2. Optoelectronic Memristor Model and Electrical Characteristics Analysis

2.1. Background of Opoelectronic Memristor

Before modeling the optoelectronic memristor, the background of the optoelectronic memristor is studied in terms of both device structure and operating principles, which leads to a better understanding of the optoelectronic effects that affect the electrical characteristics of the memristor.

The modeled optoelectronic memristor device is prepared from $ITO/MgO/HfO_2/ITO$ material, where two layers of transparent conductive oxide ITO are used as the top electrode (TE) and bottom electrode (BE) of the device [21]. The wide bandwidth oxide (MgO/HfO_2) introduces the optical functionality in the resistive switching device while maintaining the optical transparency, and the MgO layer increases the durability, retention performance, and resistive ON/OFF ratio of the device. The specific device structure is shown in Figure 1a.

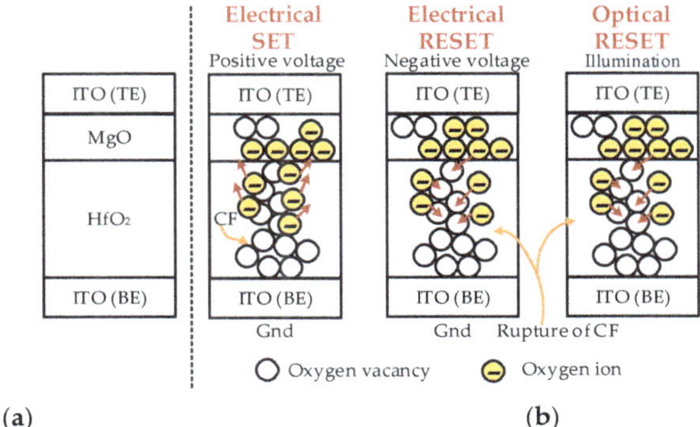

Figure 1. The schematic diagram of the optoelectronic memristor showing (**a**) device structure; and (**b**) working principle.

The working principle of the device is illustrated in Figure 1b, and it can be discussed in three cases depending on the voltage and light illumination. When a positive voltage is applied at the TE terminal and the BE terminal is grounded, the positive voltage induces the breakage of the Hf-O and Mg-O bonds, and a large number of oxygen vacancies (V_O) and oxygen ions (O^{2-}) are formed. Conductive filaments (CF) are formed inside the device, which causes the device to switch from a high resistance state (HRS) to a low resistance state (LRS). Owing to the larger amount of energy required to form V_O from MgO, most of the V_O during the setting process is produced by HfO_2. Notably, O^{2-} is not stacked near the TE interface, but at the intersection of MgO and HfO_2.

In the case that a negative voltage is applied at the TE terminal and the BE terminal is kept grounded, O^{2-} will recombine with V_O to form oxygen atoms under the negative voltage, leading to the rupture of CF. Further, the current is reset and the resistance state of the device is transformed to HRS.

In addition, the device is influenced by optical factors as well as electrical stimulation. The illumination is able to provide energy to O^{2-}, which in turn promotes O^{2-} and V_O to achieve recombination. The effects of illumination and negative voltage on the electrical characteristics of the device are similar, and both achieve the reset function.

2.2. Modelling of Opoelectronic Memristor

In order to facilitate subsequent studies on the application of the optoelectronic memristor, the device is modeled by improving the VTEAM modelling method and employing the optoelectronic effect. Its resistance can be represented by the following mathematical equation.

$$M(x) = R_{on} + \frac{R_{off} - R_{on}}{x_{off} - x_{on}} (x - x_{on}) \tag{1}$$

where M denotes the resistance of the optoelectronic memristor, which is limited to $[R_{on}, R_{off}]$. x is the state variable, with maximum and minimum values of x_{on} and x_{off}, and the dynamic function of x can be described by the following equation.

$$\frac{dx}{dt} = \begin{cases} \left[k_{on}\left(\frac{V(t)}{V_{th1}} - 1\right)^{\alpha_{on}} + \frac{I_p}{\varepsilon \cdot I_{pmax}} \right] \cdot f(x), & 0 < V_{th1} \leq V(t) \\ \frac{I_p}{\varepsilon \cdot I_{pmax}} \cdot f(x), & V_{th2} < V(t) \leq V_{th1} \\ \left[k_{off}\left(\frac{V(t)}{V_{th2}} - 1\right)^{\alpha_{off}} + \frac{I_p}{\varepsilon \cdot I_{pmax}} \right] \cdot f(x), & V(t) \leq V_{th2} < 0 \end{cases} \tag{2}$$

$$f(x) = 1 - (\beta x - 1)^{2p} \tag{3}$$

where $k_{on}(V(t)/V_{th1} - 1)^{\alpha on}$ and $k_{off}(V(t)/V_{th2} - 1)^{\alpha off}$ are electrical effect simulation terms that achieve not only a decrease in resistance when a positive voltage is applied, but also an increase in resistance for negative voltages. k_{on}, k_{off}, α_{on}, and α_{off} are all fitting parameters of the model, V_{th1} and V_{th2} denote the positive and negative threshold voltage, respectively. $V(t)$ represents the voltage loaded onto the memristor. $I_p/(\varepsilon * I_{pmax})$ is the light effect simulation term to realize the function of increasing the memristance upon illumination. ε denotes the parameter capable of regulating the effect of light and I_{pmax} is the maximum illumination power intensity. I_p indicates the optical power intensity applied to the memristor. $f(x)$ acts as a window function for binding x to $[x_{on}, x_{off}]$. β and p are fitting parameters of the model.

Accordingly, a SPICE model of optoelectronic memristor is proposed to perform the subsequent circuit simulation. The specific sub-circuit description is shown in Table 2.

Table 2. PSPICE sub-circuit of optoelectronic memristor model.

```
* Optoelectronic Memristor Model

.SUBCKT optoelectronic memristor model Plus Minus PARAMS:
+ xon=0 xoff=3E−9 Alphaon=0.1 Alphaoff=0.1 Ron=100 Roff=3E3 kon=−1 koff=1 Ip=100
+ Epsilon=0.6 Ipmax=500 p=1 Beta=6.6666E8 Vth1=2 Vth2=−2 xinit=3E−9
******* Differential equation modelling*******
Gx 0 x value={f(V(x), V(Plus, Minus), kon, koff, Alphaon, Alphaoff, Vth1, Vth2, Epsilon,
+ Beta, p, Ip, Ipmax)}
Cx x 0 1 IC={xinit}
R x 0 1 T
**********************Ohm's Law********************
Emem Plus Aux value={I(Emem)*(Roff-Ron)*(V(x)-xon)/(xoff-xon)}
Rs aux Minus {Ron}
Emx Mx 0 value={(Roff-Ron)*(V(x)-xon)/(xoff-xon)+Ron}
**********************Functions********************
.func f(x, v, kon, koff, Alphaon, Alphaoff, Epsilon, Ip, Ipmax, Beta, p)=
+ {If(v>Vth1, f1(x, v, kon, Vth1, Alphaon, Epsilon, Ip, Beta, Ipmax, p),
+ If(v<Vth2, f2(x, v, koff, Vth2, Alphaoff, Epsilon, Ip, Beta, Ipmax, p),
+ f3(x, Epsilon, Ip, Beta, Ipmax, p))}
.func f1(x, v, kon, Vth1, Alphaon, Epsilon, Ip, Beta, Ipmax, p)=
+ {(kon*(v/Vth1−1)^Alphaon+Ip/(Epsilon*Ipmax))*(1-(Beta*x−1)^(2*p))}
.func f2(x, v, koff, Vth2, Alphaoff, Epsilon, Ip, Beta, Ipmax, p)=
+ {(koff*(v/Vth2−1)^Alphaoff+Ip/(Epsilon*Ipmax))*(1-(Beta*x−1)^(2*p))}
.func f3(x, Epsilon, Ip, Beta, Ipmax, p)={Ip/(Epsilon*Ipmax)*(1-(Beta*x−1)^(2*p))}
.ENDS optoelectronic memristor
```

To evaluate the accuracy of the proposed model, we fit the model based on real experimental data, as shown in Figure 2. Figure 2a shows the $I-V$ curve of the memristor in the dark environment, where the black ball indicates the real experimental data and the red solid line indicates the fitting result of the model. Notably, the memristor is initialized to a high resistance state before conducting this experiment. Figure 2b shows the $I-V$ curves of the memristor before and after applying illumination to it, where the blue and black lines with triangles indicate the real experimental data pre- and post-illumination, respectively, and the gray and red lines show the fitting results pre- and post-illumination, respectively. Notably, the resistance of the initialized memristor is a lower resistance before carrying out this experiment and the illumination is white light (390–780 nm) with an optical power intensity of 100 W/m^2.

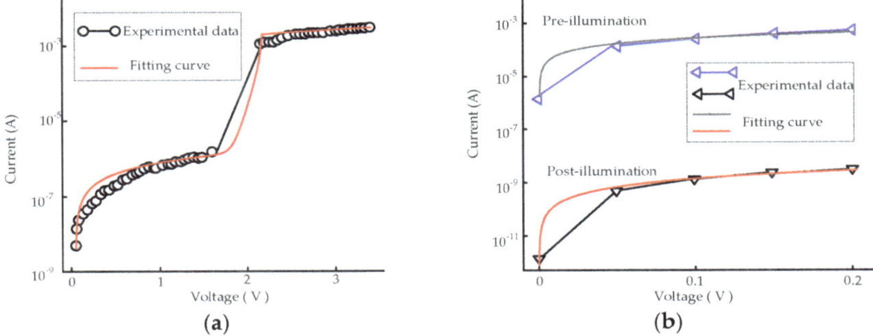

Figure 2. The fitting result of the model showing (**a**) $I-V$ curve of the memristor in the dark environment; and (**b**) $I-V$ curves of the memristor before and after applying illumination.

From Figure 2, the established memristor model has a good overlap with the actual memristor electrical characteristic curve. Root mean square error (RMSE) is introduced as an indicator for objective analysis, and its mathematical expression is shown below:

$$\text{RMSE} = \sqrt{\frac{1}{n}\left(\frac{\sum\limits_{i=1}^{n}\left(V_{\text{real},i} - V_{\text{fit},i}\right)^2}{V_{\text{fit}}^2} + \frac{\sum\limits_{i=1}^{n}\left(I_{\text{real},i} - I_{\text{fit},i}\right)^2}{I_{\text{fit}}^2}\right)} \tag{4}$$

where n is the number of samples, $V_{\text{real},i}$ and $I_{\text{real},i}$ the voltage and current on the memristor during the testing process, and $V_{\text{fit},i}$ and $I_{\text{fit},i}$ represent the voltage and current on the model during the fitting process. V_{fit} and I_{fit} denote the Euclidean criterion for the voltage and current of the memristor model. Normally, a smaller RMSE represents a better fit.

From Figure 2a, the model can achieve the electrical setting function under positive voltage, where the model can be fitted to the measured data points with the RMSE value of 1.13%. As shown in Figure 2b, the simulated I-V curves are matched with the I-V curves in the pre- and post-illumination periods with RMSE values of 1.65% and 1.98%, respectively. When illumination is applied, the resistance of the memristor increases and the current flowing through the memristor decreases. In general, the fitting results show that the constructed optoelectronic memristor model is capable of characterizing the performance of the ITO/MgO/HfO$_2$/ITO memristor.

2.3. Electrical Characteristics Analysis

To demonstrate the electrical characteristics of the proposed optoelectronic memristor model, a series of tests and analysis circuit simulations are performed, and the experimental results are shown in Figures 3–5. The specific parameters of the memristor model are detailed in Tables 2 and 3. Notably, the experiments are conducted on a desktop workstation equipped with Core i7-10700 processor, 32 GB RAM and Windows 10 operating system using Matlab2018 and PSpice software.

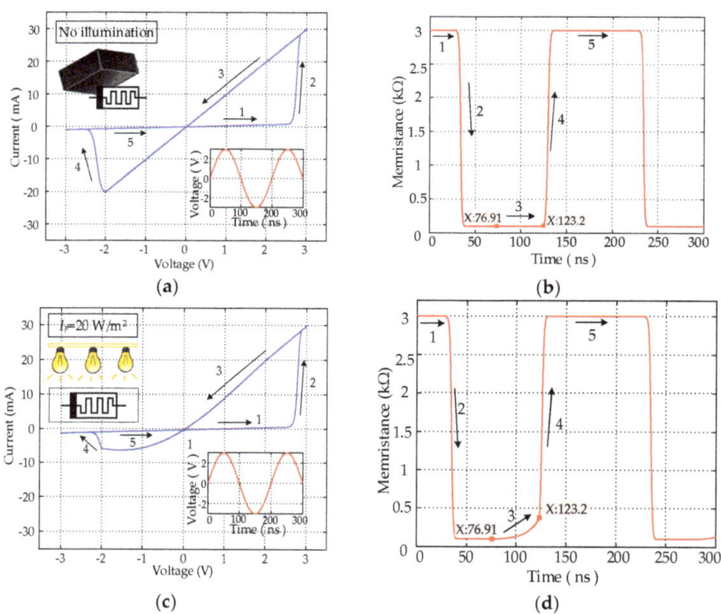

Figure 3. Simulation results under sine-wave voltage showing (**a**,**b**) $I-V$ and $M-t$ curves without illumination; and (**c**,**d**) $I-V$ and $M-t$ curves at optical radiation intensity of 20 W/m².

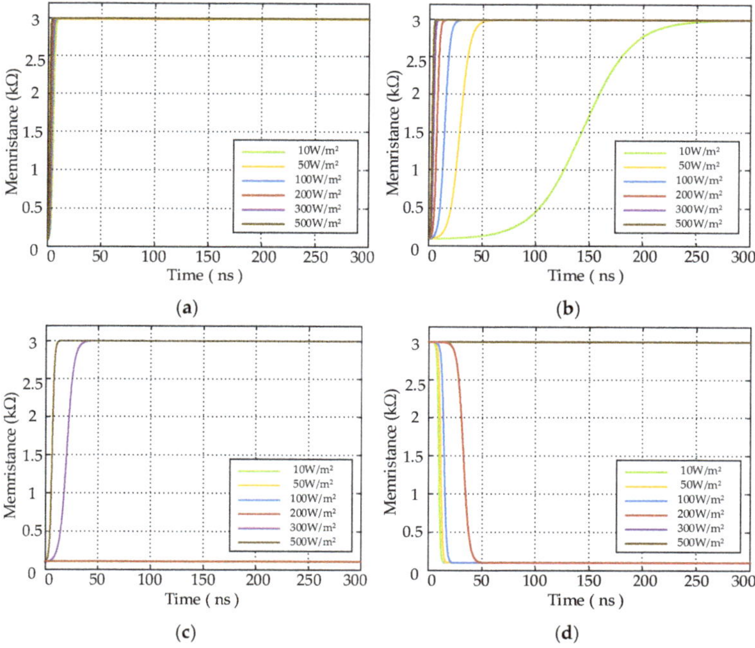

Figure 4. Simulation results at different optical power densities under constant voltages showing (**a**–**c**) $M-t$ curves of the memristor with initial resistance R_{on} at -3 V, 0 V, 3 V; and (**d**) $M-t$ curve of the memristor with initial resistance R_{off} at 3 V.

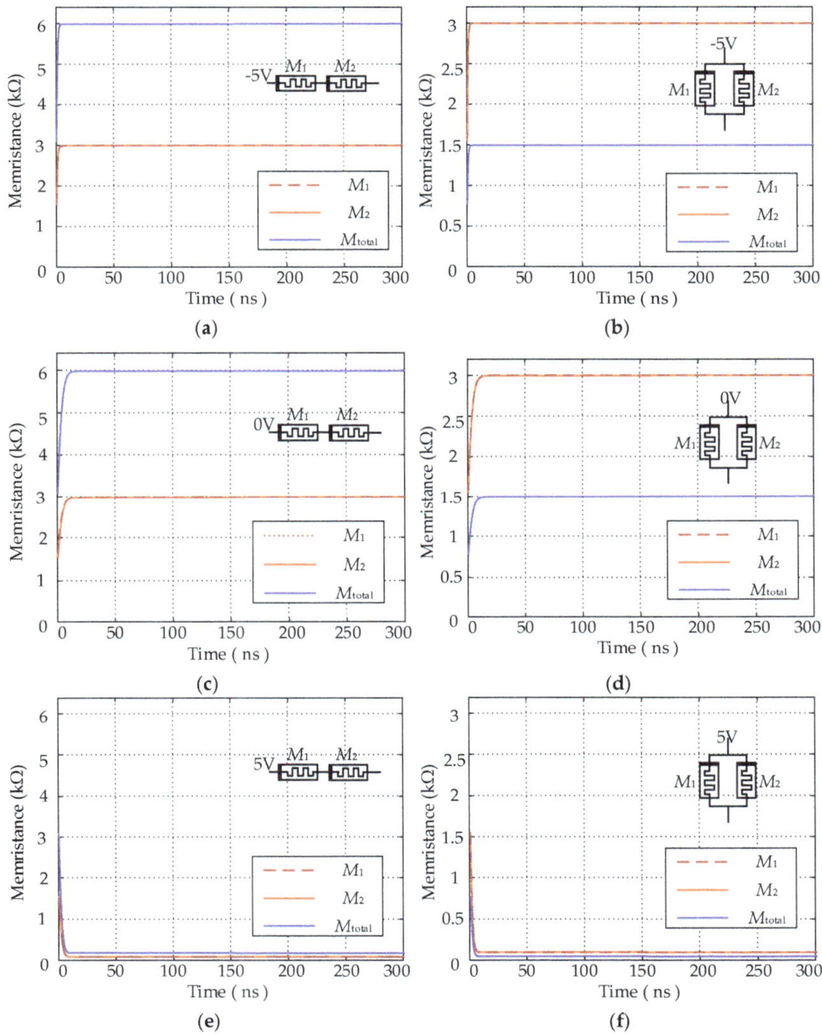

Figure 5. Electrical characteristic curves of the memristor composite circuit at the optical power density of $100\,\text{W/m}^2$ showing (**a,c,e**) $M-t$ curves of the series circuit at -5 V, 0 V, 5 V; and (**b,d,f**) $M-t$ curves of the parallel circuit at -5 V, 0 V, 5 V.

Figure 3 illustrates the electrical characteristics (volt-ampere characteristics and memristance variation rule) of the memristor under different illumination conditions. The relationship between the current flowing through the memristor model and the applied voltage is shown in Figure 3a,c, and the variation of memristance with time is shown in Figure 3b,d.

The $I-V$ characteristic curves in Figure 3a,c both show squeezed hysteresis curves at the origin, which demonstrates that the proposed memristor model conforms to the definition of a generalized memristor [37]. Notably, the model is relatively symmetrical in the positive and negative voltage regions under the condition of no illumination, whereas the area of the pinched hysteresis loop in the negative voltage region is greatly reduced upon optical stimulation, and the positive and negative regions are asymmetric. This occurs because the light stimulus increases the memristance as well as the negative voltage, weakening the effect of the negative voltage on the electrical properties of the model.

Table 3. Parameters for the electrical characteristics analysis.

	Optical Power Density (W/m^2)	Electrical Stimulation (V)	Initial Value of Memristor (kΩ)
Figure 3a,b	$I_p = 0$	$V = 3 \sin(10^7 \pi t)$	3
Figure 3c,d	$I_p = 20$	$V = 3 \sin(t)$	3
Figure 4a		$V = -3$	0.01
Figure 4b	$I_p = 10, 50, 100, 200, 300, 500$	$V = 0$	0.01
Figure 4c		$V = 3$	0.01
Figure 4d		$V = 3$	3
Figure 5a		$V = -5$	1.5, 1.5
Figure 5b		$V = -5$	1.5, 1.5
Figure 5c	$I_p = 100$	$V = 0$	1.5, 1.5
Figure 5d		$V = 0$	1.5, 1.5
Figure 5e		$V = 5$	1.5, 1.5
Figure 5f		$V = 5$	1.5, 1.5

From Figure 3b, it can be observed that the electrical characteristics of this memristor model are similar to those of the VTEAM model under conditions without illumination, and the variation of its resistance with time can be described in five stages. In stage 1, the resistance keeps R_{off} constant when the scanning voltage increases from 0 V to V_{th1}; in stage 2, the resistance decreases from R_{off} to R_{on} when applying the voltage to the memristor over V_{th1}; in stage 3, the resistance remains unchanged as the scanning voltage decreases from V_{th1} to V_{th2}; in stage 4, when the voltage is less than V_{th2}, the memristor transforms from the LRS to the HRS; and in stage 5, by the time the voltage increases from V_{th2} to 0 V, the resistance characteristics are the same as in stages 1 and 3, and the HRS is maintained. Comparing Figure 3b,d, it can be found that the optical stimulation causes an alteration of the resistance variation rule in stage 3, which is the time interval from 76.91 ns to 123.2 ns. The inclusion of the optical stimulus leads to the reversal of the soft breakdown, causing an increase in the resistance of the memristor. Notably, with the low power density of the applied optical signal here, the optical stimulus has less effect on the electrical characteristics of the memristor in comparison with the electrical stimulus and does not lead to a shift in the resistance variation pattern in the remaining phases.

Figure 4 shows the results of the effect of optical stimulation with different irradiation power densities and constant voltage of different amplitudes on the variation pattern of the memristance. Figure 4a–c shows the memristor response curves upon the application of electrical stimuli of 3 V, 0 V and −3 V to the memristor model with the initial resistance R_{on} within the time interval [0, 300]ns at different optical power densities (i.e., 10, 50, 100, 200, 300, and 500 W/m^2), respectively. Considering that the resistance state of the memristor with initial resistance R_{off} will not change with external electrical stimuli of −3 V and 0 V, thus only the effect of different optical power densities on the memristance under 3 V positive voltage is simulated. From Figure 4a,b, it is obvious that the higher the optical power density, the faster the rate of resistance enhancement. A summary analysis of Figure 4c,d shows that a sufficiently large optical stimulus can surpass the effect of positive voltage on the electrical characteristics of the memristor. Notably, the green, yellow, blue and red lines in Figure 4c are overlapped, and the purple and brown lines in Figure 4d are overlapped.

Figure 5 depicts the electrical characteristic curves of the memristor composite circuit (series- and parallel-connected configuration) at an optical power density of 100 W/m^2. Figure 5a,c,e illustrates the resistance variation curves of two optoelectronic memristors with the same initial resistance connected in series at −5 V, 0 V and 5 V electrical stimuli, respectively, while Figure 5b,d,f represents the resistance variation curves connected in parallel.

From Figure 5, since the model parameters (including the initial memristances) of the two memristors are the same, the corresponding resistance variation pattern (which can be referred to the resistance variation pattern in Figure 3d) is also the same, as shown in the overlapping of the red dashed line and the orange solid line. When two memristors are connected in series in the same direction, the resistance state of both memristors changes to the HRS under the combined effect of negative voltage and illumination, as shown in Figure 5a. In the case of illumination separately, as shown in Figure 5c, the resistances of both memristors also increase to R_{off}. Notably, compared to Figure 5a, the change (growth) curve of the resistance is relatively smooth owing to the lack of the effect of the negative voltage. Under the effect of positive voltage, as shown in Figure 5e, the resistance of both memristors decreases owing to the lower power density of illumination radiation, whose effect on the optoelectronic memristors is smaller than that of positive voltage. In addition, when two memristors are connected in parallel in the same direction, as shown in Figure 5b,d,f, the resistance variation trend of the two memristors is the same as that in series in the same direction. However, when the same voltage is applied to the input, the voltage divided by the parallel memristors is larger, so the rate of resistance variation is relatively faster. Moreover, the equivalent resistance M_{total} of the series circuit is calculated as the sum of the resistance M_1 and M_2, i.e., $M_{total} = M_1 + M_2$, and the equivalent resistance of the parallel circuit satisfies: $M_{total} = M_1 * M_2/(M_1 + M_2)$.

3. Rotation Mechanism Based Multi-Valued Logic

In this section, a composite memristor circuit (series- and parallel-connected configuration) incorporating a rotation mechanism is discussed. According to the composite circuit, a circuit capable of implementing 0–1 multiple logic functions is subsequently proposed. The specific operation procedure and simulation results are described as follows.

3.1. Rotation Mechanism Based Composite Circuit

The schematic diagram of the composite memristor circuit based on the rotation mechanism is depicted in Figure 6, where A, B, C are the ports of the connection line, and port B is the center of rotation. The rotation mechanism enables the conversion of series and parallel configuration circuits, and the circuit during the conversion contains an intermediate state in addition to the series and parallel states.

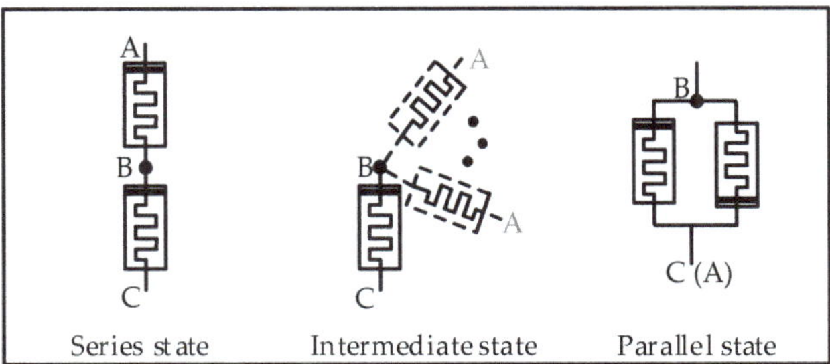

Figure 6. The circuit diagram of rotation mechanism.

3.2. Implementation of Multi-Valued Logic

Incorporating the composite circuit based on the rotation mechanism, a circuit for implementing multi-valued logic is designed (as illustrated in Figure 7). Note that the proposed circuit contains two valid states during rotation, i.e., state *I* and state *II*.

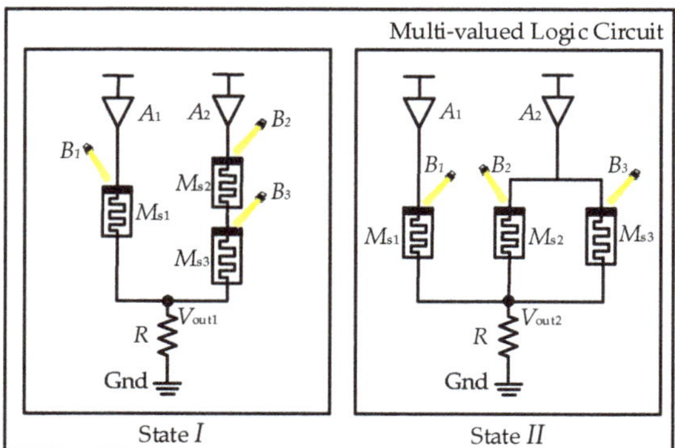

Figure 7. The diagram of multi-valued logic circuit based on rotation mechanism.

The memristors (M_{s1}, M_{s2}, and M_{s3}) employed in the circuit are the proposed optoelectronic memristors, whose resistances (R_{s1}, R_{s2}, and R_{s3}) are varied in [R_{on}, R_{off}] by the joint effect of electrical and optical stimulation. The input voltages (A_1 and A_2) and the optical power densities (B_1, B_2, and B_3) are the input state variables of the circuit. R is the regular resistor, and the voltage (V_{out1} and V_{out2}) across it indicates the output state variable of the logic circuit.

Before executing the multi-valued logic operation, the memristor M_{s1}, M_{s2}, and M_{s3} are required to be initialized to their lowest value R_{on}. The specific process of implementing multiple logic functions from 0 to 1 for the two states is described as follows.

3.2.1. State I for Multi-Valued Logic

For multi-valued logic operation, the electrical inputs (A_1 and A_2) always have two states V_{on} and V_{off}, representing the logic "1" and logic "0", respectively. Notably, the value of voltage V_{off} used in this section is 0 V. According to the principle of series voltage division and parallel current division, the node voltage V_{out1} can be calculated with the following equation.

$$V_{out1} = \begin{cases} 0 & A_1 = V_{off}, A_2 = V_{off} \\ \dfrac{R}{R+R_{s1}//(R_{s2}+R_{s3})} V_{on} & A_1 = V_{on}, A_2 = V_{on} \\ \dfrac{R//R_{s1}}{R//R_{s1}+(R_{s2}+R_{s3})} V_{on} & A_1 = V_{off}, A_2 = V_{on} \\ \dfrac{R//(R_{s2}+R_{s3})}{R//(R_{s2}+R_{s3})+R_{s1}} V_{on} & A_1 = V_{on}, A_2 = V_{off} \end{cases} \tag{5}$$

When $A_1 = A_2 = V_{off}$, the node voltage V_{out1} is always 0 V, independent of whether the state of the memristor is changed or not, thus the case is not described when the optical stimulus is applied to the memristor. The optical inputs (B_1, B_2, and B_3), which always have two states I_{ph} and I_{pl}, representing the logic "1" and logic "0", respectively, can be discussed in the following six cases.

- Case A: When $B_1 = B_2 = B_3 = I_{pl}$, the node voltage V_{out1} can be computed as:

$$V_{out1} = \begin{cases} \dfrac{R}{R+R_{on}//2R_{on}} V_{on} & A_1 = V_{on}, A_2 = V_{on} \\ \dfrac{R//R_{on}}{R//R_{on}+2R_{on}} V_{on} & A_1 = V_{off}, A_2 = V_{on} \\ \dfrac{R//2R_{on}}{R//2R_{on}+R_{on}} V_{on} & A_1 = V_{on}, A_2 = V_{off} \end{cases} \tag{6}$$

According to the resistance variation pattern of optoelectrical memristor, the low optical power density is not sufficient to trigger a change in the resistance of the optoelectrical memristor, thus the resistance of memristors (R_{s1}, R_{s2}, and R_{s3}) remains in the initial R_{on}.

- Case B: When $B_1 = I_{pl}$, $B_2 = I_{pl}$ and $B_3 = I_{ph}$ (or $B_1 = I_{pl}$, $B_2 = I_{ph}$ and $B_3 = I_{pl}$), the node voltage V_{out1} can be computed as:

$$V_{out1} = \begin{cases} \dfrac{R}{R+R_{on}//(R_{on}+R_{off})}V_{on} & A_1 = V_{on}, A_2 = V_{on} \\[2mm] \dfrac{R//R_{on}}{R//R_{on}+R_{on}+R_{off}}V_{on} & A_1 = V_{off}, A_2 = V_{on} \\[2mm] \dfrac{R//(R_{on}+R_{off})}{R//(R_{on}+R_{off})+R_{on}}V_{on} & A_1 = V_{on}, A_2 = V_{off} \end{cases} \tag{7}$$

Here, the high optical power density has a greater effect on the resistance variation of the memristor than the positive voltage, with the memristance of R_{s3} or R_{s2} increasing from R_{on} to R_{off}. The memristors M_{s1} and M_{s2} (or M_{s1} and M_{s3}) remain in the low resistance state.

- Case C: When $B_1 = I_{pl}$ and $B_2 = B_3 = I_{ph}$, the node voltage V_{out1} can be computed as:

$$V_{out1} = \begin{cases} \dfrac{R}{R+R_{on}//2R_{off}}V_{on} & A_1 = V_{on}, A_2 = V_{on} \\[2mm] \dfrac{R//R_{on}}{R//R_{on}+2R_{off}}V_{on} & A_1 = V_{off}, A_2 = V_{on} \\[2mm] \dfrac{R//2R_{off}}{R//2R_{off}+R_{on}}V_{on} & A_1 = V_{on}, A_2 = V_{off} \end{cases} \tag{8}$$

Here, the memristor M_{s1} remains in the initial low resistance state, and the resistances of M_{s2} and M_{s3} increase to the highest value R_{off}.

- Case D: When $B_1 = I_{ph}$ and $B_2 = B_3 = I_{pl}$, the node voltage V_{out1} can be computed as:

$$V_{out1} = \begin{cases} \dfrac{R}{R+R_{off}//2R_{on}}V_{on} & A_1 = V_{on}, A_2 = V_{on} \\[2mm] \dfrac{R//R_{off}}{R//R_{off}+2R_{on}}V_{on} & A_1 = V_{off}, A_2 = V_{on} \\[2mm] \dfrac{R//2R_{on}}{R//2R_{on}+R_{off}}V_{on} & A_1 = V_{on}, A_2 = V_{off} \end{cases} \tag{9}$$

Here, the memristors M_{s2} and M_{s3} remain in the initial low resistance state, and the resistance of M_{s1} increases to the highest value R_{off}.

- Case E: When $B_1 = I_{ph}$, $B_2 = I_{pl}$ and $B_3 = I_{ph}$ (or $B_1 = I_{ph}$, $B_2 = I_{ph}$ and $B_3 = I_{pl}$), the node voltage V_{out1} can be computed as:

$$V_{out1} = \begin{cases} \dfrac{R}{R+R_{off}//(R_{on}+R_{off})}V_{on} & A_1 = V_{on}, A_2 = V_{on} \\[2mm] \dfrac{R//R_{off}}{R//R_{off}+R_{on}+R_{off}}V_{on} & A_1 = V_{off}, A_2 = V_{on} \\[2mm] \dfrac{R//(R_{on}+R_{off})}{R//(R_{on}+R_{off})+R_{off}}V_{on} & A_1 = V_{on}, A_2 = V_{off} \end{cases} \tag{10}$$

Here, the memristors M_{s2} (or M_{s3}) remain in the initial low resistance state, and the resistance of M_{s1} and M_{s3} (or M_{s2}) increase to the highest value R_{off}.

- Case F: When $B_1 = B_2 = B_3 = I_{ph}$, the node voltage V_{out1} can be computed as:

$$V_{out1} = \begin{cases} \dfrac{R}{R+R_{off}//2R_{off}}V_{on} & A_1 = V_{on}, A_2 = V_{on} \\[2mm] \dfrac{R//R_{off}}{R//R_{off}+2R_{off}}V_{on} & A_1 = V_{off}, A_2 = V_{on} \\[2mm] \dfrac{R//2R_{off}}{R//2R_{off}+R_{off}}V_{on} & A_1 = V_{on}, A_2 = V_{off} \end{cases} \tag{11}$$

Here, the memristors M_{s1}, M_{s2} and M_{s3} shift to the high resistance state, and correspondingly all the memrsistances (R_{s1}, R_{s2}, and R_{s3}) increase to the highest value R_{off}.

The node voltage V_{out1} in Case A is defined as logic "1" when the electrical inputs $A_1 = A_2 = V_{on}$, and the output logic value for the rest of the cases is the ratio between the output V_{out1} and the voltage corresponding to logic "1". Assuming $R_{off} \geq 10\,R$, $R \cong 3\,R_{on}$, the truth table of state I for multi-valued logic is shown in Table 4.

Table 4. Truth table of multi-valued logic circuit.

Electronical Inputs		Optical Inputs				Output of state *I*	Output of state *II*
A_1	A_2	Cases	B_1	B_2	B_3	V_{out1}	V_{out2}
0	0	-	×	×	×	0	0
1	1	Case A	0	0	0	1	1
		Case B	0	0/1	1/0	0.9	1
		Case C	0	1	1	0.9	0.9
		Case D	1	0	0	0.8	1
		Case E	1	0/1	1/0	0.2	0.9
		Case F	1	1	1	0.2	0.3
0	1	Case A	0	0	0	0.7	0.7
		Case B	0	0/1	1/0	0.9	0.5
		Case C	0	1	1	0.9	0.1
		Case D	1	0	0	0	1
		Case E	1	0/1	1/0	0.1	0.9
		Case F	1	1	1	0.1	0.2
1	0	Case A	0	0	0	0.3	0.4
		Case B	0	0/1	1/0	0	0.5
		Case C	0	1	1	0	0.8
		Case D	1	0	0	0.7	0
		Case E	1	0/1	1/0	0.1	0
		Case F	1	1	1	0	0.1

3.2.2. State II for Multi-Valued Logic

The memristors M_{s1} and M_{s2} are rotated from series to parallel in state *I*, and the computational equation for the node voltage V_{out2} is replaced as follows:

$$V_{\text{out1}} = \begin{cases} 0 & A_1 = V_{\text{off}}, A_2 = V_{\text{off}} \\ \frac{R}{R+R_{\text{s1}}//(R_{\text{s2}}//R_{\text{s3}})} V_{\text{on}} & A_1 = V_{\text{on}}, A_2 = V_{\text{on}} \\ \frac{R//R_{\text{s1}}}{R//R_{\text{s1}}+(R_{\text{s2}}//R_{\text{s3}})} V_{\text{on}} & A_1 = V_{\text{off}}, A_2 = V_{\text{on}} \\ \frac{R//(R_{\text{s2}}//R_{\text{s3}})}{R//(R_{\text{s2}}//R_{\text{s3}})+R_{\text{s1}}} V_{\text{on}} & A_1 = V_{\text{on}}, A_2 = V_{\text{off}} \end{cases} \tag{12}$$

In the same way as state *I*, which implements multi-valued logic, the node voltage V_{out2} can be further specified into the following six cases according to the optical inputs.

- Case A: When $B_1 = B_2 = B_3 = I_{\text{pl}}$, the variation of resistance R_{s1}, R_{s2}, and R_{s3} is the same as that of Case A in state *I*, thus the node voltage V_{out2} can be computed as:

$$V_{\text{out2}} = \begin{cases} \frac{R}{R+R_{\text{on}}//(R_{\text{on}}//R_{\text{on}})} V_{\text{on}} & V_1 = V_{\text{on}}, V_2 = V_{\text{on}} \\ \frac{R//(R_{\text{on}}//R_{\text{on}})}{R//(R_{\text{on}}//R_{\text{on}})+R_{\text{on}}} V_{\text{on}} & V_1 = V_{\text{off}}, V_2 = V_{\text{on}} \\ \frac{R//R_{\text{on}}}{R//R_{\text{on}}+R_{\text{on}}//R_{\text{on}}} V_{\text{on}} & V_1 = V_{\text{on}}, V_2 = V_{\text{off}} \end{cases} \tag{13}$$

- Case B: When $B_1 = I_{\text{pl}}$, $B_2 = I_{\text{pl}}$ and $B_3 = I_{\text{ph}}$ (or $B_1 = I_{\text{pl}}$, $B_2 = I_{\text{ph}}$ and $B_3 = I_{\text{pl}}$), the variation of resistance R_{s1}, R_{s2}, and R_{s3} is the same as that of Case B in state *I*, thus the node voltage V_{out2} can be computed as:

$$V_{\text{out2}} = \begin{cases} \frac{R}{R+R_{\text{on}}//(R_{\text{on}}//R_{\text{off}})} V_{\text{on}} & V_1 = V_{\text{on}}, V_2 = V_{\text{on}} \\ \frac{R//(R_{\text{on}}//R_{\text{off}})}{R//(R_{\text{on}}//R_{\text{off}})+R_{\text{on}}} V_{\text{on}} & V_1 = V_{\text{off}}, V_2 = V_{\text{on}} \\ \frac{R//R_{\text{on}}}{R//R_{\text{on}}+R_{\text{off}}//R_{\text{on}}} V_{\text{on}} & V_1 = V_{\text{on}}, V_2 = V_{\text{off}} \end{cases} \tag{14}$$

- Case C: When $B_1 = I_{pl}$ and $B_2 = B_3 = I_{ph}$, the variation of resistance R_{s1}, R_{s2}, and R_{s3} is the same as that of Case C in state I, thus the node voltage V_{out2} can be computed as:

$$V_{out2} = \begin{cases} \dfrac{R}{R + R_{on}//(R_{off}//R_{off})} V_{on} & V_1 = V_{on}, V_2 = V_{on} \\[2ex] \dfrac{R//(R_{off}//R_{off})}{R//(R_{off}//R_{off}) + R_{on}} V_{on} & V_1 = V_{off}, V_2 = V_{on} \\[2ex] \dfrac{R//R_{on}}{R//R_{on} + R_{off}//R_{off}} V_{on} & V_1 = V_{on}, V_2 = V_{off} \end{cases} \tag{15}$$

- Case D: When $B_1 = I_{ph}$ and $B_2 = B_3 = I_{pl}$, the variation of resistance R_{s1}, R_{s2}, and R_{s3} is the same as that of Case D in state I, thus the node voltage V_{out2} can be computed as:

$$V_{out2} = \begin{cases} \dfrac{R}{R + R_{off}//(R_{on}//R_{on})} V_{on} & V_1 = V_{on}, V_2 = V_{on} \\[2ex] \dfrac{R//(R_{on}//R_{on})}{R//(R_{on}//R_{on}) + R_{off}} V_{on} & V_1 = V_{off}, V_2 = V_{on} \\[2ex] \dfrac{R//R_{off}}{R//R_{off} + R_{on}//R_{on}} V_{on} & V_1 = V_{on}, V_2 = V_{off} \end{cases} \tag{16}$$

- Case E: When $B_1 = I_{ph}$, $B_2 = I_{pl}$ and $B_3 = I_{ph}$ (or $B_1 = I_{ph}$, $B_2 = I_{ph}$ and $B_3 = I_{pl}$), the variation of resistance R_{s1}, R_{s2}, and R_{s3} is the same as that of Case E in state I, thus the node voltage V_{out2} can be computed as:

$$V_{out2} = \begin{cases} \dfrac{R}{R + R_{off}//(R_{on}//R_{off})} V_{on} & V_1 = V_{on}, V_2 = V_{on} \\[2ex] \dfrac{R//(R_{on}//R_{off})}{R//(R_{on}//R_{off}) + R_{off}} V_{on} & V_1 = V_{off}, V_2 = V_{on} \\[2ex] \dfrac{R//R_{off}}{R//R_{off} + R_{off}//R_{on}} V_{on} & V_1 = V_{on}, V_2 = V_{off} \end{cases} \tag{17}$$

- Case F: When $B_1 = B_2 = B_3 = I_{ph}$, the variation of resistance R_{s1}, R_{s2}, and R_{s3} is the same as that of Case F in state I, thus the node voltage V_{out2} can be computed as:

$$V_{out2} = \begin{cases} \dfrac{R}{R + R_{off}//(R_{off}//R_{off})} V_{on} & V_1 = V_{on}, V_2 = V_{on} \\[2ex] \dfrac{R//(R_{off}//R_{off})}{R//(R_{off}//R_{off}) + R_{off}} V_{on} & V_1 = V_{off}, V_2 = V_{on} \\[2ex] \dfrac{R//R_{off}}{R//R_{off} + R_{off}//R_{off}} V_{on} & V_1 = V_{on}, V_2 = V_{off} \end{cases} \tag{18}$$

As in the previous section, the ratio between the node voltage V_{out2} and the voltage value V_{out1} is defined as the corresponding output logic state variable. Assuming $R_{off} \geq 10\,R$, $R \cong 3\,R_{on}$, the correlation between the inputs (electrical inputs and optical inputs) and the output of state II is also shown in Table 4.

From Table 4, it can be shown that the proposed multi-valued logic circuit (including state I and state II) can implement 10 logic functions in 0–1. In addition, since the output state variables are voltages, the circuit is easy to cascade for implementing circuits with more complex functions.

3.3. Circuit Smulations and Analysis

To verify the effectiveness of the designed multi-valued logic circuit, a series of simulation experiments was performed on the same workstation as in the previous section. At the device level, the circuit uses three identical memristors, and the specific parameter configurations of the devices are detailed in Table 2. Notably, in addition to the memristors, the circuit needs to be configured with a resistor with a resistance of 0.3 kΩ, satisfying $R = 3\,R_{on}$. At the circuit level, different input variables (the electrical input variables and the optical signal variables) are necessary for implementing the multi-valued logic, configuring $V_{on} = 3$ V, $V_{off} = 0$ V, $I_{ph} = 300$ W/m^2, and $I_{pl} = 0$ W/m^2.

Figure 8 shows the simulation results of the multi-valued logic circuit (including states I and II). A_1 (the orange solid line) and A_2 (the red dashed line) indicate the input electrical signals, which are represented in the figure as the first column E(0, 0), the second column

E(1, 1), the third column E(0, 1), and the last column E(1, 0) for four electrical writing cases. B_1 (the purple dotted line), B_2 (the green dashed line), and B_3 (the pink dashed line) indicate whether or not an optical signal is applied to the memristor, written by O(i, j, z), and correspond to the optical case in the previous section. V_{out1} and V_{out2} of state *I* and *II* are indicated by the yellow solid and the blue dashed lines, respectively. The logical state variables are represented by the yellow dashed and blue solid lines, respectively, denoted as L(m, n).

As in the theoretical analysis, the simulation results are classified into Case A–Case F depending on whether or not illumination is applied to the memristors. Under Case A, Case C, Case D, and Case F, the simulation results are further divided into four cases according to whether the voltage is applied at the input ports, corresponding to the first row, the fourth row, the fifth row, and the last row in Figure 8. In addition, Case B (Case E) contains two cases, $B_2 = I_{pl}$, $B_3 = I_{ph}$ and $B_2 = I_{ph}$, $B_3 = I_{pl}$, for which the simulation results in Figure 8 contain eight small diagrams. From the first column of Figure 8, the voltage is not available at the output regardless of whether there is a light input or not, which corresponds to logic "0", since there is no voltage at the input. From Figure 8, the relationship between the input logic state variables and the output logic state variables obtained from the simulation corresponds to the truth table (Table 4), demonstrating that the circuit is capable of implementing multiple logic functions from 0–1, as well as verifying the validity of the constructed optoelectronic memristor model.

Case A

Case B

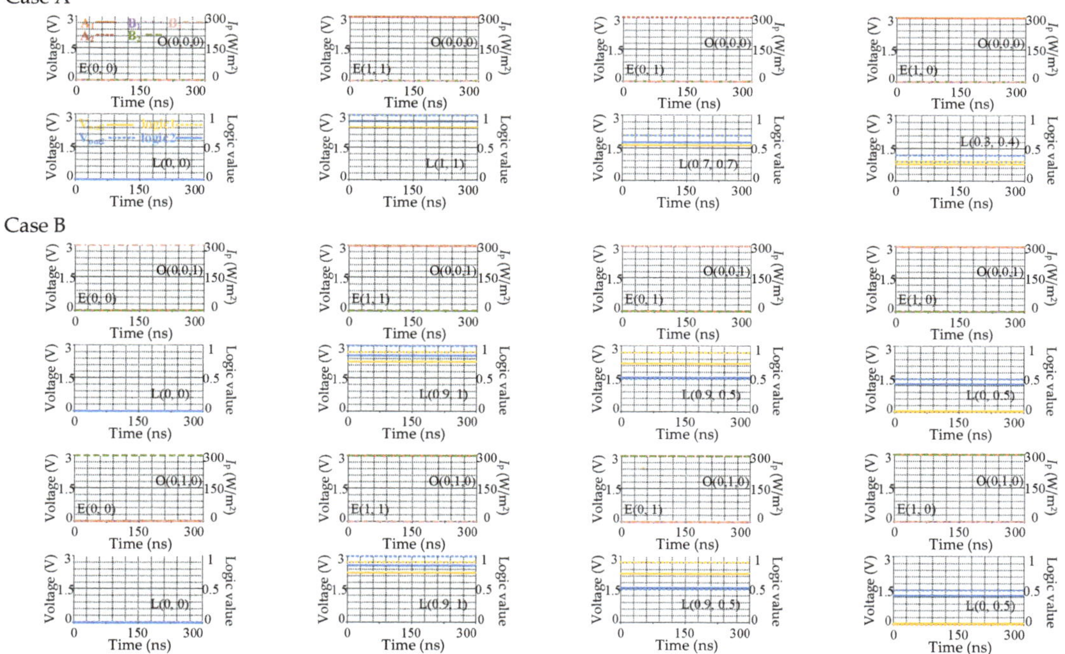

Figure 8. *Cont.*

Case C

Case D

Case E

Case F

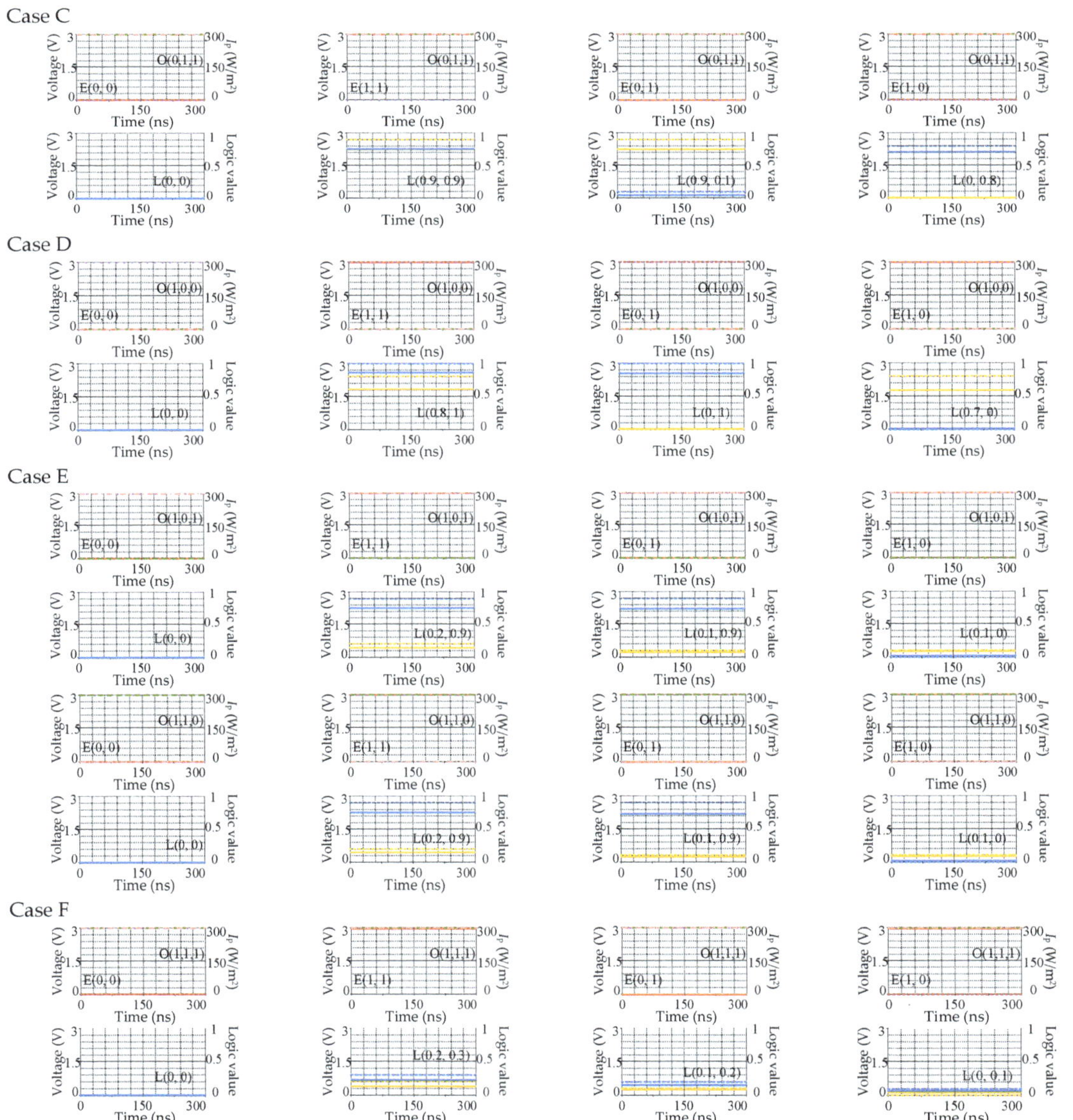

Figure 8. The simulation results of multi-valued logic circuit (including states *I* and *II*).

Then, the proposed multi-valued logic method is compared with five existing logic methods (i.e., material implication logic [29], memristor-aided logic [30], memristor ratioed logic [31], balanced ternary logic [32], and unbalanced ternary logic [33]). The comparison results including input variable, output variable, memristor type, computation form, need of resistors or transistors, initialization, cascading capacity and logic values are shown in Table 5.

Table 5. Comparison results of the proposed multi-valued logic circuits with other logic circuits.

	Proposed Logic	Material Implication Logic	Memristor-Aided Logic	Memristor Ratioed Logic	Balanced Ternary Logic	Unbalanced Ternary Logic
Input Variable	Voltage, illumination	M [1]	M [1]	Voltage	Voltage	Voltage
Output variable	Voltage	M [1]	M [1]	Voltage	Voltage	Voltage
Memristor type	Optoelectronic	HP	TEAM	VTEAM	VTEAM	Spintronic
Computation form	Parallel	Serial	Serial	Parallel	Parallel	Parallel
Need of resistors or transistors	√	√	×	√	√	√
Initialization	√	√	√	×	√	√
Cascading capacity	possible	difficult	difficult	possible	possible	possible
Logic values	Multi-valued	Binary	Binary	Binary	Ternary	Ternary

[1] M represents the memristance.

In Table 5, the proposed logic method differs from the other five logic methods in terms of input logic state variables by adding illumination variables that can affect the electrical characteristics of the device. The proposed logic circuit is easy to cascade because light is accessible and the output logic state variable is voltage. The memristor-aided logic has the simplest circuit structure and requires no additional circuit components other than the memristors. However, this method is calculated in series as in material implication logic, and the calculation process is more complicated. The proposed method requires initialization compared to the memristor ratioed logic, which increases the operation cost. In addition, the proposed method is able to implement multi-valued logic, while the other methods are restricted to binary and ternary logic.

4. Discussion

Currently, most of the research in the field of memristor logic implementation is aimed at binary logic and ternary logic. Nevertheless, relatively little research has been done to implement multi-valued logic circuits based on memristors, and there are abundant opportunities and challenges. The proposed multi-valued logic circuit cannot implement "0.5" in logic 0–1, and the circuit structure will be further improved to achieve complete logic functions in the future. Since the discrete logic output of the designed circuit cannot realize the affiliation function, continuous logic output will be explored in the future to build fuzzy systems.

5. Conclusions

This paper focuses on the modelling method of optoelectronic memristors. Specifically, the mathematical and circuit models of the optoelectronic memristor are developed using the optoelectronic effect that affects the electrical characteristics of such devices. Notably, the modelling approach is based on an improvement of the popular VTEAM modelling method. Moreover, the electrical characteristics (referring to the volt-ampere characteristics and memristance variation rule) of a single memristor and its series-parallel circuit under different illumination conditions and different voltages are tested. Furthermore, a rotation mechanism is introduced to realize the conversion between series and parallel circuits, and a multi-valued logic circuit containing two states (state *I* and state *II*) is designed. Simulation results demonstrate that the designed circuit is capable of implementing 10 logic functions from 0–1, which verifies the effectiveness of the established optoelectronic memristor model as well as providing a new approach to the circuit implementation of fuzzy logic.

Author Contributions: Methodology, J.W.; software, Y.L. and C.H.; writing—original draft preparation, J.W. and S.Z.; conceptualization, J.W. and S.G.; writing—review and editing, J.W. and Y.L.; visualization, M.Y.; supervision, Y.Y.; funding acquisition, G.M. All authors have read and agreed to the published version of the manuscript.

Funding: This research was supported by G.M. and Y.Y., and it was funded by National Natural Science Foundation of China grant numbers 62001149 and 62001416, Fundamental Research Funds for the Provincial Universities of Zhejiang grant number GK229909299001-06, and Natural Science Foundation of Zhejiang Province grant number LQ21F010009.

Data Availability Statement: Not applicable.

Acknowledgments: The authors would like to thank the editorial board and reviewers for the improvement of this paper.

Conflicts of Interest: The authors declare no conflict of interest.

References

1. Yang, X.; Taylor, B.; Wu, A.; Chen, Y.; Chua, L.O. Research progress on memristor: From synapses to computing systems. *IEEE Trans. Circuits Syst. I: Regul. Pap.* **2022**, *69*, 1845–1857. [CrossRef]
2. Ji, X.; Dong, Z.; Lai, C.S.; Qi, D. A brain-inspired in-memory computing system for neuronal communication via memristive circuits. *IEEE Commun. Mag.* **2022**, *60*, 100–106. [CrossRef]
3. Zhong, Y.; Tang, J.; Li, X.; Liang, X.; Liu, Z.; Li, Y.; Xi, Y.; Yao, P.; Hao, Z.; Gao, B.; et al. A memristor-based analogue reservoir computing system for real- time and power-efficient signal processing. *Nat. Electron.* **2022**, *5*, 672–681. [CrossRef]
4. Zhong, Y.; Tang, J.; Li, X.; Gao, B.; Qian, H.; Wu, H. Dynamic memristor-based reservoir computing for high-efficiency temporal signal processing. *Nat. Commun.* **2021**, *12*, 1–9. [CrossRef] [PubMed]
5. Dong, Z.; Ji, X.; Zhou, G.; Gao, M.; Qi, D. Multimodal neuromorphic sensory-processing system with memristor circuits for smart home applications. *IEEE Trans. Ind. Appl.* **2022**. [CrossRef]
6. Chua, L. Memristor-the missing circuit element. *IEEE Trans. Circuit Theory* **1971**, *18*, 507–519. [CrossRef]
7. Strukov, D.B.; Snider, G.S.; Stewart, D.R.; Williams, R.S. The missing memristor found. *Nature* **2008**, *453*, 80–83. [CrossRef]
8. Liao, K.; Lei, P.; Tu, M.; Luo, S.; Jiang, T.; Jie, W.; Hao, J. Memristor based on inorganic and organic two-dimensional materials: Mechanisms, performance, and synaptic applications. *ACS Appl. Mater.* **2021**, *13*, 32606–32623. [CrossRef]
9. Dong, Z.; Ji, X.; Lai, C.S.; Qi, D.; Zhou, G.; Lai, L.L. Memristor-based hierarchical attention network for multimodal affective computing in mental health monitoring. *IEEE Consum. Electr. Mag.* **2022**. [CrossRef]
10. Shen, Z.; Zhao, C.; Zhao, T.; Xu, W.; Liu, Y.; Qi, Y.; Mitrovic, I.Z.; Yang, L.; Zhao, C.Z. Artificial synaptic performance with learning behavior for memristor fabricated with stacked solution-processed switching layers. *ACS Appl. Electron. Mater.* **2021**, *3*, 1288–1300. [CrossRef]
11. Ji, X.; Lai, C.S.; Zhou, G.; Dong, Z.; Qi, D.; Lai, L.L. A flexible memristor model with electronic resistive switching memory behavior and its application in spiking neural network. *IEEE Trans. Nanobioscience* **2022**, *22*, 52–62. [CrossRef]
12. Ji, X.; Dong, Z.; Lai, C.S.; Zhou, G.; Qi, D. A physics-oriented memristor model with the coexistence of NDR effect and RS memory behavior for bio-inspired computing. *Mater. Today Adv.* **2022**, *16*, 100293. [CrossRef]
13. Khalid, M. Review on various memristor models, characteristics, potential applications, and future works. *Trans. Electr. Electron. Mater.* **2019**, *20*, 289–298. [CrossRef]
14. Li, J.; Dong, Z.; Luo, L.; Duan, S.; Wang, L. A novel versatile window function for memristor model with application in spiking neural network. *Neurocomputing* **2020**, *405*, 239–246. [CrossRef]
15. Li, T.; Duan, S.; Liu, J.; Wang, L.; Huang, T. A spintronic memristor-based neural network with radial basis function for robotic manipulator control implementation. *IEEE Trans. Syst. Man Cybern. Syst.* **2015**, *46*, 582–588. [CrossRef]
16. Kvatinsky, S.; Friedman, E.G.; Kolodny, A.; Weiser, U.C. TEAM: Threshold adaptive memristor model. *IEEE Trans. Circuits Systems I Regul. Pap.* **2012**, *60*, 211–221. [CrossRef]
17. Kvatinsky, S.; Ramadan, M.; Friedman, E.G.; Kolodny, A. VTEAM: A general model for voltage- controlled memristors. *IEEE Trans. Circuits Syst. II* **2015**, *62*, 786–790. [CrossRef]
18. Wang, X.; Li, P.; Jin, C.; Dong, Z.; Iu, H.H. General modeling method of threshold-type multivalued memristor and its application in digital logic circuits. *Int. J. Bifurcat. Chaos* **2021**, *31*, 2150248. [CrossRef]
19. Dong, Z.; Ji, X.; Lai, C.S.; Qi, D. Design and implementation of a flexible neuromorphic computing system for affective communication via memristive circuits. *IEEE Commun. Mag.* **2022**. [CrossRef]
20. Dong, Z.; Qian, Z.; Zhou, G.; Ji, X.; Qi, D.; LAI, J. Memristor-based full-function pavlov associative memory circuit design, implementation and analysis. *J. Electron. Inf. Techn* **2021**, *43*, 1–13.
21. Berco, D.; Ang, D.S.; Kalaga, P.S. Programmable photoelectric memristor gates for in situ image compression. *Adv. Intell. Syst.* **2020**, *2*, 2000079. [CrossRef]

22. Zhou, J.; Li, W.; Chen, Y.; Lin, Y.-H.; Yi, M.; Li, J.; Qian, Y.; Guo, Y.; Cao, K.; Xie, L.; et al. A monochloro copper phthalocyanine memristor with high-temperature resilience for electronic synapse applications. *Adv. Mater.* **2021**, *33*, 2006201. [CrossRef] [PubMed]
23. Zhang, X.; Zhao, X.; Shan, X.; Shan, X.; Tian, Q.; Wang, Z.; Lin, Y.; Xu, H.; Liu, Y. Humidity effect on resistive switching characteristics of the CH$_3$NH$_3$PbI$_3$ memristor. *ACS Appl. Mater. Inter.* **2021**, *13*, 28555–28563. [CrossRef] [PubMed]
24. Cao, J.; Zhang, X.; Cheng, H.; Qiu, J.; Liu, X.; Wang, M.; Liu, Q. Emerging dynamic memristors for neuromorphic reservoir computing. *Nanoscale* **2022**, *14*, 289–298. [CrossRef]
25. Liu, G.; Shen, S.; Jin, P.; Wang, G.; Liang, Y. Design of memristor-based combinational logic circuits. *Circ. Syst. Signal Pr.* **2021**, *40*, 5825–5846. [CrossRef]
26. Xu, N.; Park, T.; Yoon, K.J.; Hwang, C.S. In-memory stateful logic computing using memristors: Gate, calculation, and application. *Phys. Status Solidi Rapid Res. Lett.* **2021**, *15*, 2100208. [CrossRef]
27. Liu, B.; Zhao, Y.; Verma, D.; Wang, L.A.; Liang, H.; Zhu, H.; Li, L.-J.; Hou, T.-H.; Lai, C.-S. Bi$_2$O$_2$Se-based memristor-aided logic. *ACS Appl. Mater. Inter.* **2021**, *13*, 15391–15398. [CrossRef] [PubMed]
28. Song, Y.; Wu, Q.; Wang, X.; Wang, C.; Miao, X. Two memristors-based XOR logic demonstrated with encryption/decryption. *IEEE Electron Device Lett.* **2021**, *42*, 1398–1401. [CrossRef]
29. Sun, B.; Ngai, J.H.; Zhou, G.; Zhou, Y.; Li, Y. Voltage-controlled conversion from CDS to MDS in an azobenzene-based organic memristor for information storage and logic operations. *ACS Appl. Mater. Inter.* **2022**, *14*, 41304–41315. [CrossRef]
30. Wang, Z.; Wang, L.; Duan, S. Memristor ratioed logic crossbar-based delay and jump-key flip-flops design. *Inter. J. Circuit Theory Appl.* **2022**, *50*, 1353–1364.2. [CrossRef]
31. Dong, Z.; Qi, D.; He, Y.; Xu, Z.; Hu, X.; Duan, S. Easily cascaded memristor-CMOS hybrid circuit for high-efficiency boolean logic implementation. *Int. J. Bifurcat. Chaos* **2018**, *28*, 1850149. [CrossRef]
32. Jha, C.K.; Thangkhiew, P.L.; Datta, K.; Drechsler, R. IMAGIN: Library of IMPLY and MAGIC NOR based approximate adders for in-memory computing. *IEEE J. Explor. Solid-St. Compu. Devices Circuits* **2022**, *8*, 68–76. [CrossRef]
33. Zhang, H.; Zhang, Z.; Gao, M.; Luo, L.; Duan, S.; Dong, Z.; Lin, H. Implementation of unbalanced ternary logic gates with the combination of spintronic memristor and CMOS. *Electronics* **2020**, *9*, 542. [CrossRef]
34. Wang, X.Y.; Dong, C.T.; Wu, Z.R.; Cheng, Z.Q. A review on the design of ternary logic circuits. *Chin. Phys. B* **2021**, *30*, 128402. [CrossRef]
35. Zhang, Z.; Xu, A.; Li, C.; Liu, G.; Cheng, X. Mathematical analysis and circuit emulator design of the three-valued memristor. *Integration* **2022**, *86*, 74–83. [CrossRef]
36. Yang, J.; Lee, H.; Jeong, J.H.; Kim, T.; Lee, S.H.; Song, T. Circuit-level exploration of ternary logic using memristors and MOSFETs. *IEEE Trans. Circuits Syst. I Regul. Pap.* **2021**, *69*, 707–720. [CrossRef]
37. Dong, Z.; Lai, C.S.; Qi, D.; Xu, Z.; Li, C.; Duan, S. A general memristor-based pulse coupled neural network with variable linking coefficient for multi-focus image fusion. *Neurocomputing* **2018**, *308*, 172–183. [CrossRef]

Article

Two-Neuron Based Memristive Hopfield Neural Network with Synaptic Crosstalk

Rong Qiu [1], Yujiao Dong [2,*], Xin Jiang [2] and Guangyi Wang [2]

[1]   School of Internet of Things Engineering, Guangdong Polytechnic of Science and Technology, Guangzhou 510640, China

[2]   Institute of Modern Circuits and Intelligent Information, Hangzhou Dianzi University, Hangzhou 310018, China

*   Correspondence: yjdong@hdu.edu.cn

Abstract: Synaptic crosstalk is an important biological phenomenon that widely exists in neural networks. The crosstalk can influence the ability of neurons to control the synaptic weights, thereby causing rich dynamics of neural networks. Based on the crosstalk between synapses, this paper presents a novel two-neuron based memristive Hopfield neural network with a hyperbolic memristor emulating synaptic crosstalk. The dynamics of the neural networks with varying memristive parameters and crosstalk weights are analyzed via the phase portraits, time-domain waveforms, bifurcation diagrams, and basin of attraction. Complex phenomena, especially coexisting dynamics, chaos and transient chaos emerge in the neural network. Finally, the circuit simulation results verify the effectiveness of theoretical analyses and mathematical simulation and further illustrate the feasibility of the two-neuron based memristive Hopfield neural network hardware.

Keywords: memristor; Hopfield neural network; chaos; synaptic crosstalk; coexisting dynamics

Citation: Qiu, R.; Dong, Y.; Jiang, X.; Wang, G. Two-Neuron Based Memristive Hopfield Neural Network with Synaptic Crosstalk. *Electronics* **2022**, *11*, 3034. https://doi.org/10.3390/electronics11193034

Academic Editors: Chun Sing Lai, Zhekang Dong and Donglian Qi

Received: 5 September 2022
Accepted: 21 September 2022
Published: 23 September 2022

Publisher's Note: MDPI stays neutral with regard to jurisdictional claims in published maps and institutional affiliations.

1. Introduction

Neural systems contribute to processing information in brains, where neurons and synapses play an important role in transmitting information. It is reported that crosstalk exists between synapses because of their interaction [1]. When the neurotransmitter overflow between synapses or the diffusion of receptors between neighboring ridges emerges, the signal transmission may be affected, thereby inducing the variation in the functions of brains.

Memristor, defined by Chua in 1971 and physically realized by HP lab in 2008, can be used to emulate synaptic functions [2,3]. By adjusting the voltage across the memristor, the change in memristances can emulate the plasticity of synaptic weights, showing the activity-dependent change in neuronal connection strength. Thus, a memristor is a good choice to replace the fixed resistor-based synapses in the neural networks, which can flexibly solve different kinds of combinatorial optimization problems, such as MAX-CUT problems and TSP (Traveling Salesman Problem) [4].

Recently, memristive neural networks have attracted more and more attention [5–7]. The memristor-based cellular neural network was applied to image processing, which has nonvolatility, compactness, and programmability of synaptic weights [5]. The memristor-based pulse coupled neural network was designed to solve image fusion problems, which improves the quality of images [6]. Ma et al. established a memristor-based Hopfield neural network and emulated human emotions via the circuit simulation [7].

The Hopfield neural network (HNN) proposed by Hopfield is a well-known and typical artificial neural network [8], which has the ability to emulate complex dynamics of the human brain, such as chaos. After that, the HNN is widely applied in associative memory, image processing, combinatorial optimization, and so on [9–13]. Reference [9] proposed a novel algorithm called Teamwork Optimization Algorithm (TOA) to solve the optimization problems. Kasihmuddin et al. presented an integrated representation of *k*-satisfiability (*k*SAT) in a mutation HNN (MHNN), which overcomes the overfitting issue [10]. Citko et al. set up associative memories to retrieve images based on the HNN [11]. Reference [12] embedded the logical rule $P_{RAN3SAT}$ in HNN and optimized the ability of retrieval. Rubio-Manzano et al. created a complete explainer video about the HNN on a recognition problem [13].

Considering the great advantages of memristors, researchers established Hopfield neural networks based on memristors [14–18]. Sun et al. achieved the recognition and sequence of four characters [16]. Reference [17] explored nonlinear dynamics of a three-neuron based memristive HNN.

This paper aims to study more complex nonlinear behaviors via a simple two-neuron based HNN. In this paper, a two-neuron based memristive HNN with synaptic crosstalk is established. By analyzing the dissipation and stability of the HNN, many different types of coexisting dynamics are found. It is verified that the memristive parameter and crosstalk weight have a significant influence on the complexity of the HNN via the bifurcation diagram, basin of attraction, and so on. Finally, the circuit simulation results verify the effectiveness of theoretical analyses.

2. Simplest Hyperbolic Memristive Synapse-Coupled HNN

2.1. Hyperbolic Memristive Synapse Emulator

Neuron activation functions are used to transform the output signal of the former neuron into the input signal of the latter neuron, where the sigmoid nonlinear activation function is commonly used.

Here, we use the hyperbolic tangent function with zero-mean as the activation function, which has a higher range and greater slope than the sigmoid activation function. The mathematical description of the hyperbolic tangent function is

$$v_o = -\tanh(v_i) \tag{1}$$

The equivalent circuit of the inverting hyperbolic tangent function is shown in Figure 1, including two operational amplifiers TL084 (U_i and U_o), two transistors MPS2222 (Q_1 and Q_2), one current source $I_0 = 1.1$ mA, and several resistors ($R_1 = R_5 = R_6 = R_7 = R_8 = 10$ kΩ, $R_2 = 0.52$ kΩ, $R_3 = R_4 = 1$ kΩ), where V_{in} and V_{out} are the input and output voltage, respectively. The operational amplifiers U_i and U_o finished the inversion of the input and subtraction operation, respectively. The transistors Q_1 and Q_2 realized the exponential operation.

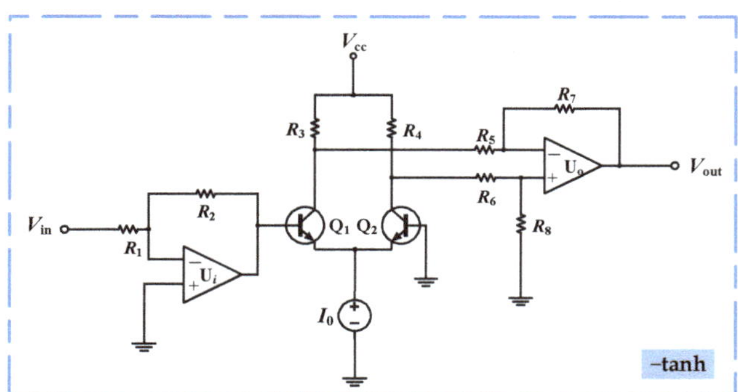

Figure 1. Circuit scheme of inverting hyperbolic tangent function.

The obtained simulated results of the inverting hyperbolic tangent circuit using Pspice are shown in Figure 2.

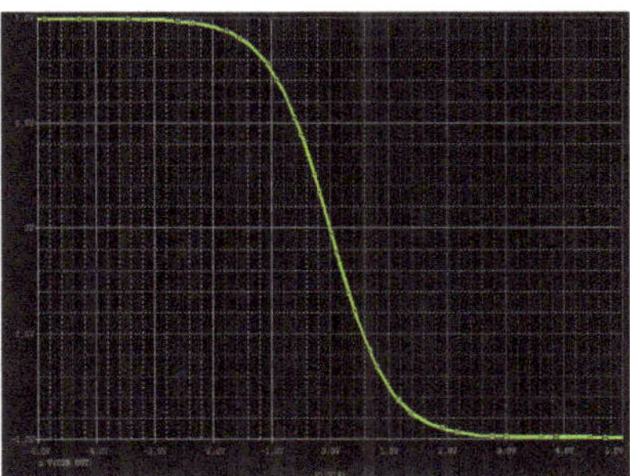

Figure 2. Simulated results of the inverting hyperbolic tangent circuit.

The generic model of hyperbolic tangent-type memristor emulator can be used to emulate synaptic weights of neurons, which is described as

$$i = W(x)v = [a - b\tanh(x)]v \tag{2}$$

where v, i, and x represent the voltage, current, and state variable of the memristor; a and b are the memristor parameters, $a > 0$, $b > 0$.

Based on Equation (2), the circuit equation of the memristor can be defined as

$$i = W(x)v = [-\frac{1}{R_a} - \frac{1}{R_b}\tanh(x)]v \tag{3}$$

From Equation (3), one can establish the hyperbolic tangent memristor emulator, as depicted in Figure 3, including an operational amplifier TL084 (U_o), a capacitor ($C = 100$ nF), four resistors ($R = 10$ kΩ, R_a and R_b are adjustable), a multiplier AD633, and a model of -tanh.

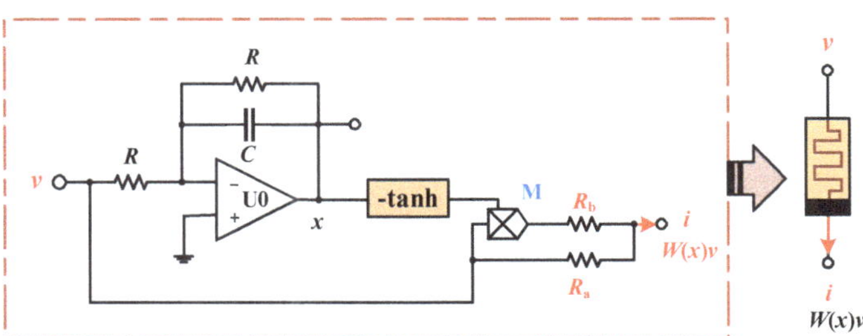

Figure 3. Hyperbolic tangent memristor emulator.

A coupled memristor emulator is obtained by coupling the two same hyperbolic tangent memristors [19], as shown in Figure 4, which can emulate the crosstalk between synapses. Thus, the synaptic weights between neurons can be described as

$$\begin{cases} W_1 = a_1 - b_1\tanh(z) + c_1\tanh(u) \\ W_2 = a_2 - b_2\tanh(u) + c_2\tanh(z) \end{cases} \tag{4}$$

where a_1, b_1, a_2 and b_2 are memristor parameters; c_1 and c_2 are crosstalk strength parameters; $a_1 = \frac{R}{R_{a1}}$, $b_1 = \frac{R}{R_{b1}}$, $c_1 = \frac{gR^2}{R_{b1}R_{c1}}$, $a_2 = \frac{R}{R_{a2}}$, $b_2 = \frac{R}{R_{b2}}$, $c_2 = \frac{gR^2}{R_{b2}R_{c2}}$.

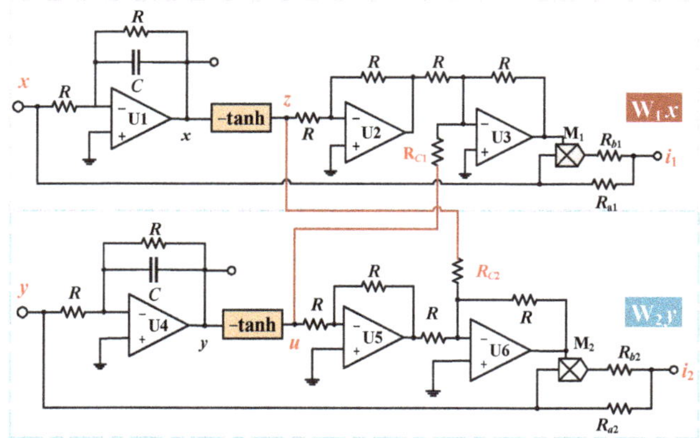

Figure 4. Hyperbolic-type memristor emulator with crosstalk.

2.2. Two Neurons-Based HNN Model

The Hopfield neural network (HNN), a fully interconnected neural network, can be used to describe dynamic behaviors of human brains [20]. An n-neuron-based HNN is defined as

$$C_i\frac{dx_i}{dt} = -\frac{x_i}{R_i} + \sum_{i=1}^{n} w_{ij}\tanh(x_i) + I_i \tag{5}$$

where C_i, R_i and x_i represent membrane capacitance, membrane resistance and voltage; $\tanh(x_i)$ is the activation function of neuron I; w_{ij} is the synaptic weight between neurons i and j; I_i is the biasing current.

Here, we design a novel two-neuron-based memristive HNN, as depicted in Figure 5, which can emulate synaptic crosstalk. The mathematical description of the established HNN is

$$\begin{cases} \dot{x} = -x + w_{11}\tanh(x) - k_2 W_2\tanh(y) \\ \dot{y} = -y + k_1 W_1\tanh(x) + w_{22}\tanh(y) \\ \dot{z} = -z + \tanh(x) \\ \dot{u} = -u + \tanh(y) \end{cases} \tag{6}$$

where $k_1 = 1, k_2 = 1$; w_{11} and w_{22} are self-connected synaptic weights; $W_1 = a_1 - b_1\tanh(z) + c_1\tanh(u)$ and $W_2 = a_2 - b_2\tanh(u) + c_2\tanh(z)$ are mutual synaptic weights between neuron 1 and neuron 2. The synaptic weight matrix is

$$W_{ij} = \begin{pmatrix} w_{11} & w_{12} \\ w_{21} & w_{22} \end{pmatrix} = \begin{pmatrix} w_{11} & -k_2 W_2 \\ k_1 W_1 & w_{22} \end{pmatrix} \tag{7}$$

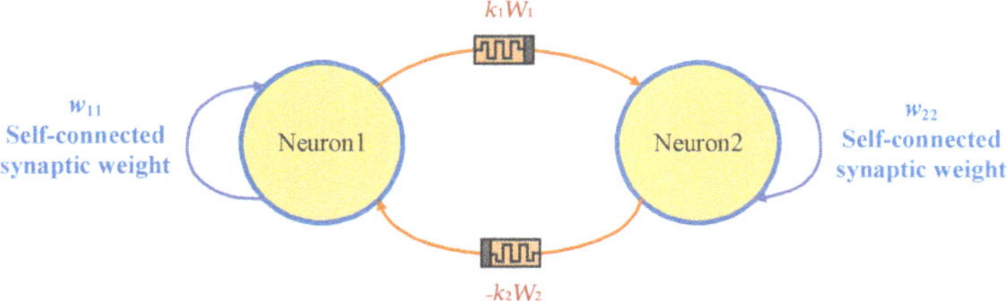

Figure 5. The simplest memristive HNN with synaptic crosstalk.

3. Dissipativity and Stability of HNN

3.1. Dissipativity Analyses

A dissipative characteristic is necessary for a system or network to generate chaos. Thus, the volumetric shrinkage rate Λ should be calculated to verify that the system in Equation (6) is dissipative based on $V(t) = Ve^{\Lambda t}$. If $\Lambda < 0$, the system is dissipative, and chaos may emerge; when $\Lambda = 0$, the system is called a conservative system; the divergent phenomenon will occur in the system when $\Lambda > 0$.

According to the method in reference [21], the Lyapunov function is introduced as

$$V(x,y,z,u) = \frac{1}{2}\left(x^2 + y^2 + z^2 + u^2\right) \tag{8}$$

whose corresponding time derivative is

$$\begin{aligned}
\dot{V}(x,y,z,u) &= x\dot{x} + y\dot{y} + z\dot{z} + u\dot{u} \\
&= -\left(x^2 + y^2 + z^2 + u^2\right) + v(x,y,z,u) \\
&= -2V(x,y,z,u) + v(x,y,z,u)
\end{aligned} \tag{9}$$

where

$$v(x,y,z,u) = (w_{11}x + W_1x + z)\tanh(x) + (w_{22}y + W_2x + u)\tanh(y) \tag{10}$$

Since $\tanh(\zeta) \in (-1,1)$ for all $\zeta = x, y, u, z$, one can obtain

$$\begin{aligned}
|W_1| &= a_1 - b_1\tanh(z) + c_1\tanh(u) \\
&\leq M_1 = \max\{|a_1 - b_1 + c_1|, |a_1 + b_1 + c_1|\} \\
|W_2| &= a_2 - b_2\tanh(u) + c_2\tanh(z) \\
&\leq M_2 = \max\{|a_2 - b_2 + c_2|, |a_2 + b_2 + c_2|\}
\end{aligned} \tag{11}$$

So we have

$$\begin{aligned}
v(x,y,z,u) &\leq |(w_{11}x + W_1y + z)\tanh(x)| + |(w_{22}y + W_2x + u)\tanh(y)| \\
&\leq (w_{11} + M_2)|x| + (M_1 + w_{22})|y| + |z| + |u|
\end{aligned} \tag{12}$$

Suppose that all state variables (x, y, z, u) satisfy $V(x, y, z, u) = D$ for $D > D_0$ ($D_0 > 0$ is a sufficiently large domain), it requires

$$\begin{aligned}
v(x,y,z,u) &< (w_{11} + M_2)|x| + (M_1 + w_{22})|y| + |z| + |u| \\
&< x^2 + y^2 + z^2 + u^2 = 2V(x,y,z,u)
\end{aligned} \tag{13}$$

where $w_{11} + M_2$ and $M_1 + w_{22}$ are positive constants. So

$$\{(x,y,z,u)|V(x,y,z,u) = D\} \tag{14}$$

Since $D > D_0$, one can obtain

$$\dot{V} = -2V(x,y,z,u) + v(x,y,z,u) < 0 \tag{15}$$

Based on Equation (15), the confined domain of the solutions in Equation (6) is given, as

$$\{(x,y,z,u)|V(x,y,z,u) \leq D\} \tag{16}$$

Thus, the memristive HNN in Equation (6) is bounded, which has the possibility to generate chaos.

3.2. Stability Analyses

From Equation (6), the equilibria $P = (\overline{x}, \overline{y}, \overline{z}, \overline{u}, \overline{w})$ can be calculated by

$$\begin{cases} 0 = -\overline{x} + w_{11}\tanh(\overline{x}) - k_2 W_2 \tanh(\overline{y}) \\ 0 = -\overline{y} + k_1 W_1 \tanh(\overline{x}) + w_{22}\tanh(\overline{y}) \\ 0 = -\overline{z} + \tanh(\overline{x}) \\ 0 = -\overline{u} + \tanh(\overline{y}) \end{cases} \tag{17}$$

where W_1 and W_2 are hyperbolic-type memristor emulator, as

$$\begin{cases} W_1 = a_1 - b_1\tanh(z) + c_1\tanh(u) \\ W_2 = a_2 - b_2\tanh(u) + c_2\tanh(z) \end{cases} \tag{18}$$

By solving Equations (17) and (18), one can obtain

$$\begin{cases} H_1(x,y) = -x + w_{11}\tanh(x) - k_2(a_2 - b_2\tanh(u) + c_2\tanh(z))\tanh(y) \\ H_2(x,y) = -y + k_1(a_1 - b_1\tanh(z) + c_1\tanh(u))\tanh(x) + w_{22}\tanh(y) \end{cases} \tag{19}$$

where the equilibria of the HNN are the intersection points between the curve $H_1(x, y)$ and $H_2(x, y)$.

Now, as an example, we set $a_1 = 1$, $a_2 = 2$, $b_1 = 0.04$, $b_2 = 0.03$, $c_1 = 5.55$ and $c_2 = 5.9$. The two curves $H_1(x, y)$ and $H_2(x, y)$ are shown in Figure 6. Based on Figure 6 and Equation (17), the obtained equilibria are P_0 (0, 0, 0, 0), P_1 (−1.7, −0.238, −0.935, 0.234), P_2 (−0.174, −1.116, −0.184, −0.806) and P_3 (1.737, −0.149, 0.940, −0.148).

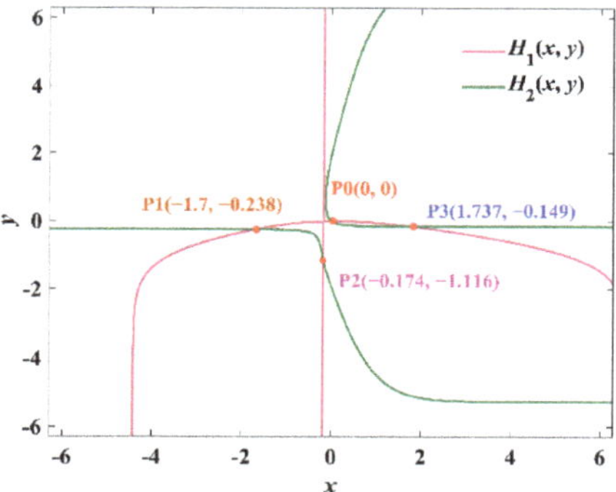

Figure 6. Equilibria of the memristive HNN, i.e., the intersection points of H_1 and H_2.

By linearizing Equation (6) at the equilibria $(\overline{x}, \overline{y}, \overline{z}, \overline{u})$, one obtains its Jacobian matrix as

$$J = \begin{bmatrix} -1 + w_{11}h_1 & -W_2 h_2 & -c_2\tanh(\overline{y})h_3 & b_2\tanh(\overline{y})h_4 \\ W_1 h_1 & -1 + w_{22}h_2 & -b_1\tanh(\overline{x})h_3 & c_1\tanh(\overline{x})h_4 \\ h_1 & 0 & -1 & 0 \\ 0 & h_2 & 0 & -1 \end{bmatrix} \tag{20}$$

where $h_1 = 1 - \tanh(\overline{x})$, $h_2 = 1 - \tanh(\overline{y})$, $h_3 = 1 - \tanh(\overline{z})$, $h_4 = 1 - \tanh(\overline{u})$.

According to the stability criterion, at least one positive eigenvalue causes an unstable equilibrium. The eigenvalues at different equilibria and their stability are listed in Table 1.

Table 1. Equilibria, eigenvalues and their stability of the HNN.

Equilibria	Eigenvalues	Stability
P_0 (0, 0, 0, 0)	$0.5000 \pm 0.8944i$, -1.0000, -1.0000	unstable
P_1 (-1.7, -0.238, -0.935, 0.234)	$-0.0760 \pm 1.6875i$, -1.2437, -0.5887	stable
P_2 (-0.186, -1.116, -0.184, -0.806)	1.6461, $-0.6511 \pm 0.2767i$, -2.6779	unstable
P_3 (1.737, -0.149, 0.940, -0.148)	2.3973, -2.4408, -1.1678, -0.7158	unstable

3.3. Chaotic Behaviors

Here, set the initial condition $(x_0, y_0, z_0, u_0) = (0.1, 0, 0, 0)$, $k_1 = 1$, $k_2 = 1$, $W_{11} = 1$, $W_{22} = 2$, $a_1 = 1$, $a_2 = 1$, $b_1 = 0.04$, $b_2 = 0.03$. The HNN system generates chaos shown in Figure 7. Figure 7a–d show chaotic attractors on the x-y phase, x-z phase, x-u phase and y-z phase, respectively. Then, we use the Poincaré map and Lyapunov exponent to verify the chaotic behaviors.

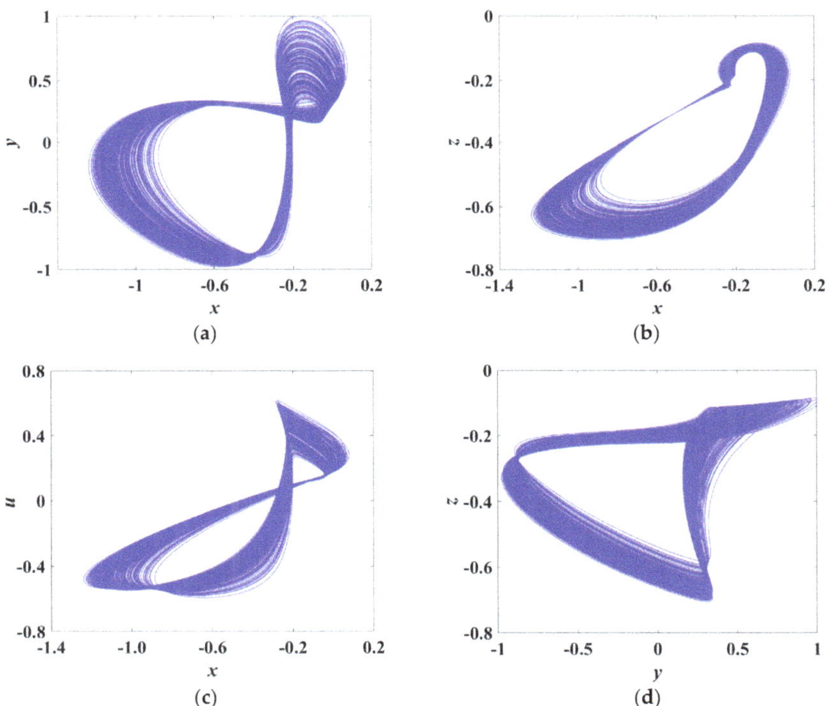

Figure 7. Chaotic attractors: (**a**) x-y phase portraits; (**b**) x-z phase portraits; (**c**) x-u phase portraits; (**d**) y-z phase portraits.

The Poincaré map is a qualitative method to verify chaotic phenomena. If there are one or several points on the Poincaré map, the system shows stable or periodic characteristics; a large number of dense points in the Poincaré map predict chaos. Figure 8 shows the Poincaré map when the cross-section is chosen as the plane $z = -0.1$. It is found that two continuous segments with dense points emerge on the x-y plane, indicating chaotic attractors.

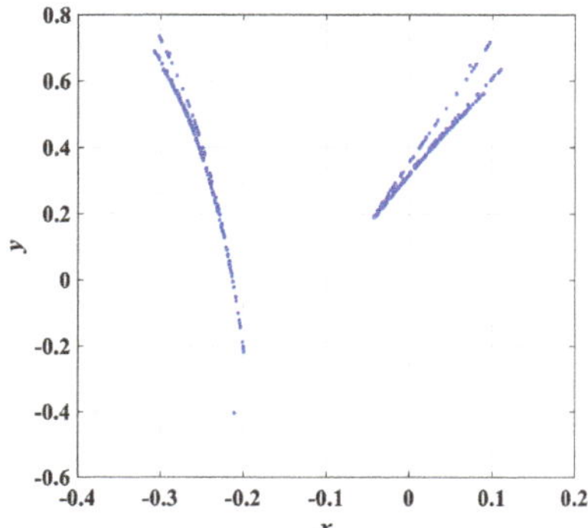

Figure 8. Poincaré map in the *x-y-z* space with the cross-section $z_0 = -0.1$.

The Lyapunov exponent is a quantitative method to judge chaotic attractors. By using the QR decomposition method to calculate the Jacobian matrix and its eigenvalues, the Lyapunov exponents of the system are calculated as $LE_1 = 0.058$, $LE_2 = 4.892 \times 10^{-4}$, $LE_3 = -0.179$, $LE_4 = -1.402$, and the corresponding Lyapunov dimension is $DL = 2.321$. Since the maximum Lyapunov exponent $LE_1 > 0$, the Hyperbolic-type memristive HNN produces chaos.

4. Dynamics Varying with Parameters

In this section, we choose two representative parameters a_2 and c_2 to study the influence on dynamics of the memristive HNN, where a_2 and c_2 are the memristive parameter and crosstalk parameter, respectively. We use the bifurcation diagram, Lyapunov exponent spectrum, and phase portraits to further explore the complex dynamics of two-neuron based HNNs varying with a_2 and c_2.

4.1. Influence of Memristive Parameter a_2 on Dynamics

The synaptic plasticity of neurons can be realized by adding memristors into neural networks, which is important for the HNN to solve many different kinds of combinatorial optimization problems including MAX-CUT problems, TSP, and so on. Changing the memristive parameter means adjusting the synaptic weight. Now, we set the initial condition $(x_0, y_0, z_0, u_0) = (0.1, 0, 0, 0)$, $W_{11} = 1.24$, $W_{22} = 0.75$, $a_1 = 1$, $b_1 = 0.03$, $b_2 = 0.02$, $c_1 = 5.7$, $c_2 = 5.9$. The bifurcation diagram of the HNN and corresponding Lyapunov exponent spectrum varying with a_2 over the range of [1.1, 1.5] are shown in Figure 9.

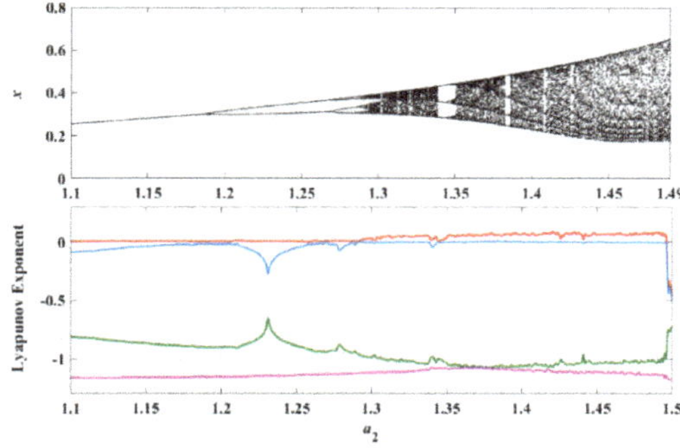

Figure 9. The bifurcation diagram and Lyapunov exponent spectrum of the HNN varying with $a_2 \in [1.1, 1.5]$.

Observe from Figure 9 that the memristive parameter a_2 has a great influence on the dynamics of the HNN. The chaotic and periodic region in the bifurcation diagram is almost consistent with that of the Lyapunov exponent spectrum. When $a_2 \in [1.1, 1.29]$, the HNN evolves from the single-period state into chaos via the period-doubling bifurcation. As a_2 gradually increases to 1.34, the HNN turns into the three-period state and then enters into the chaotic region. When $a_2 \in [1.34, 1.49]$, the HNN is mostly chaotic except for several narrow periodic windows. Interestingly, the HNN exhibits transient chaos when $a_2 = 1.5$ and finally shows non-chaotic phenomena.

The representative phase portraits of the HNN under the parameter $a_2 = 1.1, 1.2, 1.3, 1.4$ and 1.5 are depicted in Figure 10, corresponding to single-period (red), double-period (blue), quasi-period (green), chaos (purple) and transient chaos (pink), respectively.

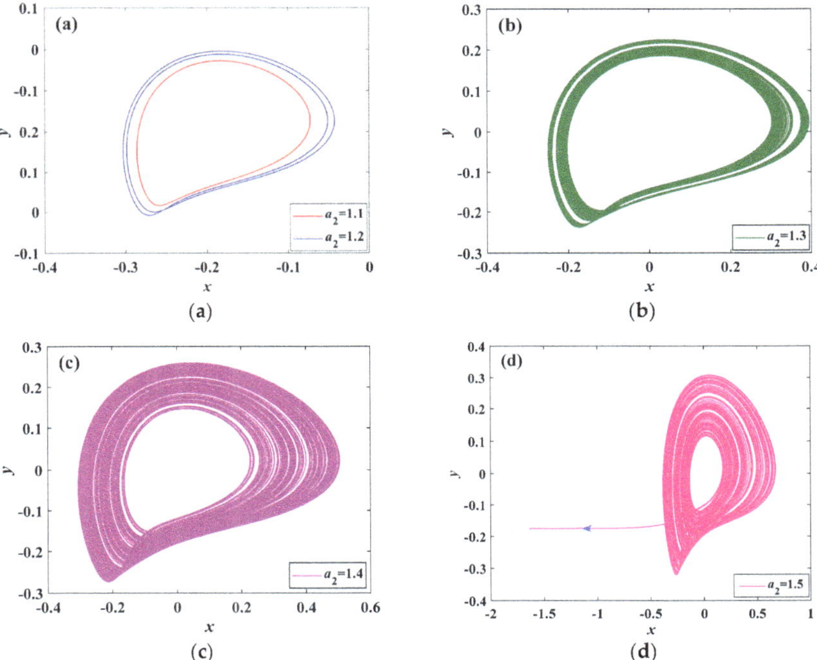

Figure 10. Phase portraits of the HNN under the parameter $a_2 = 1.1, 1.2, 1.3, 1.4, 1.5$: (**a**) $a_2 = 1.1, 1.2$; (**b**) $a_2 = 1.3$; (**c**) $a_2 = 1.4$; (**d**) $a_2 = 1.5$.

The time-domain waveforms under $a_2 = 1.4$ and $a_2 = 1.5$ are depicted in Figure 11, corresponding to red and green trajectories. Observe that the HNN shows chaotic states over the range of $t \in [200\,\mathrm{s}, 550\,\mathrm{s}]$ with $a_2 = 1.4$ and $a_2 = 1.5$. However, the HNN evolves from the chaotic state into the stable state over the range of $t \in [550\,\mathrm{s}, 800\,\mathrm{s}]$ when $a_2 = 1.5$. This special phenomenon is called transient chaos [22,23], with short-time chaotic behaviors.

Thus, changing memristive parameter a_2 can adjust synaptic weights, and finally, the dynamics of the HNN are easily controlled.

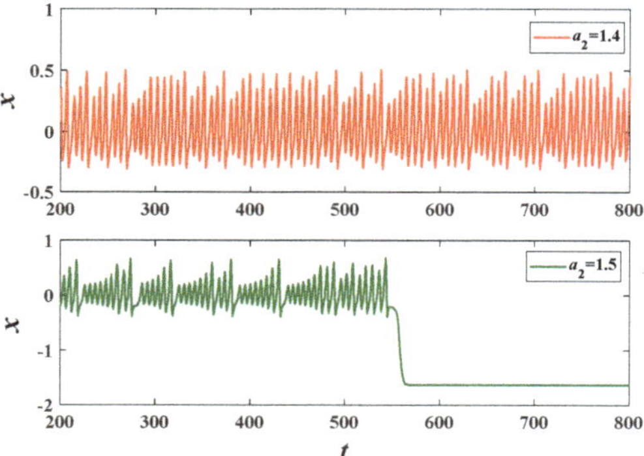

Figure 11. Time-domain waveforms with $a_2 = 1.4$ (red) and 1.5 (green).

4.2. Influence of Crosstalk Parameter c_2 on Dynamics

Here, set the parameter $c_1 = 5.56$. The bifurcation diagram of the HNN over the range of $c_2 \in [5.5, 6]$ is shown in Figure 12. Observe that the HNN exhibits stable states when $c_2 \in [5.5, 6.68]$. When $c_2 \in [5.68, 5.95]$, the system shows periodic states, transient chaos and chaos switching with each other. As c_2 increases to 5.95, the HNN always exhibits periodic states.

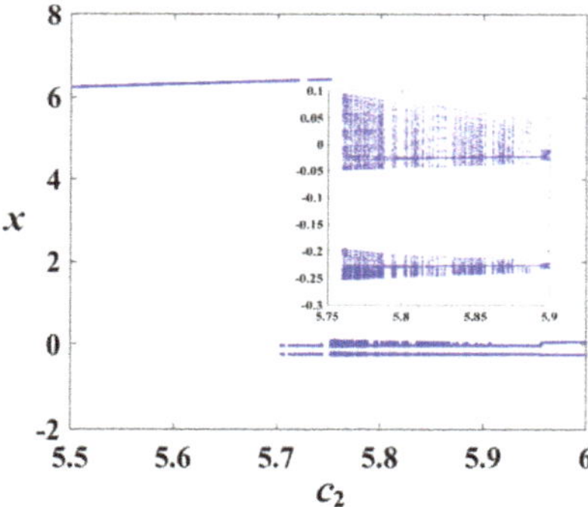

Figure 12. The bifurcation diagram of the HNN varying with crosstalk parameter c_2.

The HNN produces different attractors varying with the crosstalk parameter c_2, as shown in Figure 13. Notice that the transient chaos emerges in the HNN under $c_2 = 5.68$, as depicted in Figure 13a. In this case, the attractor is chaotic over a period of time, then switches into another nonchaotic behavior after the period of time. This phenomenon is a kind of special dynamics in neural networks, because it is difficult to find in a nonlinear system [23]. When solving the combinatorial optimization problem using the HNN, the introduction of transient chaos can help to jump from the local optimal solution to the global optimal solution. The HNN with transient chaos has stronger global search ability, so it has higher application values [24].

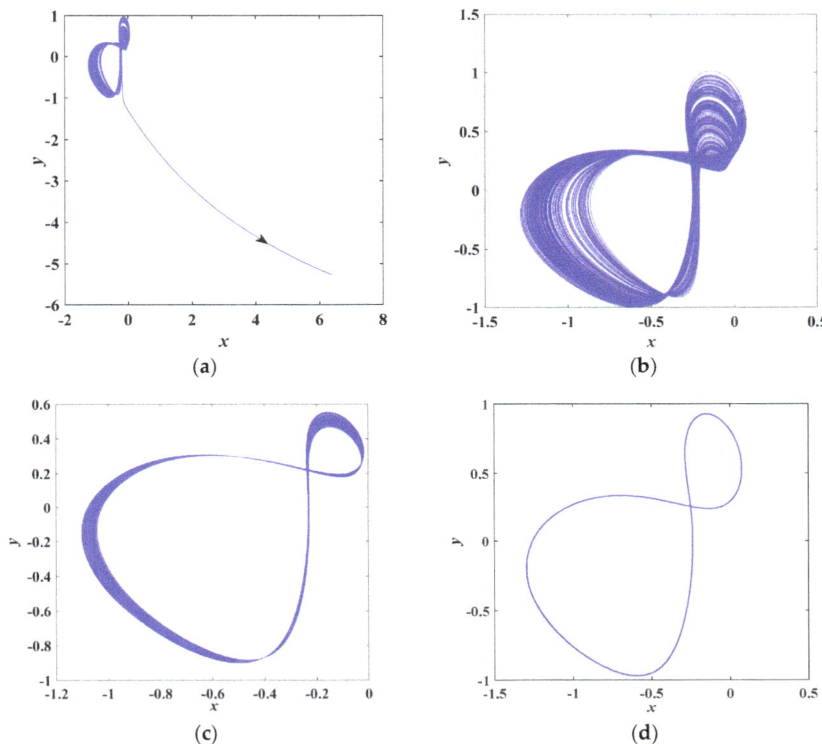

Figure 13. Phase portraits of the HNN under the parameter $c_2 = 5.68, 5.76, 5.82, 6$: (**a**) $c_2 = 5.68$; (**b**) $c_2 = 5.76$; (**c**) $c_2 = 5.82$; (**d**) $c_2 = 6$.

5. Sensitivity of Initial Conditions and Coexisting Behaviors

Chaos is sensitive to initial conditions, which is vividly described by the "butterfly effect" presented by Lorenz [25]. Small perturbations of initial conditions can eventually cause the separation of chaotic orbits. The sensitivity means the unpredictability of long-term nonlinear behaviors.

Coexisting phenomenon means different kinds of attractors emerge in a system when choosing the same system parameters and different initial values. If the obtained attractors have different dynamics, such as point attractors, periodic attractors, and chaotic attractors, these attractors are called inhomogeneous attractors. If the obtained attractors have the same dynamics but different gravity or shapes, these attractors are called homogenous attractors [26]. The generation of coexisting attractors indicates high sensitivity to initial conditions for the HNN, which also means rich dynamics.

Now, set the parameters $c_1 = 5.55$, $c_2 = 5.9$, and the initial value $(x(0), 0, z(0), 0)$. The three-dimensional bifurcation diagram of the state x varying with initial values $x(0)$ and $z(0)$ is depicted in Figure 14.

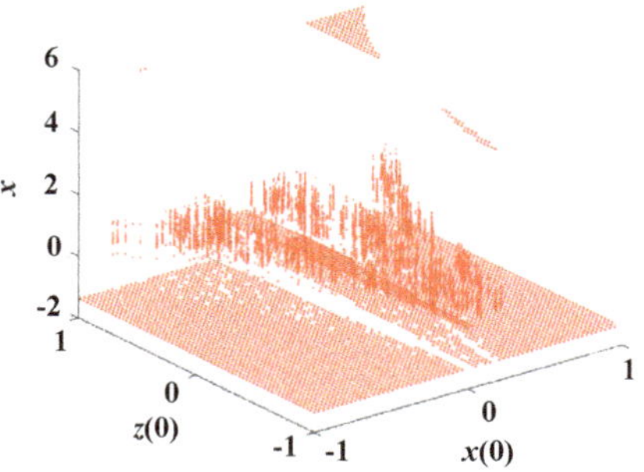

Figure 14. The three-dimensional bifurcation diagram of the HNN varying with $x(0)$ and $z(0)$.

Observe from Figure 14 that the chaotic and periodic orbits switch with each other over the range of $x(0) \in [-1,1]$ and $z(0) \in [-1,1]$, indicating that rich and complex dynamics emerge in the HNN.

When changing the initial values $x(0)$ and $z(0)$, the obtained basin of attraction and typical coexisting attractors on the x-y-z plane are shown in Figure 14.

Observe from Figure 15 that the two-neuron based HNN generates rich coexisting dynamics when changing initial values and fixing parameters, including the coexisting of periodic attractors and chaotic attractors, the coexisting of transient chaotic attractors and stable point attractors, the coexisting of chaotic attractors, periodic attractors and point attractors. The details are listed in Table 2.

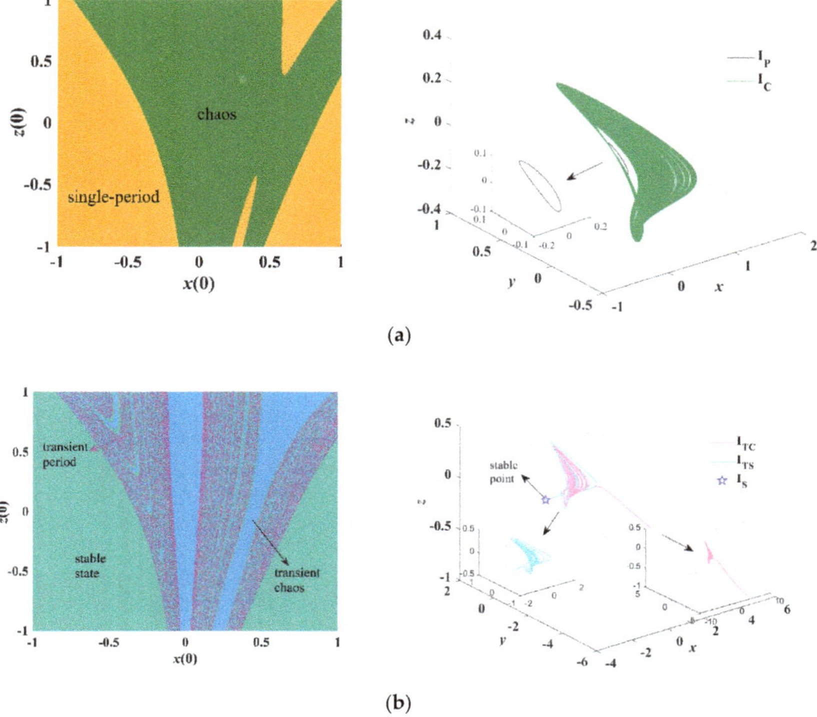

Figure 15. *Cont.*

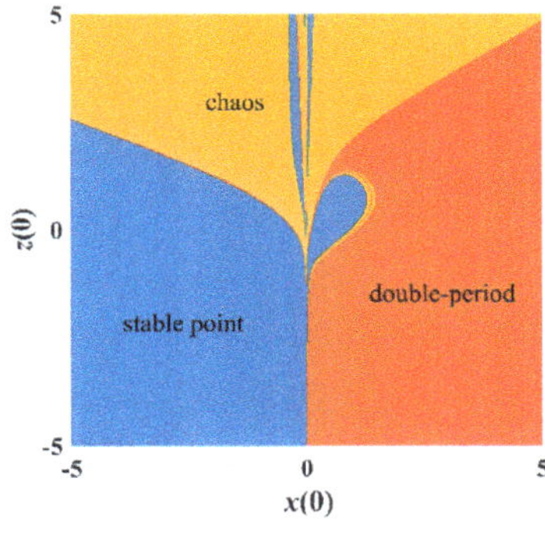
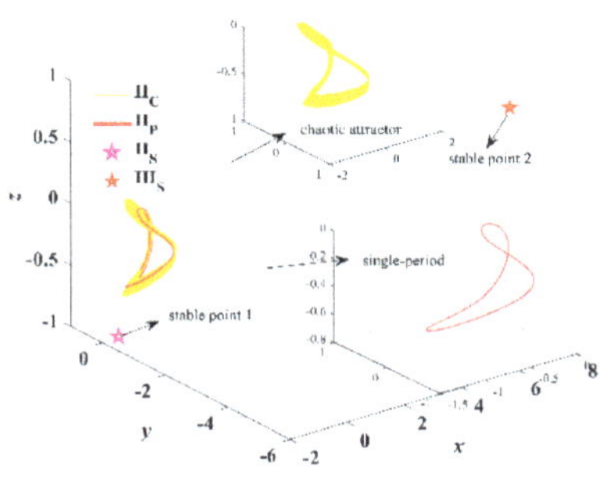

(c)

Figure 15. Basin of attraction and typical coexisting attractors under different parameters: (**a**) $a_2 = 1$, $c_1 = 5.1$, $c_2 = 3.2$; (**b**) $a_2 = 1$, $c_1 = 5.7$, $c_2 = 3.5$; (**c**) $a_2 = 1.04$, $c_1 = 5.55$, $c_2 = 5.9$.

Table 2. Characteristics of different attractors.

Color	Characteristics	Types	Initial Values
	Single-period attractor	I_P	$(-0.1, 0, 0.1, 0)$
	Single-scroll chaos	I_C	$(0.1, 0, -0.1, 0)$
	Transient chaos	I_{TC}	$(-0.2, 0, 0.2, 0)$
	Transient periodic attractor	I_{TS}	$(0.2, 0, -0.3, 0)$
	Point attractor	I_S	$(0.2, 0, -0.2, 0)$
	Double-scroll chaos	II_C	$(1, 0, -4, 0)$
	Double-periodic attractor	II_P	$(2.5, 0, 2, 0)$
	Point attractor	II_S	$(-1.5, 0, -1, 0)$
	Point attractor	III_S	$(1.5, 0, 0.5, 0, 0)$

6. Circuit Simulation

In order to verify the validity of the mathematical analyses, we made a circuit simulation by using the Pspice tool. Based on the HNN model described in Equation (6), we obtain

$$
\begin{cases}
RC_1 \frac{dv_x}{dt} = -v_x + \frac{R}{R_{11}} \tanh(v_x) - \tanh(v_y) \times \left[\frac{R}{Ra_2} - \frac{R}{Rb_2} \tanh(v_u) + \frac{R}{Rc_2} \tanh(v_z) \right] \\
RC_2 \frac{dv_y}{dt} = -v_y + \tanh(v_x) \times \left[\frac{R}{Ra_1} - \frac{R}{Rb_1} \tanh(v_z) + \frac{R}{Rc_1} \tanh(v_u) \right] + \frac{R}{R_{22}} \tanh(v_y) \\
RC_3 \frac{dv_z}{dt} = -v_z + \tanh(v_x) \\
RC_4 \frac{dv_u}{dt} = -v_u + \tanh(v_y)
\end{cases}
\tag{21}
$$

where v_x, v_y, v_z and v_u represent the voltage of capacitors C_1, C_2, C_3 and C_4, respectively.

The main circuit of two-neuron based HNN from Equation (21) is shown in Figure 16a, including four ideal operational amplifiers, four "-tanh" function models, several resistors and capacitors. Figure 16b shows memristive synaptic equivalent circuits W_1 and W_2, including four multipliers and several resistors.

Here, set the time constant as 1 ms. The parameter values of circuit components are listed in Table 3. The obtained simulation results from Pspice software are depicted in Figure 17, which are consistent with the results obtained from MatLab.

In addition, we set $R_{11} = 10$ kΩ and $R_{22} = 5$ kΩ. Different coexisting behaviors emerge in the circuit varying with the memristance Ra_2. When $Ra_2 = 7.14$ kΩ and 6.67 kΩ, the simulation results from Pspice are shown in Figure 18, which exhibit chaotic trajectory and transient chaotic trajectory, respectively. The obtained results are consistent with that of Figure 10c,d.

In conclusion, the circuit simulation results verify the feasibility of the HNN circuit, which is conductive to the hardware implementation of neural networks and studying their synaptic crosstalk.

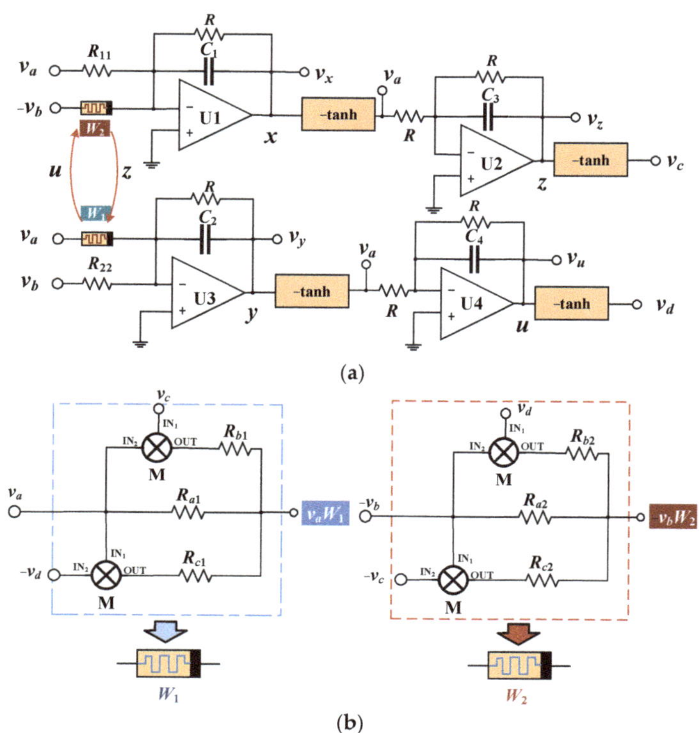

(a)

(b)

Figure 16. Circuit scheme of memristive HNN with synaptic crosstalk: (**a**) main circuit; (**b**) equivalent circuit of memristive synapse.

Table 3. Parameter values of components.

Symbol	Parameter Values	Symbol	Parameter Values
R	10 kΩ	$R_{b1} = R/b_1$	250 kΩ
C	1 μF	$R_{a2} = R/a_2$	9.52 kΩ
$R_{11} = R/W_{11}$	8.06 kΩ	$R_{b2} = R/b_2$	333.33 kΩ
$R_{22} = R/W_{22}$	13.33 kΩ	$R_{c1} = R/c_1$	1.8 kΩ
$R_{a1} = R/a_1$	10 kΩ	$R_{c2} = R/c_2$	1.69 kΩ

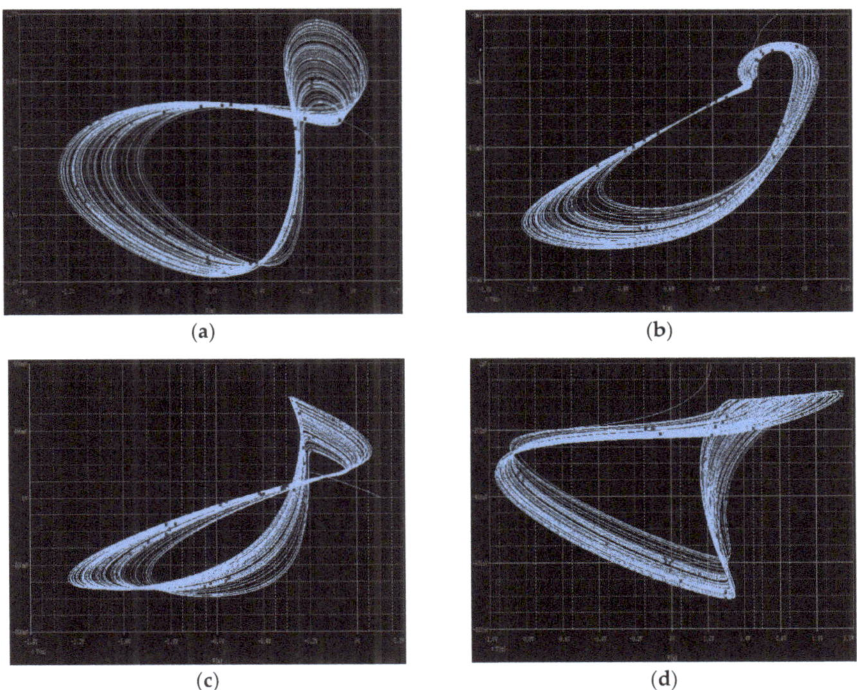

Figure 17. Simulation results of chaotic attractors: (**a**) x-y phase portrait; (**b**) x-z phase portrait; (**c**) x-u phase portrait; (**d**) y-z phase portrait.

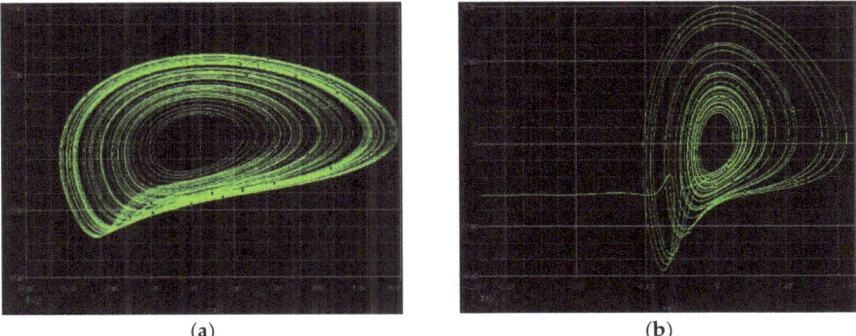

Figure 18. Simulation results of chaotic attractors and transient chaotic attractors: (**a**) $R_{a2} = 7.14$ kΩ; (**b**) $R_{a2} = 6.67$ kΩ.

7. Conclusions

Based on the synaptic plasticity and nonvolatility of the memristor, this paper presents a simple two-neuron-based Hopfield neural network, which can emulate the synaptic crosstalk of neural networks. By using the bifurcation diagram, basin of attraction and Lyapunov exponent spectrum, the dynamics of the HNN varying with memristive parameters and synaptic crosstalk weights are analyzed. Complex phenomena, including chaotic attractors, emerge in the HNN under the influence of synaptic crosstalk. In particular, a special phenomenon called transient chaos occurs in the HNN. Moreover, it is indicated that the HNN has high sensitivity and rich coexisting dynamics via the phase portraits, bifurcation diagram and basin of attraction. Finally, the circuit simulation is completed via Pspice, which is consistent with the MatLab simulation results, further verifying the implementation of the hardware of memristive HNN.

Author Contributions: Conceptualization, R.Q., Y.D. and G.W.; methodology, R.Q. and X.J.; software, X.J.; validation, G.W. and Y.D.; formal analysis, Y.D.; investigation, G.W.; resources, R.Q.; writing—original draft preparation, X.J.; writing—review and editing, R.Q., G.W. and Y.D.; visualization, Y.D.; supervision, R.Q.; project administration, G.W.; funding acquisition, G.W. All authors have read and agreed to the published version of the manuscript.

Funding: This research was funded by the National Natural Science Foundation of China, grant number 61771176.

Institutional Review Board Statement: Not applicable.

Informed Consent Statement: Not applicable.

Data Availability Statement: Not applicable.

Conflicts of Interest: The authors declare no conflict of interest.

References

1. Kawahara, M.; Kato-Negishi, M.; Tanaka, K. Cross talk between neurometals and amyloidogenic proteins at the synapse and the pathogenesis of neurodegenerative diseases. *Metallomics* **2017**, *9*, 619–633. [CrossRef]
2. Chua, L.O. Memristor—The missing circuit element. *IEEE Tran. Circuit Theory* **1971**, *18*, 507–519. [CrossRef]
3. Strukov, D.B.; Snider, G.S.; Stewart, D.R.; Williams, R.S. The missing memristor found. *Nature* **2008**, *453*, 80–83. [CrossRef]
4. Huang, Y.; Liu, J.; Harkin, J.; McDaid, L.; Luo, Y. An memristor-based synapse implementation using BCM learning rule. *Neurocomputing* **2021**, *423*, 336–342. [CrossRef]
5. Hu, X.F.; Feng, G.; Duan, S.; Liu, L. A memristive multilayer cellular neural network with applications to image processing. *IEEE Trans. Neural Netw. Learn. Syst.* **2017**, *28*, 1889–1901. [CrossRef]
6. Dong, Z.K.; Lai, C.S.; Qi, D.L.; Xu, Z.; Li, C.Y.; Duan, S.K. A general memristor-based pulse coupled neural network with variable linking coefficient for multi-focus image fusion. *Neurocomputing* **2018**, *308*, 172–183. [CrossRef]
7. Ma, D.M.; Wang, G.Y.; Han, C.Y.; Shen, Y.R.; Liang, Y. A memristive neural network model with associative memory for modeling affections. *IEEE Access* **2018**, *6*, 61614–61622. [CrossRef]
8. Hopfield, J.J. Neural networks and physical systems with emergent collective computational abilities. *Proc. Natl. Acad. Sci. USA* **1982**, *79*, 2554–2558. [CrossRef]
9. Dehghani, M.; Trojovský, P. Teamwork optimization algorithm: A new optimization approach for function minimization/maximization. *Sensors* **2021**, *21*, 4567. [CrossRef]
10. Kasihmuddin, M.S.M.; Mansor, M.A.; Basir, M.F.M.; Sathasivam, S. Discrete mutation Hopfield neural network in propositional satisfiability. *Mathematics* **2019**, *7*, 1133. [CrossRef]
11. Citko, W.; Sienko, W. Inpainted image reconstruction using an extended Hopfield neural network based machine learning system. *Sensors* **2022**, *22*, 813. [CrossRef]
12. Bazuhair, M.M.; Jamaludin, S.Z.M.; Zamri, N.E.; Kasihmuddin, M.S.M.; Mansor, M.A.; Always, A.; Karim, S.A. Novel Hopfield neural network model with election algorithm for random 3 satisfiability. *Processes* **2021**, *9*, 1292. [CrossRef]
13. Rubio-Manzano, C.; Segura-Navarrete, A.; Martinez-Araneda, C.; Vidal-Castro, C. Explainable Hopfield neural networks using an automatic video-generation system. *Appl. Sci.* **2021**, *11*, 5771. [CrossRef]
14. Njitacke, Z.T.; Isaac, S.D.; Kengne, J.; Negou, A.N.; Leutcho, G.D. Extremely rich dynamics from hyperchaotic Hopfield neural network: Hysteretic dynamics, parallel bifurcation branches, coexistence of multiple stable states and its analog circuit implementation. *Eur. Phys. J. Spec. Top.* **2020**, *229*, 1133–1154. [CrossRef]
15. Chen, C.; Chen, J.; Bao, H.; Chen, M.; Bao, B. Coexisting multi-stable patterns in memristor synapse-coupled Hopfield neural network with two neurons. *Nonlinear Dyn.* **2019**, *95*, 3385–3399. [CrossRef]
16. Sun, J.; Xiao, X.; Yang, Q.; Liu, P.; Wang, Y. Memristor-based Hopfield Network Circuit for Recognition and Sequencing Application. *AEU-Int. J. Electron. Commun.* **2021**, *134*, 1536984. [CrossRef]
17. Njitacke, Z.T.; Kengne, J.; Fotsin, H.B. A plethora of behaviors in a memristor based Hopfield neural networks (HNNs). *Int. J. Dyn. Control* **2019**, *7*, 36–52. [CrossRef]
18. Bao, B.; Qian, H.; Xu, Q.; Chen, M.; Wang, J.; Yu, Y. Coexisting behaviors of asymmetric attractors in hyperbolic-type memristor based Hopfield neural network. *Front. Comput. Neurosci.* **2017**, *11*, 81. [CrossRef]
19. Leng, Y.; Yu, D.; Hu, Y.; Yu, S.S.; Ye, Z. Dynamic behaviors of hyperbolic-type memristor-based Hopfield neural network considering synaptic crosstalk. *Chaos* **2020**, *30*, 033108. [CrossRef]

20. Hopfield, J.J. Neurons with graded response have collective computational properties like those of 2-state neurons. *Proc. Natl. Acad. Sci. USA* **1984**, *81*, 3088–3092. [CrossRef]
21. Parks, P.C. A new proof of the Routh-Hurwitz stability criterion using the second method of lyapunov. *Math. Proc. Camb. Philos. Soc.* **1962**, *58*, 694–702. [CrossRef]
22. Margielewicz, J.; Gaska, D.; Opasiak, T.; Litak, G. Multiple solutions and transient chaos in a nonlinear flexible coupling model. *J. Braz. Soc. Mech. Sci. Eng.* **2021**, *43*, 471. [CrossRef]
23. Yang, X.S.; Xu, Q. Chaos and transient chaos in simple Hopfield neural networks. *Neurocomputing* **2005**, *69*, 232–241. [CrossRef]
24. Yang, K.; Duan, Q.; Wang, Y.; Zhang, T.; Yang, Y.; Huang, R. Transiently chaotic simulated annealing based on intrinsic nonlinearity of memristors for efficient solution of optimization problems. *Sci. Adv.* **2020**, *6*, eaba9901. [CrossRef]
25. Lorenz, E.N. Deterministic nonperiodic flow. *J. Atmos. Sci.* **1963**, *20*, 130–141. [CrossRef]
26. Faradja, P.; Qi, G. Analysis of multistability, hidden chaos and transient chaos in brushless DC motor. *Chaos Solitons Fractals* **2020**, *132*, 1884–2022. [CrossRef]

Article

A Passive but Local Active Memristor and Its Complex Dynamics

Fupeng Li [1], Jingbiao Liu [2], Wei Zhou [1], Yujiao Dong [1], Peipei Jin [1], Jiajie Ying [1] and Guangyi Wang [1,*]

[1] Institute of Modern Circuits and Intelligent Information, Hangzhou Dianzi University, Hangzhou 310018, China; lfp_99@hdu.edu.cn (F.L.); zhouwei1995@hdu.edu.cn (W.Z.); yujiao_dong@hdu.edu.cn (Y.D.); peipjin@163.com (P.J.); jiajieying@hdu.edu.cn (J.Y.)
[2] School of Mechanical Engineering, Hangzhou Dianzi University, Hangzhou 310018, China; ab@hdu.edu.cn
* Correspondence: wanggyi@163.com

Abstract: This paper proposes a globally passive but locally active memristor, which has three stable equilibrium points and two unstable equilibrium points, exhibiting two stable locally active regions and four unstable locally active regions. We find that when the memristor operates in a stable local active region, the memristor-based second-order circuit with a parallel capacitor or a series inductor can produce periodic oscillation. Moreover, the memristor-based third-order circuit with two energy storage elements, a capacitor and an inductor, can produce complex chaotic oscillation, forming the simplest chaotic circuit.

Keywords: memristor; local activity; chaos

Citation: Li, F.; Liu, J.; Zhou, W.; Dong, Y.; Jin, P.; Ying, J.; Wang, G. A Passive but Local Active Memristor and Its Complex Dynamics. *Electronics* **2022**, *11*, 1843. https://doi.org/10.3390/electronics11121843

Academic Editor: Kiat Seng Yeo

Received: 12 May 2022
Accepted: 6 June 2022
Published: 9 June 2022

1. Introduction

According to Chua's theory of local activity [1], all memristors can be classified into either locally passive memristors or locally active memristors. A subset of locally active memristors have the ability to exhibit fascinating phenomena and attributes, such as periodic oscillation, chaotic oscillation, and even action potentials and artificial intelligence. Relevant research on locally active memristors may help us to promote the applications of memristors in artificial neural networks, memristor oscillation circuits, chaos circuits, and so on.

Chua proposed a locally active memristor model named Chua Corsage memristor (CCM) [2], which has two stable equilibrium points and a locally active domain, and is, therefore, a nonvolatile local active memristor. Based on the CCM, a second-order periodic oscillation circuit was carried out via the small-signal circuit analysis method [3]. It was found that the circuit parameters and bias voltage can cause Hopf bifurcation, which, in turn, leads to periodic oscillation. Subsequently, a four-lobe (a section of the DC *V–I* curve of the memristor that looks like a lobe) Chua Corsage memristor [4] and a six-lobe Chua Corsage memristor [5] were proposed respectively.

As the state equations of the three locally active CCMs are based on the piecewise linear functions, they are globally passive but locally active memristors. The number of lobes of CCM increases with its stable equilibrium points. However, due to the nondifferentiable points of the state function, there is only one locally active region obtained. Moreover, the correlation between multiple stability and local active properties has not been discussed.

References [6,7] proposed two voltage-controlled locally active memristor models and found that chaotic oscillations occurred in the constructed memristor circuits. Reference [8] combined a piecewise function with a polynomial to construct a globally passive but locally active memristor, which has two locally active regions, and found that chaotic oscillation occurred in its locally active region. Reference [9] constructed a current-controlled S-type locally active memristor model based on the existing memristor physical devices.

To identify the correlation between the local active characteristics and the complex dynamic behaviors of a memristor, this paper designs a globally passive but locally active memristor model with a continuous state function and three-valued stability, and explores the correlation between the multi-stability and the locally active regions of the memristor. Based on this, we construct two chaotic circuits using the quantitative method rather than trial-and-error, and analyze their complex dynamics, illustrating the mechanism of the Hopf bifurcation caused by passive parameters connected to the memristor.

2. A Six-Lobe Locally Active Memristor with a Continuous and Derivable State Equation

Based on Chua's unfolding theorem [10], we propose a six-lobe locally active memristor with a continuous and derivable state equation, which is a voltage-controlled extended memristor. The state-dependent Ohm's law and its state equation related to the internal state variable x of the memristor are written as follows:

$$\begin{cases} i = G(x)v \\ \frac{dx}{dt} = f(x,v) = k_2(\gamma(x) + \alpha(x)\beta(v)) \end{cases} \tag{1}$$

where $G(x) = k_1(ax^2 + bx + c), \alpha(x) = a_1 + a_2x + a_3x^2, \beta(v) = v$ and

$$\gamma(x) = \begin{cases} -k_3 \sum\limits_{n=1}^{\infty} \frac{(-1)^n}{(2n+1)(2n)(2n-1)} x^{2n-2}, \ 0 \le |x| \le 1 \\ -k_3 \sum\limits_{n=1}^{\infty} \frac{(-1)^n}{(2n+1)(2n)(2n-1)} \left(\frac{1}{x}\right)^{2n-2}, \ |x| > 1 \end{cases}.$$

To make the memristor model globally passive, the function $G(x)$ must be nonnegative for all real numbers x. Therefore, the parameters need to satisfy $a > 0$ and $b^2 - 4ac < 0$, where we set the parameters: $a = 1$, $b = 0$, and $c = 0.1$. We adjust the order of magnitude of the memristance with the parameter k so that it matches the actual memristor, where we take $k = 10^{-3}$, while parameter k_2 is used to control the change rate of state variable x. If $k_2 = 10^3$, the variable rate agrees with the actual memristors. As there is no solution if $\alpha(x) = 0$ when calculating stable equilibrium points of the memristor on its DC V–I curve, we set $\alpha(x) \neq 0$, i.e., $a_3 \neq 0$ and $a_2^2 - 4a_1a_3 < 0$, where we set $a_1 = 3$, $a_2 = 0.2$, and $a_3 = 0.1$.

When the memristor is power-off, the rate of change of the state variable x is $\gamma(x)$, which is only related to the memristor itself and is called "internal force". When a voltage is applied to the memristor, function $f(x, v)$ is increased by an "external force", $\alpha(x)\beta(v)$. We also introduce parameter k_3 for balancing the effect of internal force and external force, where $k_3 = 0.05$. In addition, $\gamma(x)$ can be written as a continuous function $\gamma(x) = -k_3((48x^5 - 480x^3 + 240x)/(1 + x^2)^5)$.

2.1. Dynamic Route Map of the Memristor

The dynamic route map is one of the important methods to explore the dynamic properties of nonlinear equations [11], which can be used to analyze memory characteristics and switch characteristics of memristors. Figure 1 shows five dynamic routes of the proposed memristor, each parametrized by a value of the memristor voltage v, where v = −0.6 V, −0.3 V, 0 V, 0.3 V, and 0.6 V, respectively.

In Figure 1, the dynamic route with $v = 0$ is called the Power Off Plot (POP), which has five intersections with the horizontal axis (Q1 to Q5), in which the state variable x is stable at Q1, Q3, and Q5 but unstable at Q2 and Q4. When the memristor is power-off, the state variable x gradually stabilizes to Q1 $(-3.0777, 0)$ or Q3 $(0, 0)$ or Q5 $(3.0777, 0)$. Observe that when the absolute value of the state variable x increases, the rate of change of the state variable x, i.e., dx/dt, tends to zero asymptotically.

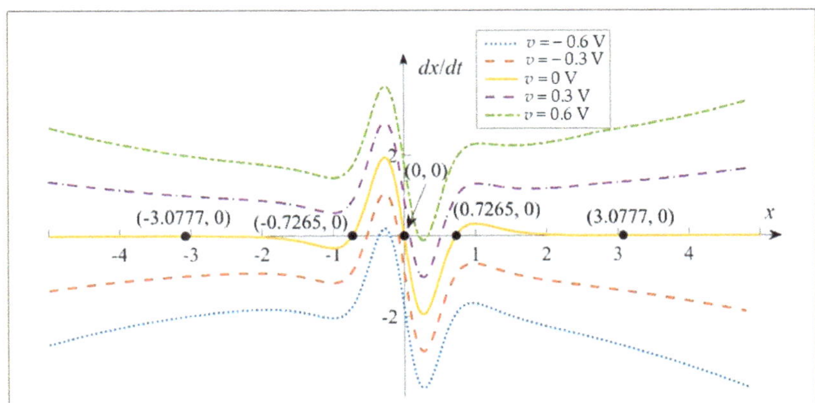

Figure 1. Dynamic route map of the memristor.

2.2. Hysteresis Characteristics of Memristor

Driven by the zero-bias AC power, the relationship between the voltage and the current through the memristor has hysteresis characteristics and pinches with the increase in the frequency of the AC power [12]. Applying the voltage $v(t) = 2A\sin(2\pi ft)$ with $A = 1$ V and different frequencies f at both ends of the memristor, if the initial state value $x(0) = 0$, the current response of the memristor is shown in Figure 2, which is a hysteresis loop and located in the first and third quadrants of the plane coordinate. If the frequency of the power is greater than 2 kHz, the curve coincides with a straight line with a slope of 10^{-4}, and the memristor is equivalent to a linear resistor with a resistance of 10^4 Ohms. Therefore, the model is consistent with the characteristics of a memristor [13].

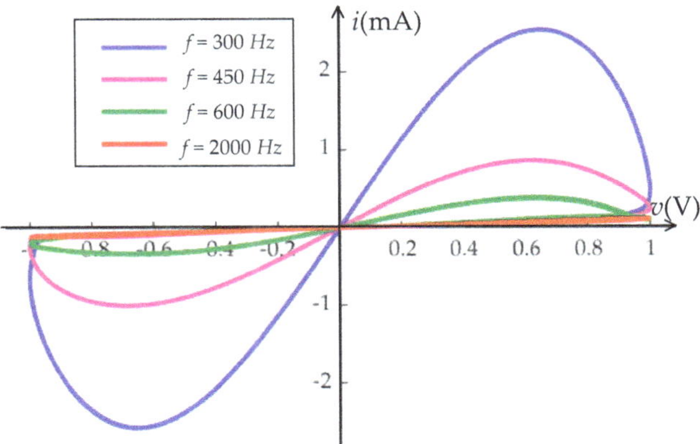

Figure 2. Memristor-pinched hysteresis loop.

2.3. Local Activity

The local activity of the memristor is analyzed to determine whether the memristor can amplify small signals at some DC operating points. Based on a given DC voltage V, we solve the equation $f(x,V) = 0$ to obtain the solution of x and calculate the corresponding current I of the memristor. Then, the DC V–I curve of the memristor can be obtained by drawing the point set of the voltage V and the corresponding current I on the V–I coordinate plane, as shown in Figure 3.

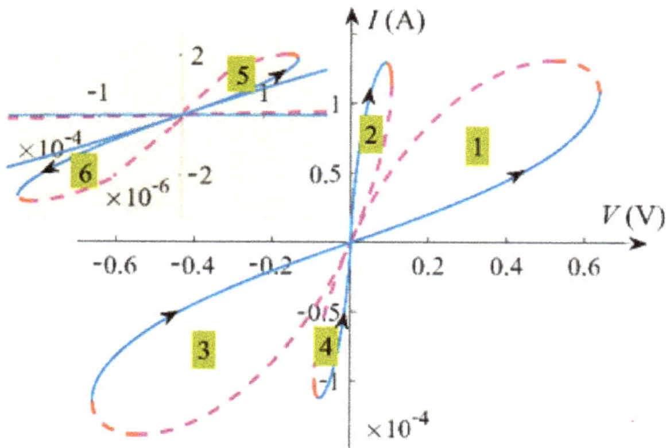

Figure 3. DC *V–I* curve of the memristor.

Observe from Figure 3 that the DC *V–I* curve has six unique, smooth, and continuous lobes, and each lobe has a locally active region. All lobes are marked with numbers from 1 to 6, in which the red curve represents the negative differential resistance region of the memristor. The arrow indicates the increasing direction of state variable *x*, the real line segment indicates that the operating points on it are stable, and the dotted line segment indicates that the operating points on it are unstable. Observe also that the DC *V–I* curve of the memristor is distributed in the first and third quadrants of the coordinate plane, which shows that the memristor is globally passive but locally active. Table 1 shows the value range of six locally active regions.

Table 1. Locally active regions of the memristor.

Range of x	Lobe	Corresponding Voltage (V)	Corresponding Current I (mA)	Stability
$(-3.979, -3.667)$	6	$(-1.816 \times 10^{-4}, -1.989 \times 10^{-4})$	$(-2.894 \times 10^{-6}, -2.694 \times 10^{-6})$	Unstable
$(-1.169, -1.000)$	2	$(0.08855, 0.1034)$	$(1.298 \times 10^{-4}, 1.138 \times 10^{-4})$	Stable
$(-0.3963, -0.2702)$	3	$(-0.5395, -0.6600)$	$(-1.387 \times 10^{-4}, -1.140 \times 10^{-4})$	Unstable
$(0.2649, 0.3909)$	1	$(0.6369, 0.5206)$	$(1.084 \times 10^{-4}, 1.316 \times 10^{-4})$	Unstable
$(0.9921, 1.152)$	4	$(-0.09095, -0.07837)$	$(-9.862 \times 10^{-5}, -1.119 \times 10^{-4})$	Stable
$(3.653, 3.954)$	5	$(1.414 \times 10^{-4}, 1.295 \times 10^{-4})$	$(1.902 \times 10^{-6}, 2.038 \times 10^{-6})$	Unstable

3. Small-Signal Analysis for Locally Active Region

Locally active regions have the potential to amplify infinitely small signals and therefore can generate complex phenomena, such as periodic oscillation, chaotic oscillation, and even action potentials after assembling passive circuit components [14,15]. Small-signal analysis of a locally active region is helpful to explore its complex dynamic behaviors.

3.1. Zero-Pole Analysis of Memristor

Small-signal analysis is used to approximate the local dynamic behavior of the nonlinear memristor via its associated linearized equations. Let the DC voltage and current at an operating point Q be *V* and *I*, respectively. Assume that there is a voltage increment δv at the operating point Q(*V*, *I*), and the resulting rate of current change di/dt obtained from Equation (1) is

$$\frac{di}{dt} = \frac{\partial i}{\partial x} \cdot \frac{dx}{dt} + \frac{\partial i}{\partial v} \cdot \frac{dv}{dt} = a_{11}\frac{dx}{dt} + a_{12}\frac{dv}{dt}, \tag{2}$$

where $a_{11}(Q) = \partial i/\partial x = 2xv/1000$ and $a_{12}(Q) = \partial i/\partial v = (x^2 + 0.1)/1000$. Furthermore, the operating point $Q(V, I)$ on the DC V–I curve must meet the condition $dx/dt = g(x,v)$. Through differential expansion of the state equation $dx/dt = g(x,v)$ in Equation (1), we obtain

$$\frac{d\left(\frac{dx}{dt}\right)}{dt} = b_{11}(Q)\frac{dx}{dt} + b_{12}(Q)\frac{dv}{dt} \tag{3}$$

where $b_{11}(Q) = \left.\frac{\partial g(x,v)}{\partial x}\right|_Q = \frac{\partial \gamma(x)}{\partial x} + \frac{\partial \alpha(x)}{\partial x} \times \beta(v)$ and $b_{12}(Q) = \left.\frac{\partial g(x,v)}{\partial v}\right|_Q = \alpha(x)$.

Applying the Laplace transformation to (2) and (3), we have

$$\begin{cases} \hat{i}(s) = a_{11}(Q)\hat{x}(s) + a_{12}(Q)\hat{v}(s) \\ s\hat{x}(s) = b_{11}(Q)\hat{x}(s) + b_{12}(Q)\hat{v}(s) \end{cases} \tag{4}$$

Solving Equation (4), the small-signal-equivalent admittance function $Y(s, Q)$ about $Q(V, I)$ is obtained as

$$Y(s, Q) = \hat{i}(s)/\hat{v}(s) = a_{11}(Q)b_{12}(Q)/(s - b_{11}(Q)) + a_{12}(Q) \tag{5}$$

By rearranging (5), the admittance function can be equivalent to

$$Y(s, Q) = 1/(sL_x + R_x) + 1/R_y \tag{6}$$

where $Lx = 1/(a_{11}(Q)b_{12}(Q))$, $Rx = b_{11}(Q)/(a_{11}(Q)b_{12}(Q))$, and $Ry = a_{12}(Q)$.

Figure 4 shows the equivalent circuit of the memristor at operating point $Q(V, I)$. We find by calculation that $R_x < 0$ and $R_y > 0$ in the locally active regions, and $L_x < 0$ for the stable locally active points, while $L_x > 0$ for the unstable locally active points.

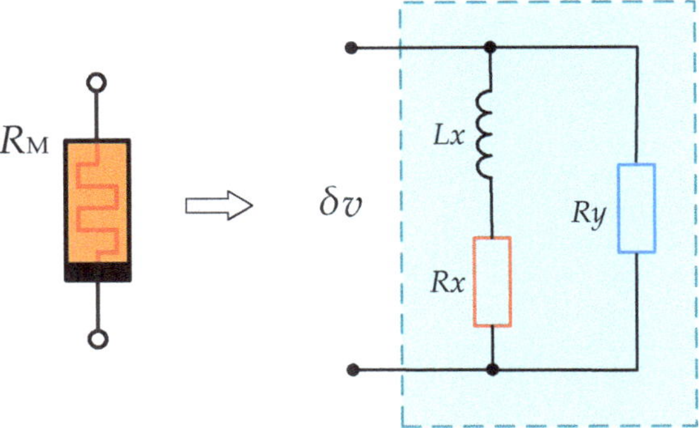

Figure 4. Small-signal-equivalent circuit of the memristor.

Through the zero-pole simplification of admittance function $Y(s, Q)$, the small-signal admittance $Y(s, Q)$ in terms of the pole $s = P$ and the zero $s = Z$ can be recast as

$$Y(s, Q) = K(s - Z)/(s - P) \tag{7}$$

where $K = a_{12}(Q)$, $Z = (a_{12}(Q)b_{11}(Q) - a_{11}(Q)b_{12}(Q))/a_{12}(Q)$, and $P = b_{11}(Q)$.

Figure 5 shows the pole and the zero trajectories of the admittance function $Y(s, Q)$ with respect to the voltage V, where the solid and the dotted lines represent the poles and zero, respectively. The purple dotted line segment on the pole curve and the blue dotted line segment on the zero curve represent the zero and pole values of the memristor in the locally active regions, respectively. Obviously, $P > 0$ and $Z < 0$ for the stable locally active regions, while $P < 0$ and $Z > 0$ for the unstable locally active regions.

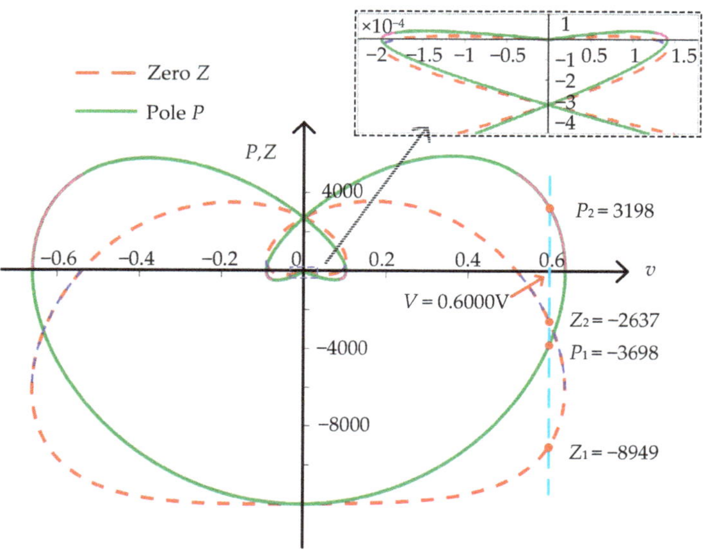

Figure 5. Zeros and poles of memristor.

3.2. Frequency Response of the Memristor

From Equation (7), the frequency response of the memristor can be written as:

$$Y(i\omega, Q) = K(i\omega - Z)/(i\omega - P) = \left(K\left(\omega^2 + PZ\right) + iK\omega(Z - P)\right)/\left(\omega^2 + P^2\right) \quad (8)$$

At equilibrium point $V = V_0$, the real part and imaginary part of the memductance function are $\mathrm{Re}Y(i\omega, V) = K(\omega^2 + pz)/(\omega^2 + p^2)$ and $\mathrm{Im}Y(i\omega, V) = iK\omega(z - p)/(\omega^2 + p^2)$, respectively. To make the locally active system generate oscillation, there should be a pair of complex conjugate poles (Hopf bifurcation points) on the imaginary axis for the admittance function $Y(s, Q)$ of the locally active memristor. In other words, it is necessary to add an energy storage element (capacitance or inductance) in series or in parallel with the memristor to form a locally active system. The selection of capacitance or inductance depends on the frequency response of the memristor. Figure 6 shows the frequency response of the memristor where the voltage at both ends of the memristor is 0.6000 V. In order to construct an oscillation system, if $Lx < 0$ ($Lx > 0$) in the equivalent circuit of the memristor shown in Figure 4, an inductance $L^* = 1/(\omega \mathrm{Im}Y(i\omega, V))$ (capacitor $C^* = \mathrm{Im}Y(i\omega, V)/\omega$) in series (parallel) with the memristor is required.

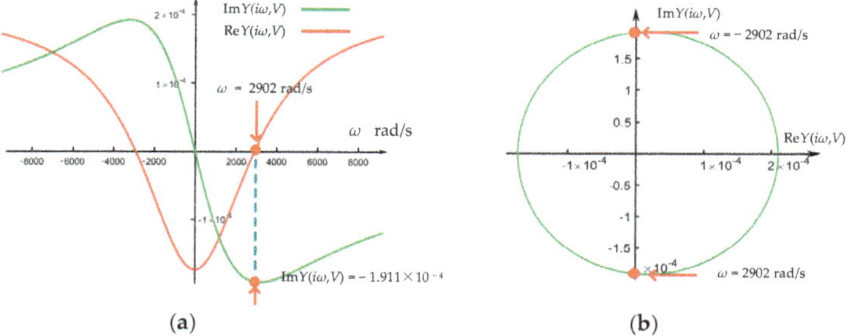

Figure 6. (**a**) Frequency response and (**b**) Nyquist diagram of the memristor at $V = 0.6$ V.

Figure 6 shows the frequency response of $\mathrm{Re}Y(i\omega,V)$ and $\mathrm{Im}Y(i\omega,V)$, where $V = 0.6$ V. Observe that $\mathrm{Re}Y(i\omega^*,V) = 0$ and $\mathrm{Im}Y(i\omega^*,V) = -1.911 \times 10^{-4}$ S at $\omega^* = 2902$ rad/s, indicating that the memristor is inductive and therefore a positive capacitance C^* in parallel with the memristor is needed to compensate the $\mathrm{Im}Y(i\omega^*,V)$, as well as to make the total impedance of the C^*-augmented memristive circuit equal to zero at operating point $V = 0.6000$ V. The compensated capacitance can be obtained using the following formula:

$$C^* = -\mathrm{Im}Y(i\omega^*,\mathrm{V})/\omega^* = 65.84 \text{ nF}$$

Figure 7 shows the frequency response of $\mathrm{Re}Y(i\omega,V)$ and $\mathrm{Im}Y(i\omega,V)$, where $V = 0.1031$ V. Observe that $\mathrm{Re}Y(i\omega^*,V) = 0$ and $\mathrm{Im}Y(i\omega^*,V) = -1.911 \times 10^{-4}$ S at $\omega^* = 200.7$ rad/s, indicating that the memristor is capacitive and therefore a positive inductance L^* in series with the memristor is needed to compensate the $\mathrm{Im}Y(i\omega,V)$, as well as to make the total impedance of the L^*-augmented memristive circuit equal to zero at operating point $V = 0.1031$ V. The compensated inductance can be obtained using the following formula:

$$L^* = 1/(\omega^*\mathrm{Im}Y(i\omega,V)) = 1.972 \text{ H}$$

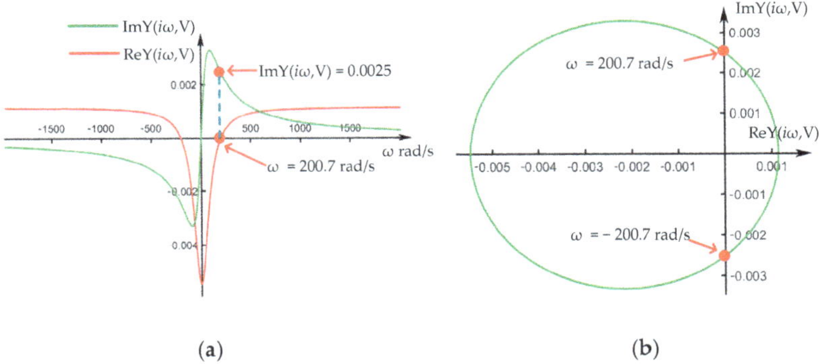

(a)   (b)

Figure 7. (**a**) Small-signal frequency response of memristor and (**b**) Nyquist diagram of memristor when it operates under $V = 0.1031$ V, $x = -1.022$, $L^* = 1.972$ H.

Using the above small-signal-equivalent circuit method, we analyze the properties of the six local active regions of the memristor and obtain the following results shown in Table 2, which includes the small-signal-equivalent inductance, zero, pole, and the compensating energy storage element to cause the memristor oscillation.

Table 2. Properties of locally active regions.

Lobe	Equivalent Circuit Lx	P	Z	Energy Storage Element
1	> 0	> 0	< 0	Parallel capacitance
2	< 0	< 0	> 0	Series inductance
3	> 0	> 0	< 0	Parallel capacitance
4	< 0	< 0	> 0	Series inductance
5	> 0	> 0	< 0	Parallel capacitance
6	> 0	> 0	< 0	Parallel capacitance

4. Second-Order Periodic Circuit of Memristor

The above analysis shows that the memristor has two different types of locally active regions with deferent small-signal circuits, namely $L_x < 0$ and $L_x > 0$. For the two cases, we design two second-order memristive circuits, as shown in Figure 8.

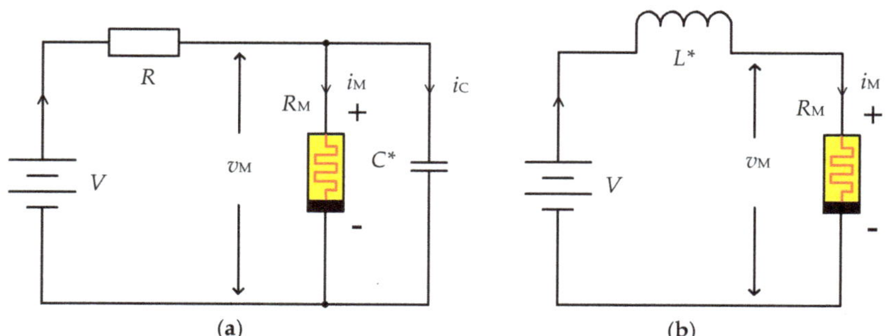

(a)   (b)

Figure 8. (**a**) Parallel capacitance and (**b**) series resistance of locally active memristor oscillation circuit.

4.1. Properties of the Memristive Circuit in the Unstable Locally Active Region of the Memristor

The memristive circuit shown in Figure 8a corresponds to the first unstable locally active region of the memristor, where R is a segregation resistor or load resistor, which provides an AC path for the memristor and also stabilizes the memristor at a certain operating point Q_2, as shown in Figure 9.

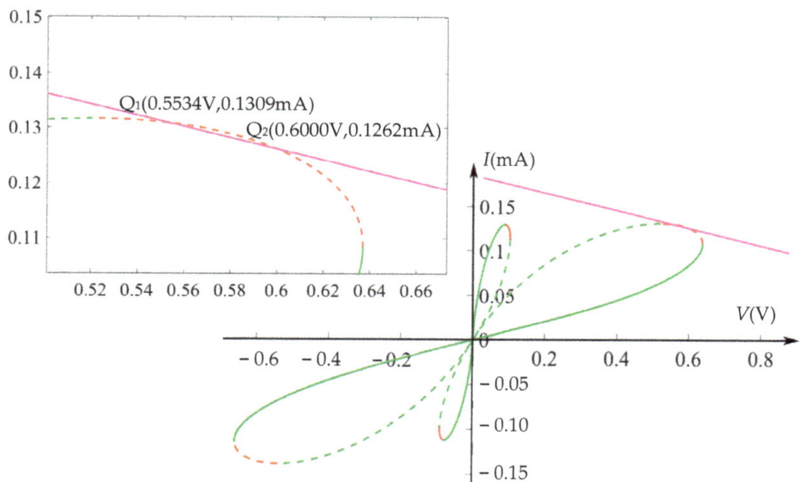

Figure 9. DC load line is superimposed on the *V–I* plane.

Through small-signal analysis of the memristor oscillation circuit in Figure 8a, the composite admittance $Y_C(s, Q)$ of the circuit satisfies that $1/Y_C(s, Q) = 1/(Y(s, Q) + Y_{C^*}) + R$, where $Y_{C^*} = sC^*$. Therefore, $Y_C(s, Q)$ can been written as follows:

$$Y_C(s, Q) = \frac{Cs^2 + (k - pC)s - kz}{RCs^2 - (pRC - 1 + kR)s - p - kzR} = \frac{(s - s_{z_1})(s - s_{z_2})}{(s - s_{p_1})(s - s_{p_2})R} \tag{9}$$

The zeroes and the poles of the composite admittance $Y_C(s, Q)$ of the circuits are obtained from Equation (9) as follows:

$$\begin{cases} p_1 = \dfrac{-a_1 + \sqrt{a_1{}^2 - 4a_2}}{2}, p_2 = \dfrac{-a_1 - \sqrt{a_1{}^2 - 4a_2}}{2} \\ z_1 = \dfrac{-b_1 + \sqrt{b_1{}^2 - 4b_2}}{2}, z_2 = \dfrac{-b_1 - \sqrt{b_1{}^2 - 4b_2}}{2} \end{cases} \tag{10}$$

where $a_1 = (kR + 1 - pRC)/RC$, $a_2 = -(p + kzR)/RC$, $b_1 = (k - pC)/RC$, and $b_2 = -kz/RC$; p, z, and k are the pole, the zero, and the coefficient of admittance function $Y(s, Q)$ of the memristor in Equation (7), respectively.

Figure 10 shows the loci of the real parts versus imaginary parts of the poles (P_1 and P_2) of $Y_C(s, Q)$ with respect to the capacitance C, in which $V = 0.6$ V and the state variable x = 0.3322. Observe from Figure 10 that $Y_C(s, Q)$ has a pair of complex conjugate poles on the imaginary axis at $C = 97.13$ nF and Im $p_{1,2} = \pm1556$, which are the Hopf bifurcation parameters. When $C < 97.13$ nF, such as $C = 65.84$ nF, the real parts of the complex conjugate poles are less than zero, and the circuit gradually stabilizes to an equilibrium point, as shown in Figure 10a. However, when $C = 97.13$ nF, the periodic oscillation shown in Figure 10b,c occurs in the circuit. Moreover, the oscillation amplitude increases with the initial value, as described in Figure 10d. If $C > 97.13$ nF, the system enters the unstable right half plane of the complex plane and may oscillate.

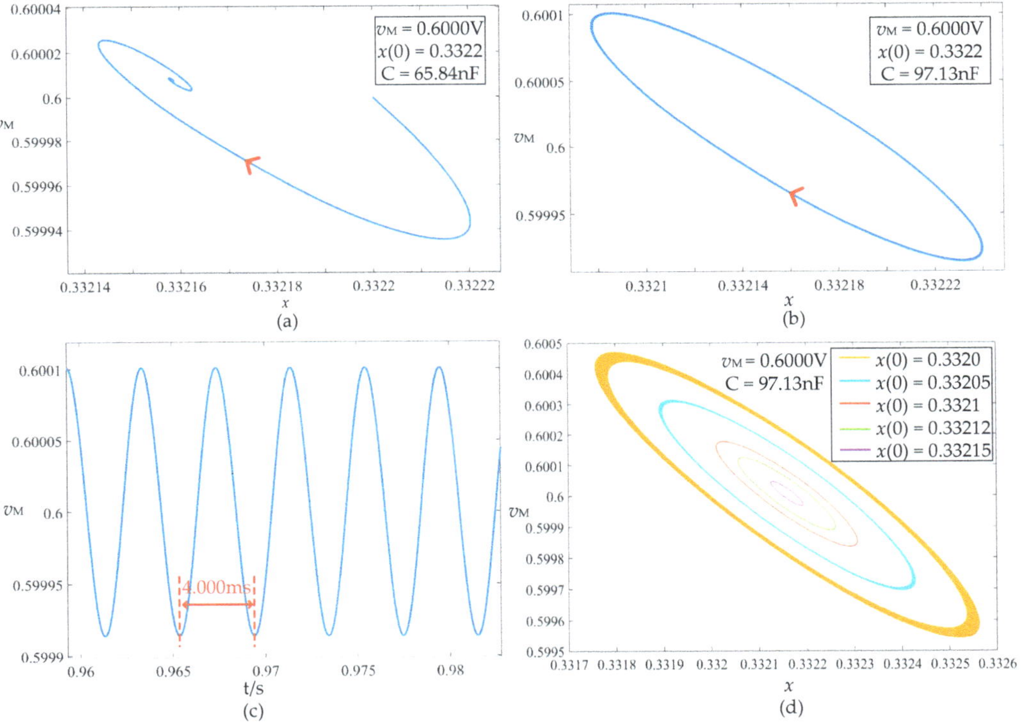

Figure 10. Simulation of memristor oscillation circuit: (**a–c**) system initial value $V = 0.6000$ V and $x = 0.3322$ and (**d**) increasing oscillation amplitude, if the initial value is far from the equilibrium point.

Figure 11 shows the loci of the real parts versus imaginary parts of the poles (p_1 and p_2) of $Y_C(s, Q)$ with respect to the voltage V, where $C = 97.13$ nF. Observe that $Y_C(s, Q)$ has a pair of complex conjugate poles on the imaginary axis at the Hopf bifurcation parameters: $V = 0.6$ V and Im $p_{1,2} = \pm1556$.

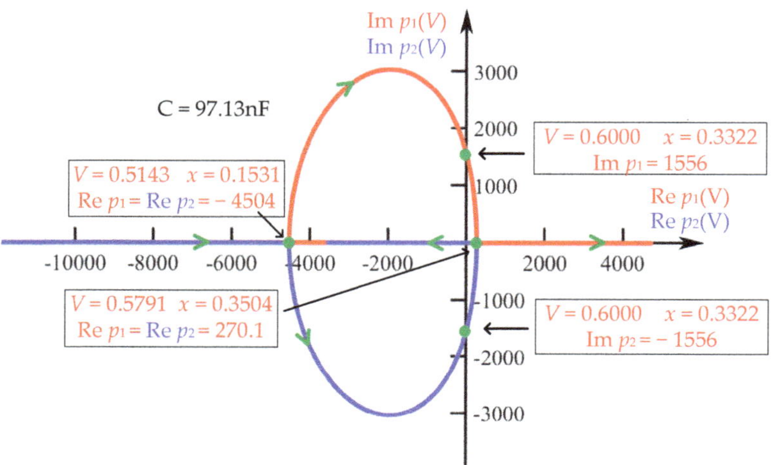

Figure 11. Variation in the real and imaginary parts of the poles of the admittance function in the oscillation circuit with the DC voltage (if C = 97.13 nF).

From Figures 11 and 12, we find that the Hopf bifurcation frequency ω_H = 1556 rad/s and the system oscillation frequency $\omega_C = 2\pi/4.000$ ms = 1570 rad/s through numerical calculation. It follows that ω_C is consistent with the expected oscillation frequency ω_H.

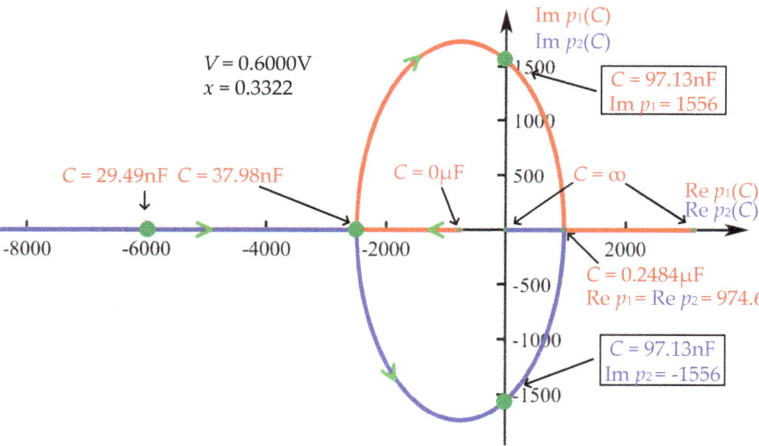

Figure 12. Variation in the real and imaginary parts of the poles of the admittance function in the oscillation circuit with the external capacitance (if V = 0.6000 V).

4.2. Properties of the Memristive Circuit in the Stable Locally Active Region of the Memristor

For Figure 8b, the composite admittance $Y_C(s, Q)$ of the circuit satisfies

$$\frac{1}{Y_L(s,Q)} = \frac{1}{Y(s,Q)} + \frac{1}{Y_{L*}}$$

where $Y_{L*} = 1/sL*$. Therefore, $Y_C(s, Q)$ can be written as follows:

$$Y_L(s,Q) = \frac{K(s-Z)}{LKs^2 - (KZL-1)s - P} = \frac{(s-Z)}{(s-s_{p_1})(s-s_{p_2})L} \tag{11}$$

where P, Z, and K are the pole, the zero, and the coefficient of admittance function $Y(s, Q)$ of the memristor in Equation (7), respectively.

The zero and the poles of the composite admittance of the circuit are obtained from Equation (11) as follows:

$$\begin{cases} s_{p_1} = \dfrac{-a_3 + \sqrt{a_3{}^2 - 4a_4}}{2}, s_{p_2} = \dfrac{-a_3 - \sqrt{a_3{}^2 - 4a_4}}{2} \\ s_z = z \end{cases} \tag{12}$$

where $a_3 = (1 - KZL)/KL$ and $a_4 = -P/KL$.

Figure 13 depicts the loci of the real and imaginary parts of the poles (P_1 and P_2) of the admittance $Y_L(s, Q)$ versus the inductance L, where $V = 0.1031$ V and $x = -1.022$. Observe that $L = 1.972$ H is the Hopf bifurcation point of the memristive circuit where Im $p_{1,2} = \pm 200.7$. If $L > 1.972$ H, the system enters the right half plane of the complex plane and may oscillate; for example, for $L = 2.021$ H, a periodic oscillation appears as shown in Figure 14.

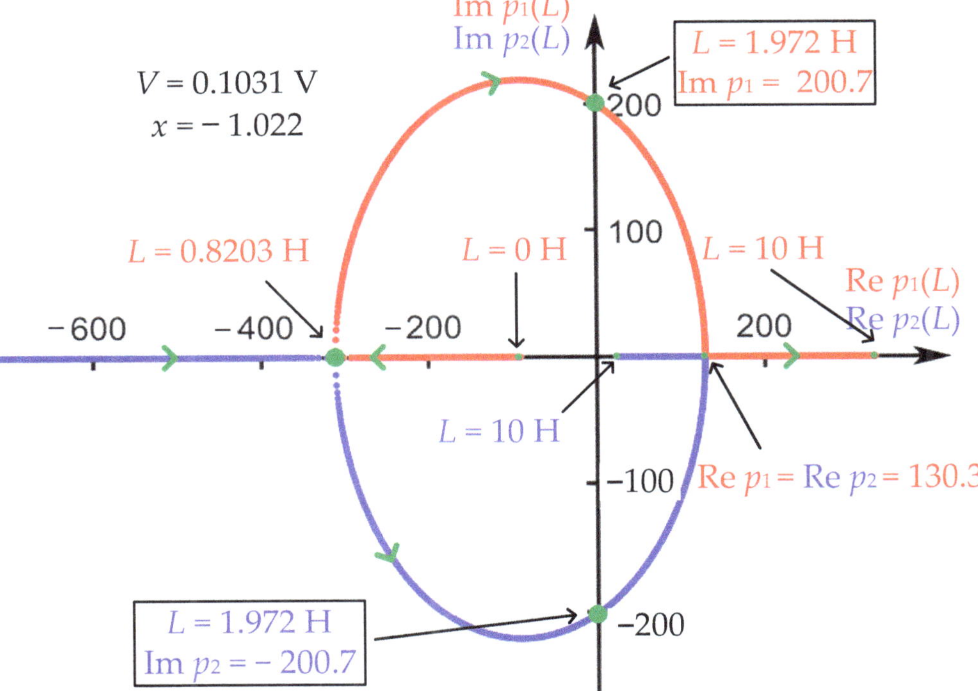

Figure 13. Variation in the real and imaginary parts of the poles of the admittance function in the oscillation circuit with the external capacitance (if $V = 0.1031$ V).

Figure 15 shows the loci of the real and imaginary parts of the poles (p_1 and p_2) of the admittance $Y_L(s, Q)$ versus the voltage V, where $L = 1.972$ H.

Based on the Hopf bifurcation frequency $\omega_H = 200.7$ rad/s in Figures 13 and 15, the system oscillation frequency can be calculated as $\omega_L = 2\pi/0.031$ s $= 202.5$ rad/s through circuit simulation, where ω_L is consistent with the expected oscillation frequency ω_H.

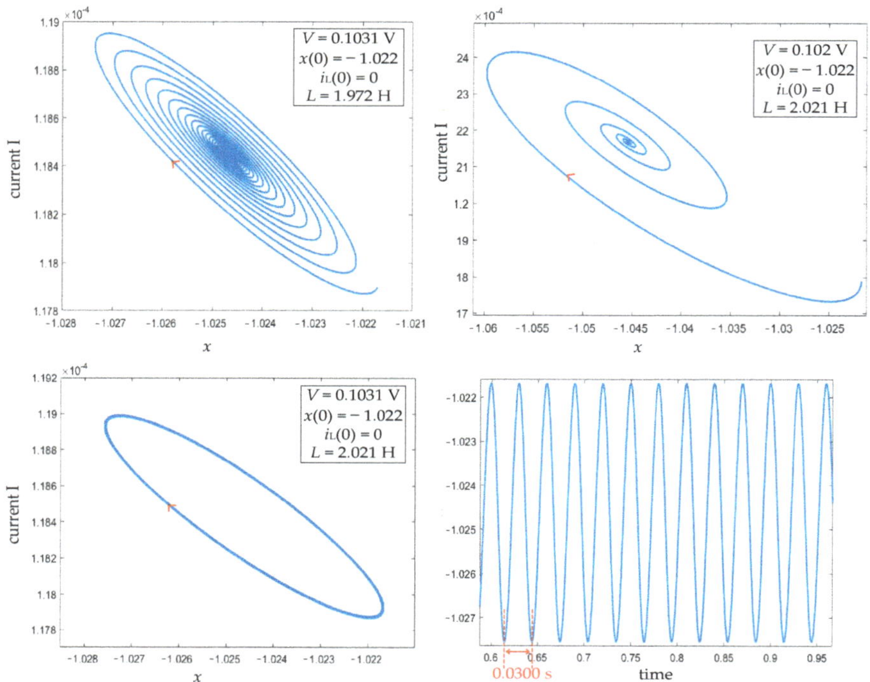

Figure 14. Simulation of memristor oscillation circuit (if $V = 0.1031$ V and $x = -1.022$).

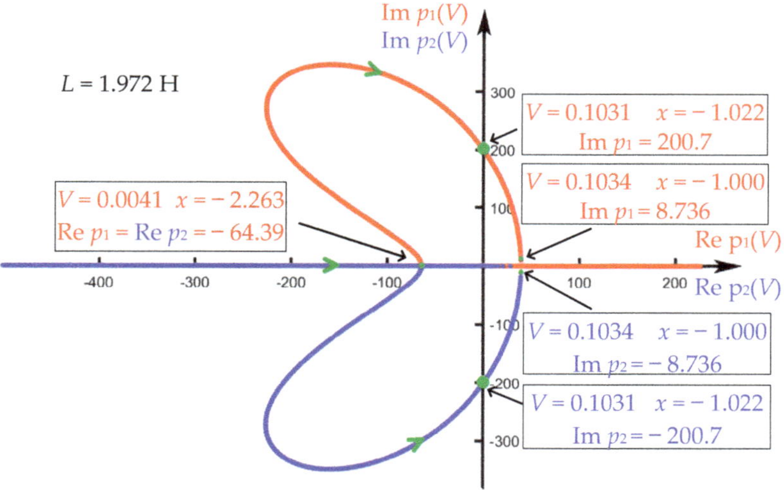

Figure 15. Variation in the real and imaginary parts of the poles of the admittance function in the series inductance oscillation circuit with the DC voltage.

5. Memristor-Based Third-Order Chaotic Circuit

Based on the second-order memristive circuit shown in Figure 8b, we design the simplest third-order chaotic circuit by connecting a capacitor in parallel with the memristor that operates at the stable locally active regions of lobe 2 or 4 of the memristor's DC *V–I* curve in Figure 3, where the small-signal-equivalent inductance $L_x < 0$, as shown in Figure 16.

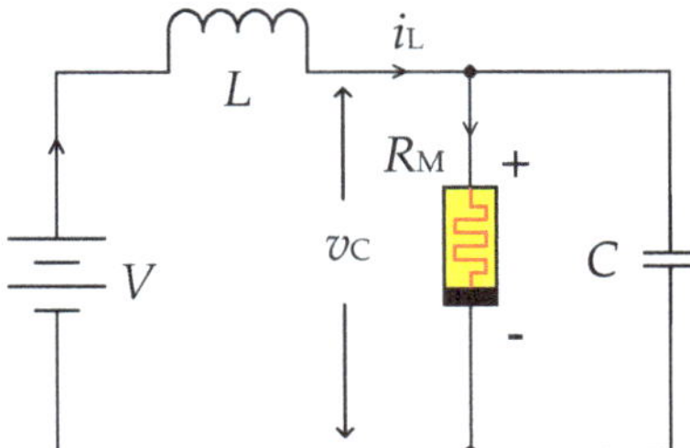

Figure 16. Chaotic oscillation circuit.

According to Kirchhoff's law, the state equations of the circuit can be written as follows:

$$\begin{cases} \frac{dx}{dt} = 1000\left(-0.05 \times \frac{48x^5-480x^3+240x}{(1+x^2)^5} + 3v_C + 0.2xv_C + 0.1x^2v_C\right) \\ \frac{dv_C}{dt} = \frac{1}{C}(i_L - G(x)v_C) \\ \frac{di_L}{dt} = \frac{1}{L}(V - v_C) \end{cases} \tag{13}$$

where v_C is the voltage across the memristor, i_L is the current through the inductance, x is the state variable of the memristor, and v is the supply voltage.

5.1. System Equilibrium Points

Let $dx/dt = 0$, $dv_c/dt = 0$, and $di_L/dt = 0$ in Equation (13); the following four equilibrium points of the system can be obtained: E_1 (−1.022, 0.1031 V, 0.179 mA), E_2 (−0.98, 0.1031 V, 0.109 mA), E_3 (0.026, 0.1031 V, 0.104 mA), E_4 (0.633, 0.1031 V, 0.516 mA). The four equilibrium points of the system (13) happens to be the equilibrium points of the memristor. Equilibrium E1 is located in the locally active region.

It can be known from the system state equations that the equilibrium points of the system is only related to the internal properties of the memristor and the applied voltage, and has nothing to do with the inductance L and the capacitance C in the circuit. The Jacobian matrix is obtained at an equilibrium point as follows:

$$J = \begin{bmatrix} \frac{\partial(g(x,v))}{\partial x} & 1000 \times \left(3 + 0.2x + 0.1x^2\right) & 0 \\ -\frac{xv_C}{500C} & -\frac{x^2+0.1}{1000C} & \frac{1}{C} \\ 0 & -\frac{1}{L} & 0 \end{bmatrix} \tag{14}$$

where

$$\frac{\partial(g(x,v))}{\partial x} = 1000 \times \left(\frac{-0.05 \times (-240\,x^6+3600x^4-3600x^2+240)}{(1+x^2)^6} + 0.2v_C + 0.2v_Cx\right)$$

Table 3 shows the characteristic roots of the Jacobian matrix obtained at the four equilibrium points, in which it has three Saddle focuses (E1, E2, and E4) and one stable focus (E3).

Table 3. Characteristic roots of the Jacobian matrix *J*.

Equilibrium Point	Characteristic Value			Equilibrium Points Types
	λ_1	λ_2	λ_3	
E_1	-157.01	$9.55 + 156.13i$	$9.55–156.13i$	Saddle focus
E_2	144.14	$-45.92 + 159.2i$	$-45.92–159.2i$	Saddle focus
E_3	-11810	$-2 + 204i$	$-2–204i$	Stable focus
E_4	4273.2	$-8.1 + 204.2i$	$-8.1–204.2i$	Saddle focus

Simulation analysis finds that the circuit has the phenomenon of coexisting attractors, i.e., when the system parameters are fixed, the circuit produces different dynamic characteristics with the different initial values. For example, let the parameters C = 26 μF and L= 0.96 H be fixed; the circuit exhibits a chaotic attractor and a stable equilibrium point under the conditions of initial values E1 and E2, respectively, as shown in Figure 17.

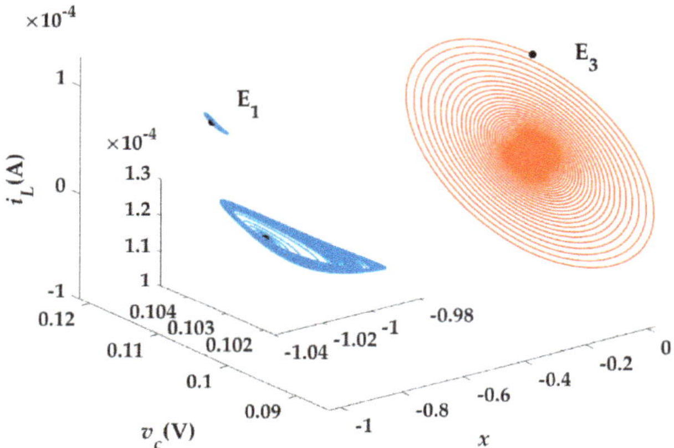

Figure 17. Two kinds of coexisting attractors of the system with different initial values.

5.2. Influence of Parameters L and C on System Dynamics

Let us fix the voltage V = 0.1031 V and the initial values E_1 (−1.022, 0.1031 V, 0.179 mA); the variation in inductance L and capacitance C can cause the system to bifurcate. Figure 18a shows the variation in the system Lyapunov exponent spectrum [16] with the capacitance C within the interval of 10 μF–30 μF, where inductance L = 0.96 H. Figure 18b shows the bifurcation of the state variable x with the capacitance C within the interval of 21 μF–26.2 μF.

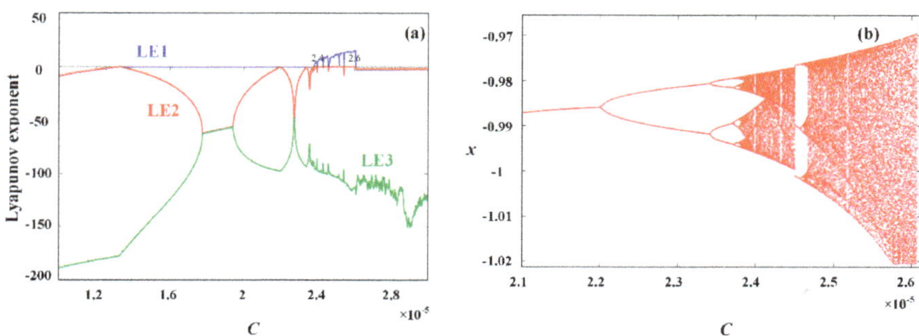

Figure 18. (**a**) Lyapunov exponent of the system, and (**b**) bifurcation of variable x with the capacitance C (if L = 0.96 H).

Observe from Figure 18 that as C increases, the system enters into chaotic oscillation through period-doubling bifurcation and finally enters into a stable state rapidly. In this doubling bifurcation process, the system oscillates with two limit cycles of period 1 (yellow cycle in Figure 18a) and period 2 (Figure 19b) when $C = 21.5$ µF and $C = 21.5$ µF, respectively; when $C > 23.8$ µF, the system enters the chaotic region (Figure 19d shows a chaotic attractor with $C = 26$ µF), where there is a period 3 window, whose corresponding phase diagram is shown in Figure 19c. When the capacitance $C \geq 26.2$ µF continues to increase, the system will rapidly stabilize from chaotic oscillation to a stable point attractor.

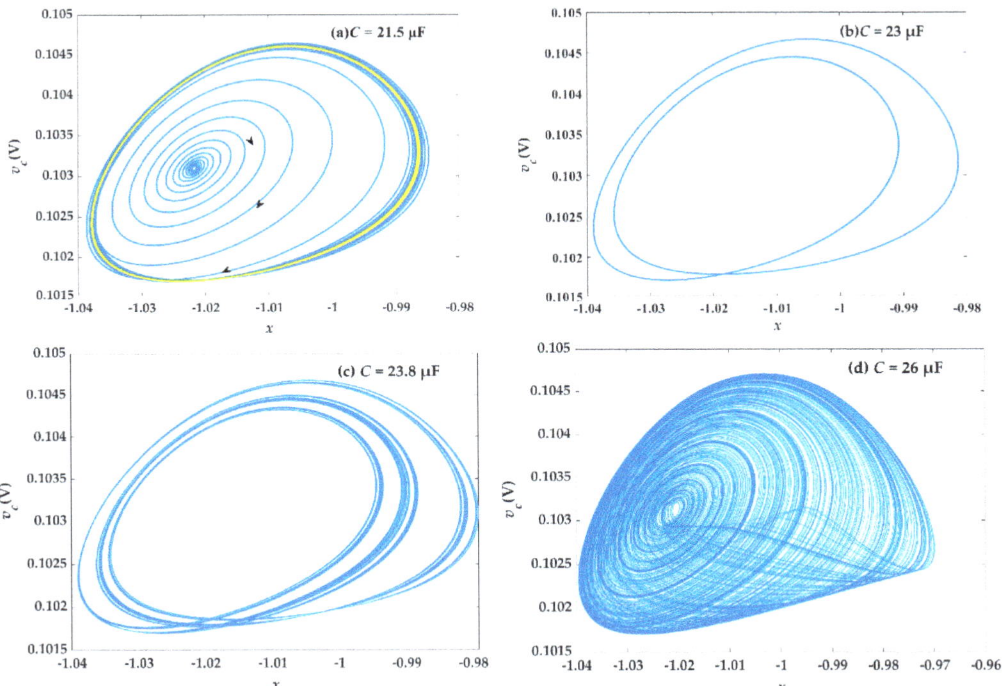

Figure 19. Variation in system attractor with capacitance C, where $V = 0.1031$ V, $L = 0.96$ H. (**a**) $C = 21.5$ µF, (**b**) $C = 23$ µF, (**c**) $C = 23.8$ µF, (**d**) $C = 26$ µF.

Figure 20 shows the waveforms and attractors of the system as capacitance $C = 26.2$ µF, where the system gradually stabilizes to E_3 (0.026, 0.1031 V, 0.104 mA).

Figure 21 shows the Lyapunov exponent spectrum and the bifurcation of the system with respect to the inductance L, where the applied voltage $V = 0.1031$ V, the initial value of the memristor is E_1 (−1.022, 0.1031 V, 0.179 mA), and the capacitance $C = 26$ µF. Observe from Figure 21 that for 0.925 H $\leq L \leq 0.962$ H, the system generates chaotic oscillation. With the increase in inductance L, the system bifurcates from period doubling to chaos by period-doubling bifurcation. Obviously, a Period 3 window can be observed from Figure 22b. If $L > 0.962$ H, the system gradually stabilizes to the system equilibrium point E3.

Figure 23 shows the system dynamics map with respect to both inductance L and capacitance C, where the system operating voltage $V = 0.1031$ V. Obviously, the dynamics map looks like a rainbow pattern, in which the areas labeled P1, P2, P3, C, and E represent period 1, period 2, period 3, chaos, and the stable point, respectively. The typical phase diagrams of those statuses are shown in Figure 24.

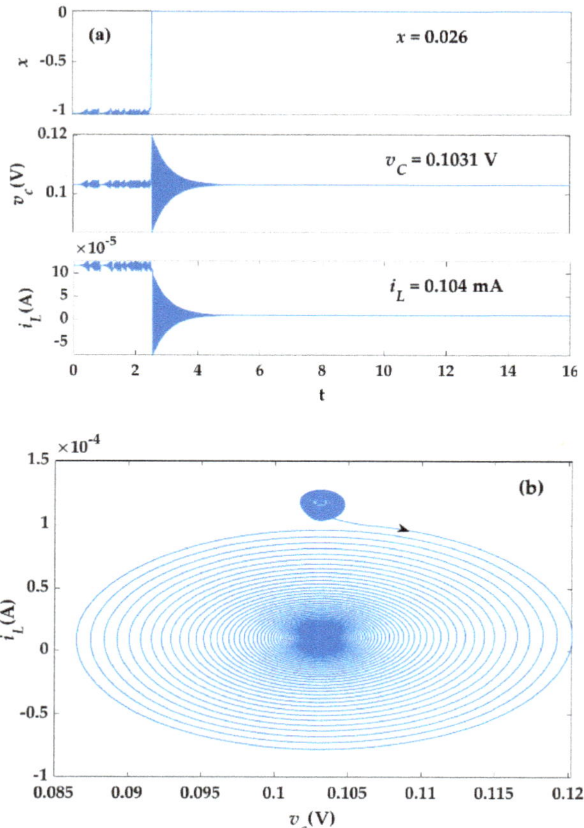

Figure 20. Variation in system attractor (**a**) waveforms and (**b**) v_c-i phase with $C = 26.2$ μF.

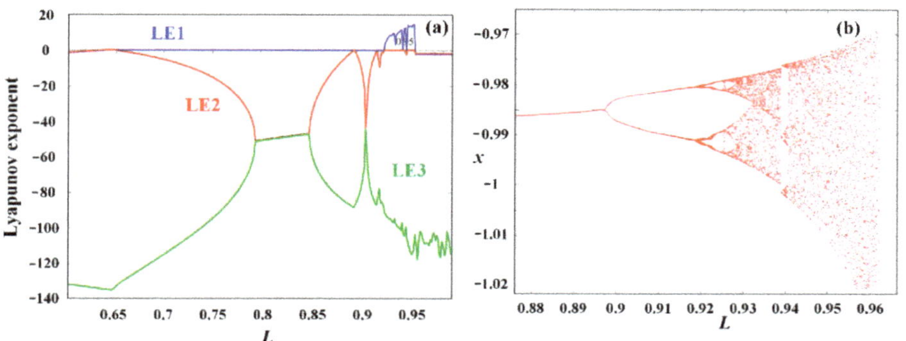

Figure 21. (**a**) Lyapunov exponent spectrum and (**b**) bifurcation of system with the inductance L (if $C = 26$ μF).

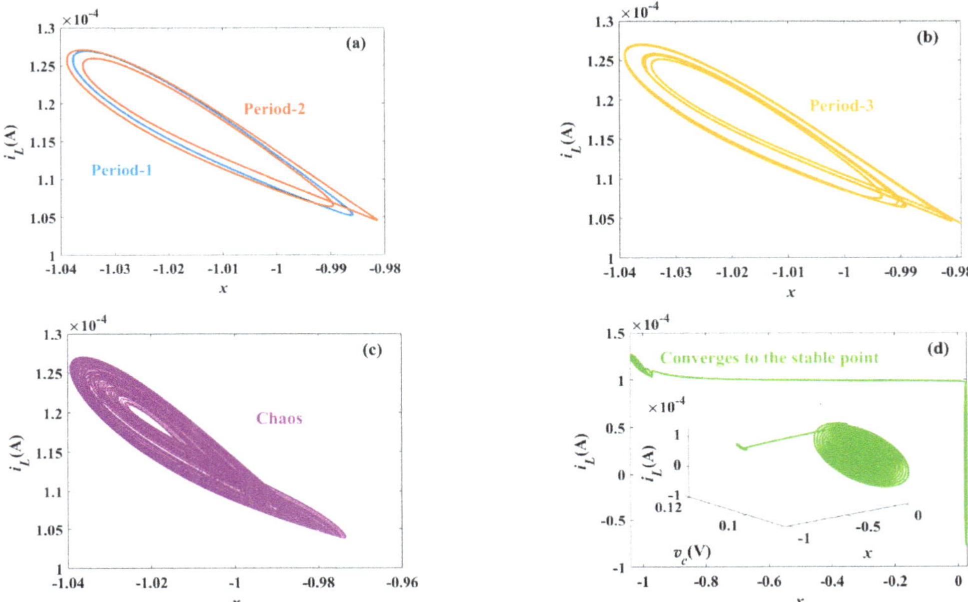

Figure 22. Variation in system attractor with inductance L, $v = 0.1031$ V, $C = 26$ μF. (**a**) $L = 0.88$ H, $L = 0.91$ H, (**b**) L = 0.925 H, (**c**) $L = 0.95$ H, (**d**) $L = 0.963$ H.

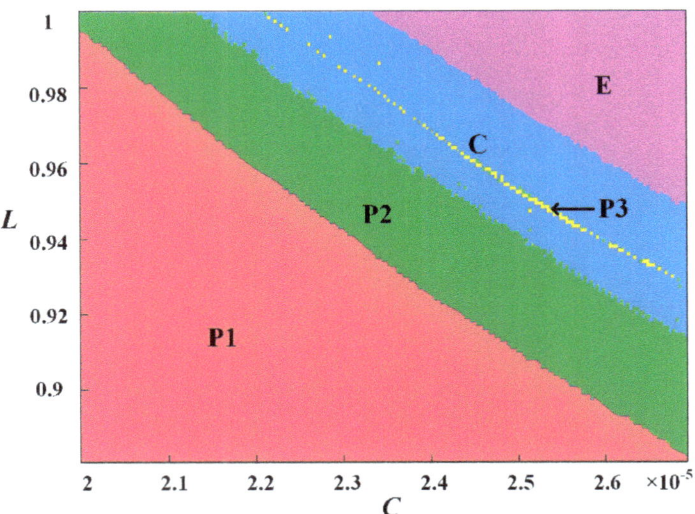

Figure 23. Dynamic map with $V = 0.1031$ V.

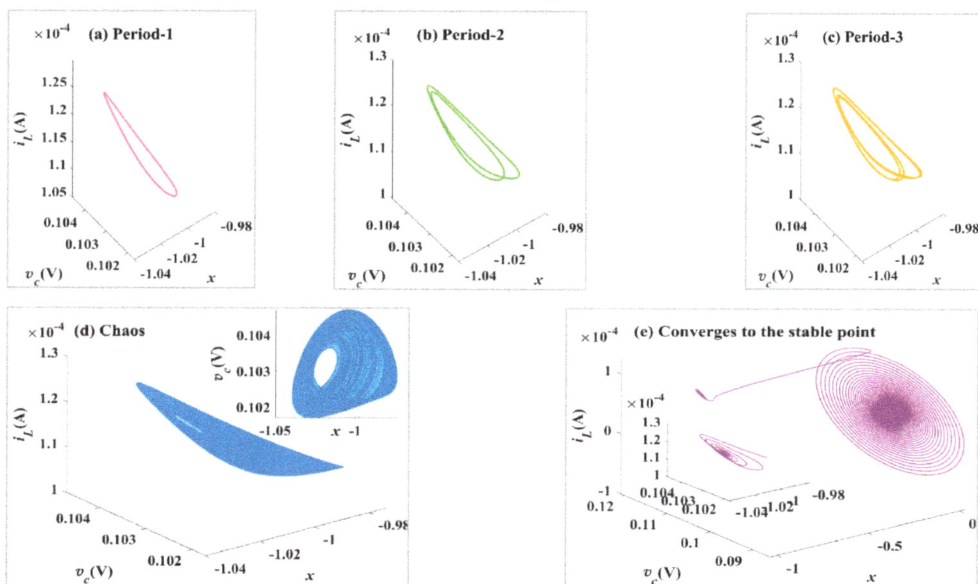

Figure 24. Typical phase diagrams. (**a**) C = 21.5 μF, L = 0.9 H; (**b**) C = 24 μF, L = 0.94 H; (**c**) C = 23.5 μF, L = 0.96 H; (**d**) C = 24 μF, L = 0.98 H; (**e**) C = 26 μF, L = 0.98 H.

6. Conclusions

We design a locally active memristor model whose state function is continuous and derivable within the whole real number interval. Its basic property has been analyzed via the dynamic route map, the pinched v–i hysteresis curve, and the DC V–I curve. It is found that the memristor is globally passive but locally active and contains two stable locally active regions called the edge of chaos and four unstable locally active regions.

In the stable locally active region, the pole of the admittance function is negative and the zero of the admittance function is positive, while in the unstable locally active region, the pole of admittance function is negative, and the zero of admittance function is positive.

At the two different locally active regions, two kinds of second-order memristive circuits have been built by connecting the memristor in series with a positive inductance L or in parallel with a passive capacitor C. Small-signal analysis and simulations show that the built circuits can generate period oscillation signals.

Adding another passive energy storage component to the second-order circuit, we construct the simplest third-order chaotic circuit, whose equilibrium points and stability are discussed, and the influence of the capacitance and the inductance on system dynamics is further studied. With the change in system parameters, the circuit exhibits various characteristics such as periodic oscillation and chaos, as well as coexisting attractors.

Author Contributions: Conceptualization, F.L., J.L. and G.W.; methodology, F.L., W.Z. and Y.D.; formal analysis, J.L., P.J., J.Y. and G.W.; writing—original draft preparation, F.L., W.Z. and P.J.; writing—review and editing, F.L., Y.D. and J.Y.; supervision, J.L. and G.W.; funding acquisition G.W. All authors have read and agreed to the published version of the manuscript.

Funding: This work was supported in part by the National Natural Science Foundation of China under Grant 62171173, Grant 61771176, and Grant 61801154.

Conflicts of Interest: The authors declare that they have no conflicts of interest.

References

1. Chua, L. Local activity is the origin of complexity. *Int. J. Bifurcat. Chaos* **2011**, *15*, 3435–3456. [CrossRef]
2. Mannan, Z.I.; Choi, H.; Kim, H. Chua corsage memristor oscillator via Hopf bifurcation. *Int. J. Bifurcat. Chaos* **2016**, *26*, 1630009. [CrossRef]
3. Dong, Y.; Wang, G.; Chen, G. A Bistable Nonvolatile Locally-active Memristor and its Complex Dynamics. *Commun. Nonlinear Sci. Numer. Simul.* **2020**, *84*, 105203. [CrossRef]
4. Mannan, Z.I.; Yang, C.; Kim, H. Oscillation with 4-lobe Chua corsage memristor. *IEEE Circuits Syst. Mag.* **2018**, *18*, 14–27. [CrossRef]
5. Mannan, Z.I.; Yang, C.; Adhikari, S.P. Exact Analysis and Physical Realization of the 6-Lobe Chua Corsage Memristor. *Complexity* **2018**, *2018*, 8405978. [CrossRef]
6. Jin, P.; Wang, G.; Iu, H.H.C. A Locally-Active Memristor and its Application in Chaotic Circuit. *IEEE Trans. Circuits Syst. II Express Briefs* **2017**, *65*, 246–250. [CrossRef]
7. Gu, W.; Wang, G.; Dong, Y.; Ying, J. Nonlinear dynamics in non-volatile locally-active memristor for periodic and chaotic oscillations. *Chin. Phys. B* **2020**, *29*, 110503. [CrossRef]
8. Ying, J.; Liang, Y.; Wang, J.; Dong, Y.; Wang, G.; Gu, M. A tristable locally-active memristor and its complex dynamics. *Chaos Solitons Fractals* **2021**, *148*, 111038. [CrossRef]
9. Liang, Y.; Wang, G.; Chen, G.; Dong, Y.; Yu, D.; Iu, H.H.C. S-Type Locally active memristor-based periodic and chaotic oscillators. *IEEE Trans. Circuits Syst. I Regul. Pap.* **2020**, *67*, 5139–5152. [CrossRef]
10. Ascoli, A.; Slesazeck, S.; Tetzlaff, R.; Maehne, H.; Mikolajick, T. Unfolding the local activity of a memristor. In Proceedings of the 2014 14th International Workshop on Cellular Nanoscale Networks and their Applications (CNNA), Notre Dame, IN, USA, 29–31 July 2014; IEEE: Piscataway, NJ, USA, 2014; pp. 1–2.
11. Chua, L.O. Five non-volatile memristor enigmas solved. *Appl. Phys. A Mater. Sci.* **2018**, *124*, 563. [CrossRef]
12. Adhikari, S.P.; Sah Kim, M.P.H.; Chua, L.O. Three fingerprints of memristor. *IEEE Trans. Circuits Syst. I Regul. Pap.* **2013**, *60*, 3008–3021. [CrossRef]
13. Chua, L. If it's pinched it's a memristor. *Semicond. Sci. Technol.* **2014**, *29*, 1–42. [CrossRef]
14. Mannan, Z.I.; Choi, H.; Rajamani, V.; Kim, H.; Chua, L.O. Chua Corsage Memristor: Phase Portraits, Basin of Attraction, and Coexisting Pinched Hysteresis Loops. *Int. J. Bifurc. Chaos* **2017**, *27*, 1730011. [CrossRef]
15. Kumar, S.; Strachan, J.P.; Williams, R.S. Chaotic dynamics in nanoscale NbO_2 Mott memristors for analogue computing. *Nature* **2017**, *548*, 318–321. [CrossRef] [PubMed]
16. Sandri, M. Numerical calculation of Lyapunov exponents. *Math. J.* **1996**, *6*, 78–84.

 *electronics*

Article

Complex Oscillations of Chua Corsage Memristor with Two Symmetrical Locally Active Domains

Jiajie Ying, Yan Liang, Fupeng Li, Guangyi Wang * and Yiran Shen

Institute of Modern Circuits and Intelligent Information, Hangzhou Dianzi University, Hangzhou 310018, China; jiajieying@hdu.edu.cn (J.Y.); liangyan@hdu.edu.cn (Y.L.); lfp_99@hdu.edu.cn (F.L.); yrshen@hdu.edu.cn (Y.S.)
* Correspondence: wanggyi@hdu.edu.cn

Abstract: This paper proposes a modified Chua Corsage Memristor endowed with two symmetrical locally active domains. Under the DC bias voltage in the locally active domains, the memristor with an inductor can construct a second-order circuit to generate periodic oscillation. Based on the theories of the edge of chaos and local activity, the oscillation mechanism of the symmetrical periodic oscillations of the circuit is revealed. The third-order memristor circuit is constructed by adding a passive capacitor in parallel with the memristor in the second-order circuit, where symmetrical periodic oscillations and symmetrical chaos emerge either on or near the edge of chaos domains. The oscillation mechanisms of the memristor-based circuits are analyzed via Domains distribution maps, which include the division of locally passive domains, locally active domains, and the edge of chaos domains. Finally, the symmetrical dynamic characteristics are investigated via theory and simulations, including Lyapunov exponents, bifurcation diagrams, and dynamic maps.

Keywords: memristor; chaos; local activity; the edge of chaos

Citation: Ying, J.; Liang, Y.; Li, F.; Wang, G.; Shen, Y. Complex Oscillations of Chua Corsage Memristor with Two Symmetrical Locally Active Domains. *Electronics* **2022**, *11*, 665. https://doi.org/10.3390/electronics11040665

Academic Editor: Christos Volos

Received: 25 January 2022
Accepted: 18 February 2022
Published: 21 February 2022

Publisher's Note: MDPI stays neutral with regard to jurisdictional claims in published maps and institutional affiliations.

1. Introduction

Local activity is considered to be the origin of complexity, which can amplify infinitesimal fluctuations to generate oscillations [1,2]. The complex behaviors and rich dynamics only appear in the locally active systems [3]. A locally active memristor-based circuit can generate complex oscillations such as limit cycles, chaos, or neuromorphic behaviors [4].

The locally active memristor exhibits negative differential memductance or memristance in its locally active domain of the DC V–I plot [5]. The edge of chaos domain of the memristor, a subset of the locally active domain, satisfies the asymptotically stable and locally active characteristics, where chaos, intelligence, and creativity may emerge [6,7]. Many hardware implementations of memristor have been reported, such as NbO_x, VO_2, and TaO_x devices, which are passive but local activity to be locally active memristors [8–11]. A bistable and a tristable locally active memristors are applied to construct chaotic circuits with rich dynamics, respectively [12,13], whose basic characteristics, coexisting dynamics, and oscillation mechanisms are analyzed. Locally active memristors can generate complex dynamical behaviors with potential application in many fields, including neurobiology [14,15] and nonlinear dynamics [16–18].

Chua Corsage Memristor (CCM), proposed by Chua, is a typical locally active memristor [5]. Chua provides analysis tools for analyzing the characteristics of memristors, including power-off plot, DC V–I plot, dynamic route map, quasi DC V–I plot, and small-signal equivalent circuit, etc., laying the foundation for the research of memristors [5,6,19]. The CCM family with rich nonlinear dynamics, including 2-Lobe CCM [19], 4-Lobe CCM [20], and 6-Lobe CCM [21], is built by designing various state equations with multiple stable states. Based on the theory of edge of chaos, the periodic oscillation mechanism of the CCM family is analyzed [22].

When the neural network operates in an edge of chaos domain, it may exhibit complexity, learning efficiency, adaptability [16], which is essential for solving global optimization problems and is more effective [23,24]. In addition, many researchers have investigated the robust H-infinity performance, exponential synchronization, and stability problems of memristor-based neural networks with time-varying delays [25,26]. Locally active memristors, with nonlinear and non-volatile, hold great potential to simulate neuromorphic behavior and apply to neural networks. An isolated third-order nanocircuit element is reported in [27], which is the first time to realize an integrated circuit element to express neuromorphic nonlinear dynamics. A vanadium dioxide VO_2 locally active memristor is used to design a two-channel neuron, which possesses most of the known biological neuronal dynamics [28].

The CCM exhibits complex behaviors of biological neurons when it operates at the edge of chaos domain [29,30]. In [29], circuits are constructed through a CCM and passive elements, verifying that action potentials emerge near the edge of chaos domain. It has been shown that CCM with two locally active domains can be used to model neurons to simulate some action potentials, and the chaos emerges in one of the locally active domains [30]. However, the original CCM has only one locally active domain and has a vast kiloamp level of current under standard operating voltage, which greatly limits its practical applications.

Since the Chua Corsage Memristor (CCM) is proposed in 2010 [5], many researchers have studied complex dynamics of CCM-based circuits, but there are still some mysteries to be explored. To further explore the complex dynamics and reveal the oscillation mechanisms of the CCM family, this paper proposes a symmetrical Chua Corsage Memristor (SCCM) model with two locally active domains. The parameter k is added to the state equation of the SCCM model to make its operating current in the milliampere level, which is more applicable for the practical circuit. It is found that the SCCM exhibits capacitive characteristics by analyzing the frequency response of the admittance function, so it can connect with a passive inductor to form a second-order nonlinear system. Using the Nyquist plot of the poles of the admittance function, this paper analyzes the transition from the stable state to the unstable state via the Hopf bifurcation point. It is obtained that the periodic oscillations of the second-order circuit only occur on the right half-plane pole domain. The third-order circuit is obtained by adding a passive capacitor to the second-order circuit, which can generate chaotic oscillation. A domains distribution map in the V–L plane is drawn, through the Nyquist plot of the poles of the admittance function, including the locally active domain, the locally passive domain, the edge of chaos domain, and the RHP pole domain. The third-order circuit has symmetric domains distribution map, which has symmetric oscillations at positive and negative voltage. It is demonstrated that the complex oscillations emerge either on or near the edge of chaos domain, which is speculated by Chua [31]. Furthermore, the rich symmetric dynamics of the SCCM-based circuits are explored in this paper with Lyapunov exponents, bifurcation diagrams, and dynamic maps [31–33].

In this paper, a novel CCM with two symmetrical locally active domains is proposed and its basic characteristics are analyzed in Section 2. In Section 3, the small-signal admittance function is used to analyze its edge of chaos characteristic. In Section 4, a second-order circuit is constructed by adding an inductor to the SCCM. The oscillation mechanism and symmetrical dynamic behaviors admitted by the third-order circuit are expounded in Section 5.

2. Symmetrical Chua Corsage Memristor

2.1. Mathematical Model

The CCM is a typical locally active memristor, which is a first-order memristor, described by

$$\begin{cases} i = x^2 v \\ \frac{dx}{dt} = 30 - x + |x - 20| - |x - 40| + v \end{cases} \tag{1}$$

where x, v, and i represent the state variable, voltage, and current of the CCM, respectively. Based on this CCM, we proposed a modified CCM with symmetrical locally active domains, called symmetrical Chua Corsage Memristor (SCCM). The mathematical model is as follows:

$$\begin{cases} i = G_0\left(x^2 + 0.1\right)v \\ \frac{dx}{dt} = k\left(30 - x + |x - 20| - |x - 40| + m|v|\right) \end{cases} \tag{2}$$

The parameter k is equal to 1000 for reducing its operating current in the milliampere level. The parameters G_0 and m are equal to 0.01 and -10, respectively.

2.2. Pinched Hysteresis Loops

Chua proposed that the pinched hysteresis loop in the v–i plane is the only required fingerprint for determining a memristor [34]. When input voltage signals $v(t) = \sin(2\pi ft)$, with different frequencies of 100 Hz, 300 Hz, 1 kHz, and 10 kHz, are applied to the SCCM, the v–i pinched hysteresis loops are depicted in Figure 1.

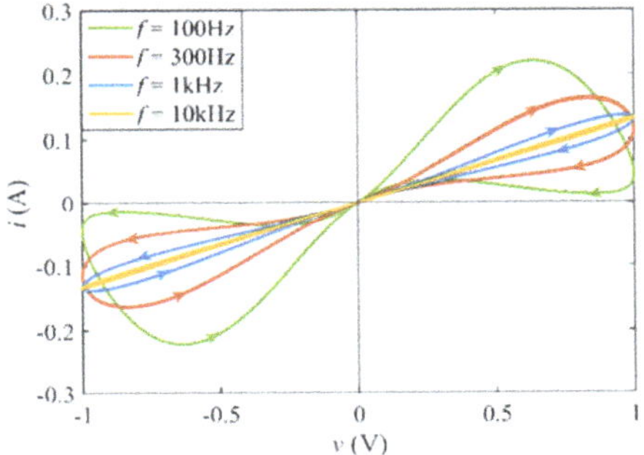

Figure 1. Pinched hysteresis loops of the SCCM when the input voltage signals $v(t) = \sin(2\pi ft)$.

In addition, Figure 1 indicates the pinched hysteresis loops vary with the frequencies. As the frequency increases, the pinched hysteresis loop area decreases monotonically, and then the area shrinks to approximately zero at 10 kHz, where the pinched hysteresis loop approximates a straight line [35].

2.3. Local Activity

DC V–I plot is an effective method to determine the locally active domains of memristor, which is proposed by Chua [5]. The negative differential resistance (NDR) domains on its DC V–I plot are the locally active domains with potential instability, where complex oscillations may emerge. When the SCCM is driven by an array of DC voltages V_k, respectively, we measure the corresponding collection of DC currents I_k flow through the SCCM, and then draw points (V_k, I_k) on the V–I plane to form a DC V–I plot.

When the differential equation dx/dt in Equation (2) is set as zero, the equation of DC voltage is derived as

$$|V| = 0.1\left(30 - X + |X - 20| - |X - 40|\right) \tag{3a}$$

Obviously, $|V| > 0$, which means $0.1(30 - X + |X - 20| - |X - 40|) > 0$. So, the ranges of the state variable X are calculated as $X < 20$ and $30 < X < 50$. Then, if we substitute the Equation (3a) into the equation $i = G_0(x^2 + 1)v$, the equation of DC current is derived. Hence, the equations of the DC voltage and DC current are given by

$$\begin{cases} V = \pm 0.1\left(30 - X + |X - 20| - |X - 40|\right), X < 20 \; or \; 30 < X < 50 \\ I = G_0\left(X^2 + 0.1\right)V \end{cases} \tag{3b}$$

The DC *V–I* plot of the SCCM for $-20 < X < 20$ and $30 < X < 50$ are depicted in Figure 2, where the NDR region is marked with the red line, and the other region is marked with the blue line.

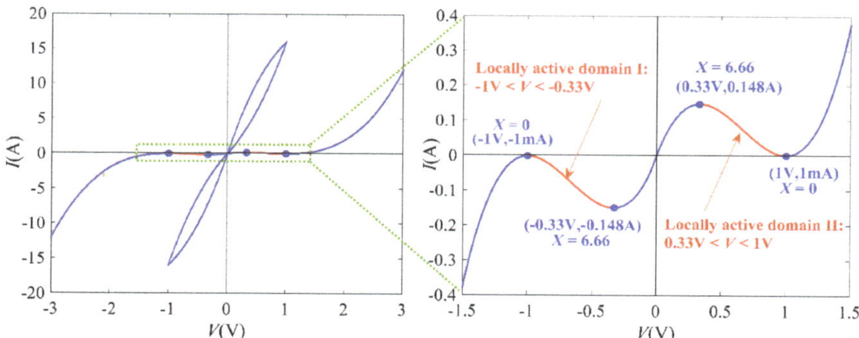

Figure 2. DC *V–I* plot of the SCCM.

It is observed from Figure 2 that the SCCM exhibits negative differential conductance when the state variable X ranges from 0 to 6.66. Therefore, the SCCM has two symmetrical locally active domains of $-1\ \text{V} < V < -0.33\ \text{V}$ and $0.33\ \text{V} < V < 1\ \text{V}$.

3. Edge of Chaos of the SCCM

3.1. Small-Signal Equivalent Circuit

When the SCCM operating point Q is in the locally active domain, the small-signal equivalent admittance can be obtained from [22]:

$$Y_M(s, Q) = \frac{a_{11}b_{12}}{s - b_{11}} + a_{12} \tag{4a}$$

where

$$a_{11} = v\frac{\partial\left(G_0\left(x^2 + 0.1\right)\right)}{\partial x}\bigg|_Q = 0.02XV$$
$$a_{12} = G_0\left(x^2 + 0.1\right)\frac{\partial v}{\partial v}\bigg|_Q = 0.01\left(X^2 + 0.1\right)$$
$$b_{11} = \frac{\partial(k\,(30 - x + |x - 20| - |x - 40| + m|v|))}{\partial x}\bigg|_Q = 1000(sign(X - 20) - sign(X - 40) - 1) \tag{4b}$$
$$b_{12} = \frac{\partial(k\,(30 - x + |x - 20| - |x - 40| + m|v|))}{\partial v}\bigg|_Q = -10^4 sign(V)$$

Equation (4a) can be rewritten as:

$$Y_M(s, Q) = \frac{1}{sL_x + R_x} + \frac{1}{R_y} \tag{5}$$

where $L_x = 1/(a_{11}b_{12})$, $R_x = -b_{11}/(a_{11}b_{12})$, and $R_y = 1/a_{12}$. The small-signal equivalent circuit of the SCCM operated in the locally active domains is described in Figure 3.

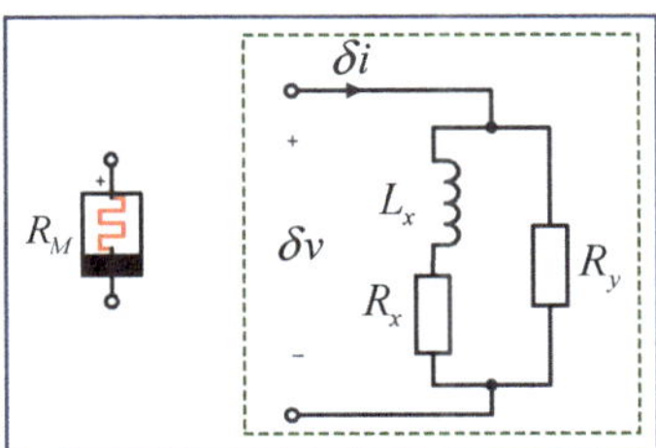

Figure 3. The small-signal equivalent circuit of the SCCM.

3.2. Edge of Chaos

When the memristor-based circuits operate at the edge of chaos domain, the operating point must be in the locally active domain and asymptotically stable [1]. According to Equation (4a), the SCCM operates at the locally active domain, i.e., $Re\ Y_M\ (i\omega, Q) < 0$ for some $\omega \in (-\infty, \infty)$. The SCCM is asymptotically stable only if the real part of the pole of the admittance function is less than zero [2].

The pole and zero of the small-signal admittance in Equation (4a) are calculated as $p = b_{11}$, and zero $z = -(a_{11}b_{12} - a_{12}b_{11})/a_{12}$, respectively.

From Equation (4a), the frequency response $Y_M\ (i\omega, Q)$ of the SCCM is derived, as

$$Y_M(i\omega, Q) = \frac{a_{11}b_{12}}{i\omega - b_{11}} + a_{12} \tag{6a}$$

where the real part of $Y_M\ (i\omega, Q)$ is

$$\mathrm{Re}Y_M(i\omega, Q) = \frac{-b_{11}(a_{11}b_{12} - a_{12}b_{11}) + \omega^2 a_{12}}{b_{11}^2 + \omega^2} = \frac{a_{12}(pz + \omega^2)}{b_{11}^2 + \omega^2} \tag{6b}$$

and the imaginary part is

$$\mathrm{Im}Y_M(i\omega, Q) = \frac{-\omega a_{11}b_{12}}{b_{11}^2 + \omega^2} \tag{6c}$$

Obviously, $a_{12} = 0.01(X^2 + 0.1) > 0$ in Equation (4b). Therefore, if the product of the pole and the zero is less than 0, i.e., $pz < 0$, the Re $Y_M\ (i\omega, Q)$ is negative for the frequency range $-\sqrt{-pz} < \omega < \sqrt{-pz}$.

Both local activity and asymptotic stability are satisfied simultaneously, i.e., $pz < 0$ and $p < 0$, calculated $z > 0$ and $p < 0$. Therefore, the SCCM operates at the edge of chaos domain if and only if the zero $z > 0$ and pole $p < 0$.

The pole and zero of the SCCM are shown in Figure 4, where the edge of chaos domains satisfying zero $z > 0$ and pole $p < 0$ are -1 V $< V < -0.33$ V and 0.33 V $< V < 1$ V. Because the pole is always less than 0, the locally active domain of SCCM is consistent with the edge of chaos domain of SCCM.

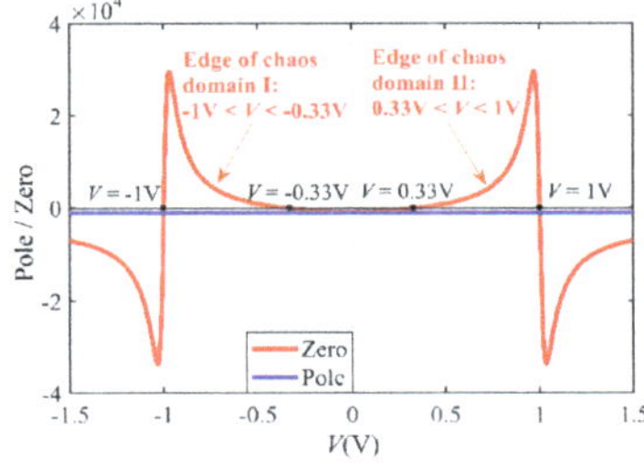

Figure 4. Zero and pole diagrams of $Y_M\ (s, Q)$ of the SCCM.

4. SCCM-Based Second-Order Circuit

4.1. Second-Order Circuit

This section will connect an inductor or capacitor with the SCCM to construct a second-order oscillator. The connected inductor or capacitor is determined by judging whether the memristor has capacitive or inductive characteristics.

According to Equations (6b) and (6c), the real and imaginary diagrams of $Y_M (i\omega, Q)$ of the SCCM over the range -3×10^3 rad/s $< \omega < 3 \times 10^3$ rad/s with $V = 0.5$ V, are shown in Figure 5.

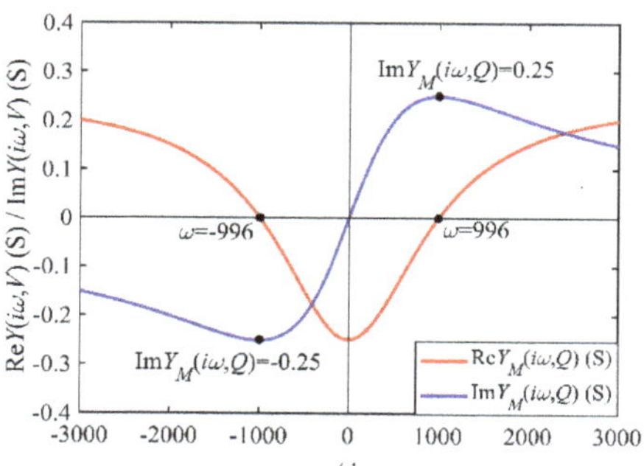

Figure 5. The real and imaginary diagrams of $Y_M (i\omega, Q)$ of the SCCM versus frequency ω at $V = 0.5$ V.

Figure 5 shows that Re $Y_M (i\omega, Q) = 0$ S and Im $Y_M (i\omega, Q) = \pm 0.25$ S at $\omega = \pm 996$ rad/s at the locally active domain I with $V = 0.5$ V, which indicates that the SCCM exhibits capacitive characteristic. Hence, an inductor is required to connect with the SCCM to form the oscillator. Based on Chua's theory [22], the inductance is calculated by $L = 1/(\omega \times$ Im $Y_M (i\omega, Q)) = 4.02$ mH.

Therefore, the second-order circuit is constructed by connecting the SCCM and an inductor, shown in Figure 6.

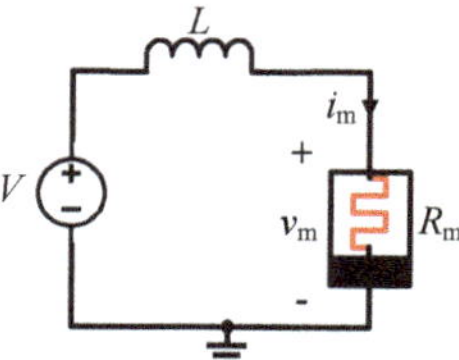

Figure 6. The SCCM-based second-order circuit.

According to Kirchhoff's laws, the state equations of the nonlinear system are:

$$\begin{cases} \frac{dx}{dt} = 1000 \left(30 - x + \left| x - 20 \right| - \left| x - 40 \right| - 10 \frac{i_L}{0.01(x^2+0.1)} \right) \\ \frac{di_L}{dt} = \frac{1}{L} \left(V - \frac{i_L}{0.01(x^2+0.1)} \right) \end{cases} \tag{7}$$

where x, i_L, and V represent the state variable, the current, and the bias voltage, respectively.

4.2. Complexity Mechanism

From Equation (6a), the composite admittance function $Y_S(s, Q)$ in Figure 6 is described as

$$Y_S(s, Q) = \frac{Y_L Y_M(s, Q)}{Y_L + Y_M(s, Q)} = \frac{a_{12}s + a_{11}b_{12} - a_{12}b_{11}}{a_{12}Ls^2 + (a_{11}b_{12}L - a_{12}b_{11}L + 1)s - b_{11}} \tag{8}$$

with poles

$$p_1 = \frac{-(L(a_{11}b_{12} - a_{12}b_{11}) + 1) + \sqrt{(L(a_{11}b_{12} - a_{12}b_{11}) + 1)^2 + 4a_{12}b_{11}L}}{2a_{12}L} \tag{9a}$$

and

$$p_2 = \frac{-(L(a_{11}b_{12} - a_{12}b_{11}) + 1) - \sqrt{(L(a_{11}b_{12} - a_{12}b_{11}) + 1)^2 + 4a_{12}b_{11}L}}{2a_{12}L} \tag{9b}$$

When the circuit operates at DC voltage with a frequency of 0 Hz, the inductor is a short circuit. Therefore, the inductor does not affect the DC characteristics of the system. The second-order circuit has the identical DC V–I plot as the SCCM, so their locally active domains are exactly the same.

Then, if the bias voltage V is within the ranges of $-1\,\text{V} < V < -0.33\,\text{V}$ and $0.33\,\text{V} < V < 1\,\text{V}$, the system in Equation (7) is on the locally active domain. The system is asymptotically stable when the real part of two poles p_1 and p_2 of $Y_S(s, Q)$ are less than zero. If these two conditions of local activity and asymptotical stability are satisfied simultaneously, the system is on the edge of chaos domain.

When the bias voltage is 0.5 V on the locally active domain, the Nyquist plots of the poles are shown in Figure 7a over the range 0.6 mH $< L <$ 100 mH.

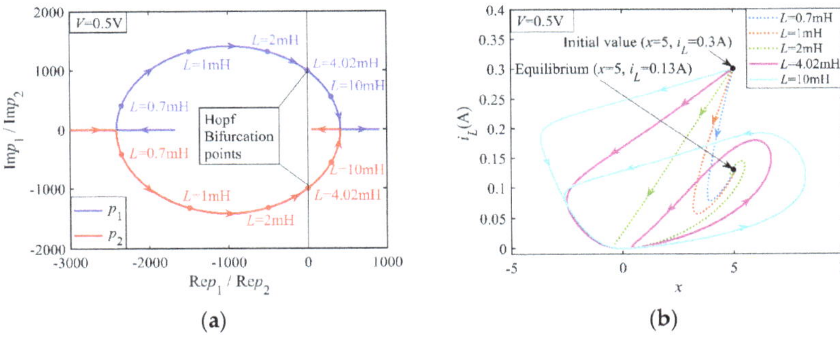

Figure 7. (**a**) The Nyquist plots of the poles over the range 0.6 mH $< L <$ 100 mH. (**b**) The $x - i_L$ phase diagrams with different inductances of 0.7 mH, 1 mH, 2 mH, 4.02 mH, and 10 mH.

The Hopf bifurcation point is the intersection of the Nyquist plot with the imaginary axis, referring to the pole where the real part is 0. The stability of the system changes when crossing the Hopf bifurcation point. The system is stable on the left half-plane (LHP) of the Nyquist plot, and the system is unstable on the right half-plane (RHP). In Figure 7a, Hopf bifurcation point is L = 4.02 mH. Therefore, the edge of chaos domain is $L <$ 4.02 mH, the open RHP domain is $L >$ 4.02 mH.

The initial value is set as (5, 0.3). The $x - i_L$ phase diagrams with different inductances of 0.7 mH, 1 mH, 2 mH, 4.02 mH, and 10 mH, are shown in Figure 7b, where dotted trajectories tend to point attractors, and solid trajectories tend to periodic attractors.

Observed that the periodic oscillation generated via the Hopf bifurcation on the RHP Pole domain.

The inductance is set as 4.02 mH. The Nyquist plots of the poles are shown in Figure 8a over the range 0 V < V < 0.92 V. Observed that Hopf bifurcation points are V = 0.5 V and 0.83 V, and the edge of chaos domain is 0 V < V < 0.5 V and 0.83 V < V < 0.92 V.

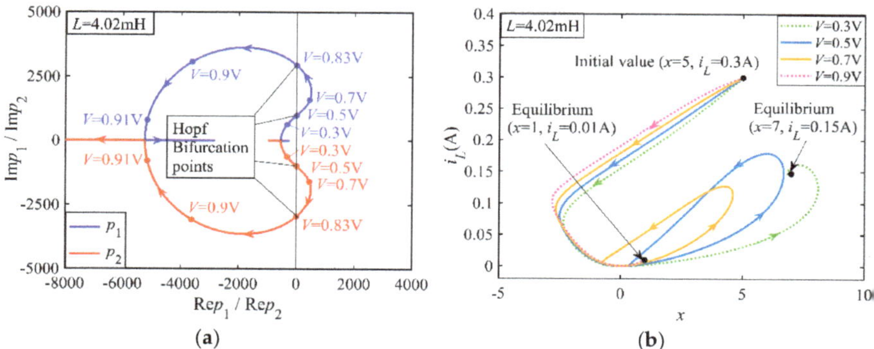

(a)   (b)

Figure 8. (a) The Nyquist plots of the poles over the range 0 V < V < 0.92 V. (b) The $x - i_L$ phase diagrams with different bias voltages of 0.3 V, 0.5 V, 0.7 V, and 0.9 V.

The initial value is set as (5, 0.3). The $x - i_L$ phase diagrams with different bias voltages of 0.3 V, 0.5 V, 0.7 V, and 0.9 V, are shown in Figure 8b, where dotted trajectories tend to point attractors, and solid trajectories tend to periodic attractors. It follows that the periodic oscillations only occur on the RHP pole domains and the Hopf bifurcation point.

Due to the symmetry, the analysis of the oscillation mechanism of the locally active domains I and II is similar, so the analysis of the locally active domain I is omitted.

According to the above analysis, any parameter domain can be divided into the following three cases:

(i) Edge of chaos domain (Locally active domain): $Re\ Y(i\omega, Q) < 0$ for some $\omega \in (-\infty, \infty)$, and real part of all poles are less than zero.

(ii) RHP pole domain (Locally active domain): $Re\ Y(i\omega, Q) < 0$ for some $\omega \in (-\infty, \infty)$, and there are poles with real parts less than zero.

(iii) Locally passive domain: $Re\ Y(i\omega, Q) > 0$ for all $\omega \in (-\infty, \infty)$.

Then, the domains distribution map of these three domains in the V–L plane is shown in Figure 9.

In Figure 9, the locally active domains are located in the bias voltage ranges -1 V < V < -0.33 V and 0.33 V < V < 1 V, in which the edge of chaos domains are painted green areas and the RHP pole domains are painted yellow areas. The locally passive domains are painted to blue areas with no oscillation. The black lines are the Hopf bifurcation lines, which are the dividing line of the edge of chaos domains and the RHP pole domains. Observed that the symmetrical oscillation occurs on the RHP pole domain of the locally active domain.

Observe from Figure 9 that the oscillation can only appear either on or near the edge of chaos domain in the locally active domains.

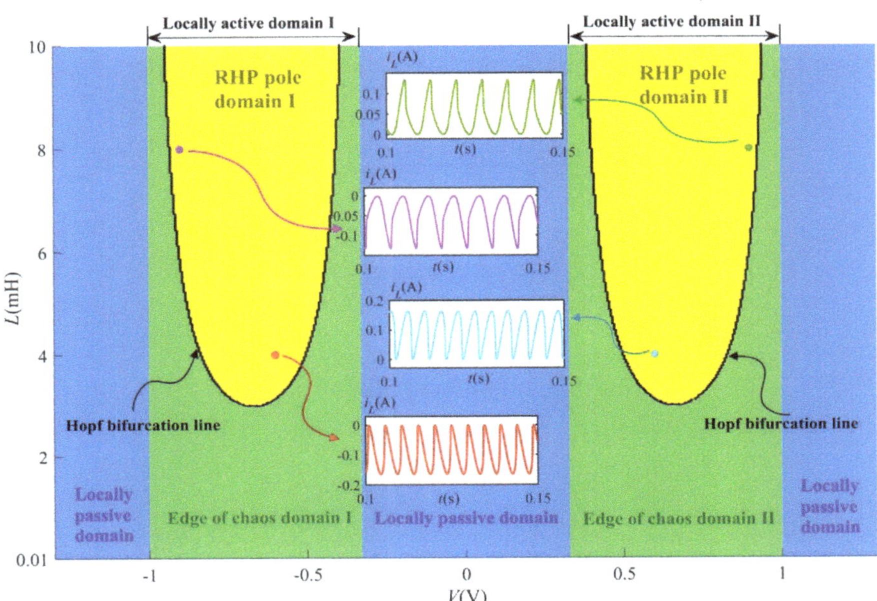

Figure 9. Domains distribution map in the *V*–*L* plane over the ranges $-1.3\text{ V} < V < 1.3\text{ V}$ and $0.01\text{ mH} < L < 10\text{ mH}$.

4.3. Symmetric Dynamics

The SCCM-based second-order circuit with two symmetrical locally active domains will generate symmetric oscillation. The inductance L is still set as 4.02 mH. The bias voltages are chosen to be ±0.4 V, ±0.5 V, and ±0.6 V located at the edge of chaos domain, Hopf bifurcation line, and RHP pole domain, respectively.

When $V = \pm0.4$ V, ±0.5 V, and ±0.6 V, respectively, the time-domain waveforms and the symmetric phase orbit diagrams are shown in Figure 10, where the blue and red orbit diagrams represent $V < 0$ V and $V > 0$ V, respectively. The second-order circuit generates the point attractor oscillation with $V = \pm0.4$ V, and the periodic attractor oscillation with $V = \pm0.5$ V and ±0.6 V.

In Section 4.1, we calculated that the Hopf bifurcation point is inductance $L = 4.02$ mH with bias voltage $V = 0.4$ V and frequency $f = 996$ rad/s. The corresponding time-domain waveforms and phase orbit diagrams are shown in Figure 10c,d, in which the oscillation frequency is 997 rad/s, verifying the prediction of Hopf bifurcation.

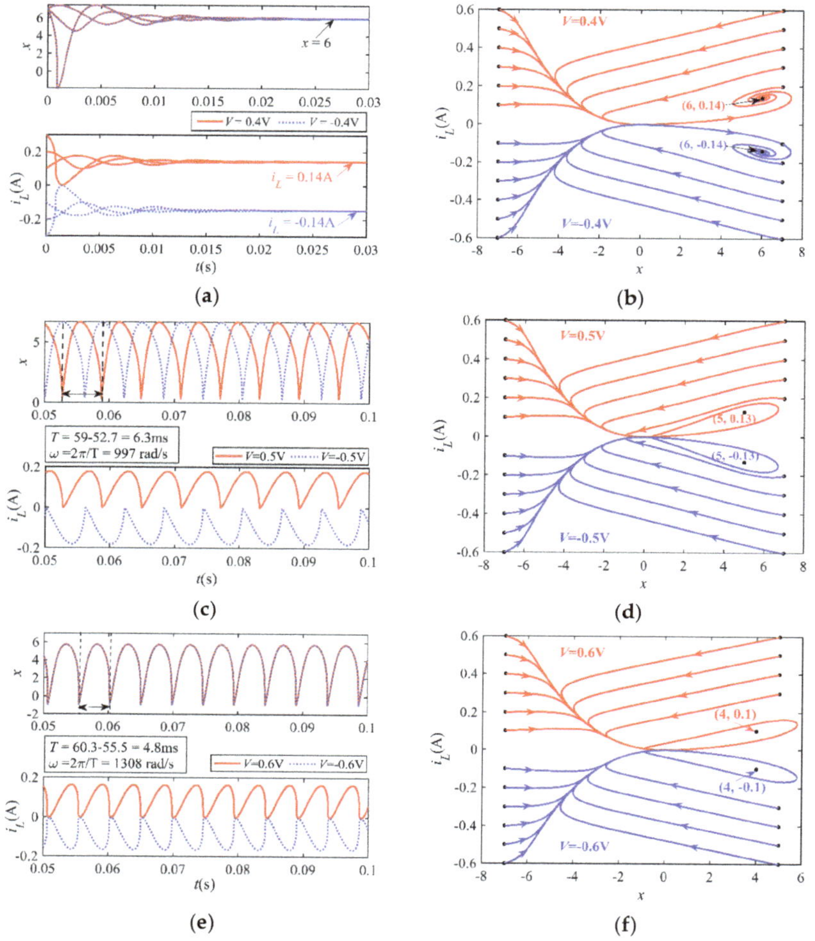

Figure 10. (**a**) Symmetric oscillations of the second-order circuit. (**a**,**c**,**e**). The time-domain waveforms with $V = \pm 0.4$ V, ± 0.5 V, and ± 0.6 V, respectively. (**b**,**d**,**f**). The $x - i_L$ phase diagram with $V = \pm 0.4$ V, ± 0.5 V, and ± 0.6 V, respectively.

5. SCCM-Based Third-Order Circuit

To reveal the oscillation mechanism of chaos [36], an SCCM-based third-order circuit is built by paralleling a capacitor to the SCCM in the second-order circuit, as shown in Figure 11.

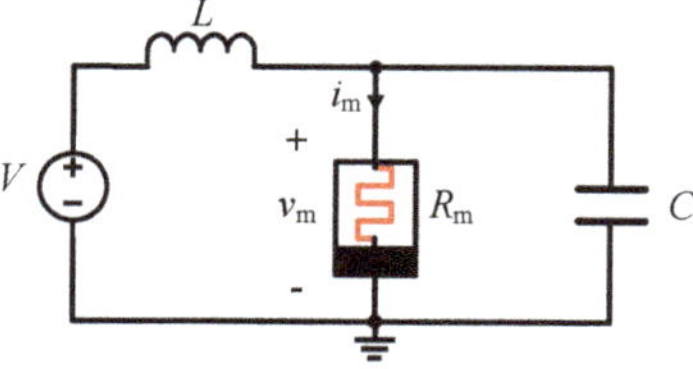

Figure 11. The SCCM-based third-order circuit.

According to Kirchhoff's laws, the state equations of the system in Figure 11 are:

$$\begin{cases} \frac{dx}{dt} = 1000\left(30 - x + |x - 20| - |x - 40| - 10v_C\right) \\ \frac{di_L}{dt} = \frac{1}{L}(V - v_C) \\ \frac{dv_C}{dt} = \frac{1}{C}\left(i_L - 0.01\left(x^2 + 0.1\right)\right) \end{cases} \tag{10}$$

where x, i_L, v_C, and V represent the state variable, inductor current, capacitor voltage, and the bias voltage, respectively.

The inductance is set as 7.2 mH, and the capacitance is set as 10 μF. The initial value (x, i_L, v_C) is (0, 0, 0). The phase diagrams of the system (10) with $V = \pm0.92$ V are shown in Figure 12, which are chaotic oscillations with Lyapunov value $LE_1 = 204.5$, $LE_2 = 0.123$, and $LE_3 = -4260.3$.

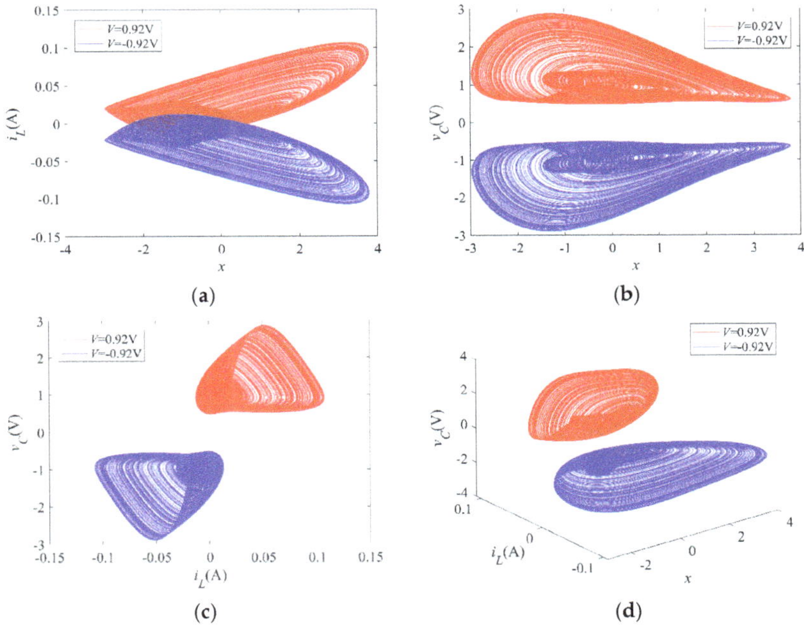

Figure 12. Chaotic oscillation of the third-order circuit with $V = \pm\,0.92$ V. (**a**) $x - i_L$ plane. (**b**) $x - u_C$ plane. (**c**) $i_L - u_C$ plane. (**d**) $x - i_L - u_C$ plane.

5.1. Complexity Mechanism

From Equation (8), the composite admittance function $Y_T(s, Q)$ of the third-order circuit shown in Figure 11 is derived as

$$\begin{aligned} Y_T(s, Q) &= \frac{Y_L(Y_C + Y_M(s,Q))}{Y_L + Y_C + Y_M(s,Q)} \\ &= \frac{Cs^2 + (a_{12} - b_{11}C)s + (a_{11}b_{12} - a_{12}b_{11})}{CLs^3 + (a_{12} - b_{11}C)Ls^2 + (1 + a_{11}b_{12}L - a_{12}b_{11}L)s - b_{11}} \\ &= \frac{k_1(s - z_1)(s - z_2)}{k_2(s - p_1)(s - p_2)(s - p_3)} \end{aligned} \tag{11}$$

where the z_1 and z_2 are the zeros of the composite admittance; the p_1, p_2, and p_3 are the poles of the composite admittance; and the k_1 and k_2 are parameter.

The poles p_1, p_2, and p_3 of the composite admittance function $Y_T(s, Q)$ cannot be derived from the formula, but it can be calculated by MATLAB software.

When the circuit operates at DC voltage with a frequency of 0 Hz, the inductor is equivalent to a short circuit, and the capacitor is equivalent to an open circuit. Therefore, the inductor and capacitor do not affect the DC characteristics of the system. The third-order circuit has the identical DC V–I plot as the SCCM, so their locally active domains are exactly the same.

Then, if the bias voltage V is within the ranges of -1 V $< V < -0.33$ V and 0.33 V $< V < 1$ V, the system is on the locally active domain. When these all poles p_1, p_2, and p_3 of $Y_T(s, Q)$ are located in the LHP of Nyquist plot, the system is asymptotically stable. If these two conditions of local activity and asymptotical stability are satisfied simultaneously, the system is on the edge of chaos domain.

When the capacitance is set as 10 μF, the parameter plane V–L of the third-order circuit over the parameter ranges -1.3 V $< V < 1.3$ V and 0.01 mH $< L < 10$ mH is divided to three locally passive domains, two symmetrical RHP pole domains, and two symmetrical edge of chaos domains, as shown in Figure 13.

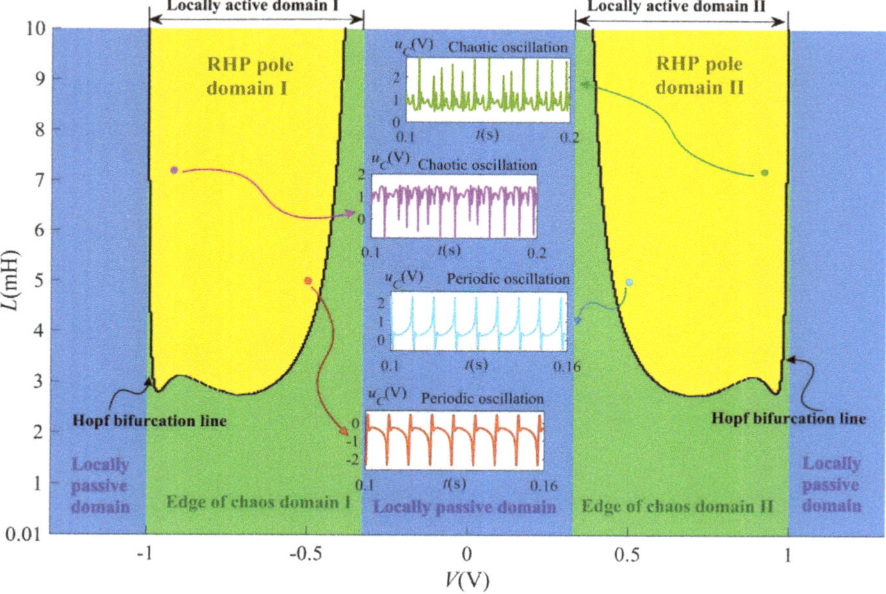

Figure 13. Domains distribution map in the V–L plane over the ranges -1.3 V $< V < 1.3$ V and 0.01 mH $< L < 10$ mH.

In Figure 13, the locally active domains are located in the bias voltage ranges -1 V $< V < -0.33$ V and 0.33 V $< V < 1$ V, in which the edge of chaos domains are painted green areas and the RHP pole domains are painted yellow areas. Observed that the symmetrical periodic oscillation and symmetrical chaos occur on the RHP pole domains. The locally passive domains are painted to blue areas, and the black lines are the Hopf bifurcation lines.

Observe from Figure 13 that the oscillation can only appear either on or near the edge of chaos domain in the locally active domains.

5.2. Symmetric Dynamics

The parameters are set as $L = 7.2$ mH and $C = 10$ μF. The initial value (x, i_L, v_C) is set as $(0, 0, 0)$. The bifurcation diagrams and Lyapunov exponent spectrums over the bias voltage ranges -0.96 V $< V < -0.86$ V and 0.86 V $< V < 0.96$ V, are shown in Figure 14 a–d, respectively. The minimum Lyapunov exponent value LE_3 is too small and omitted.

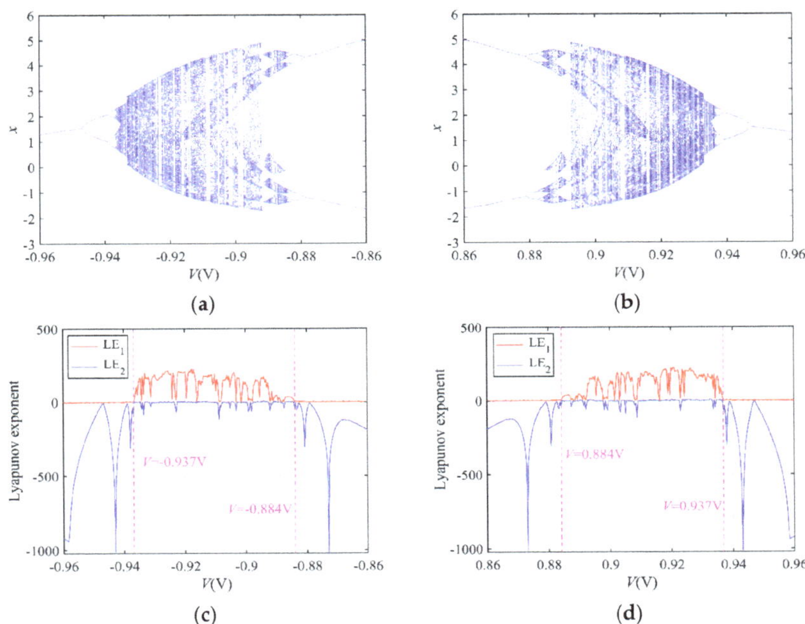

Figure 14. (**a**,**b**). The bifurcation diagrams over the ranges -0.96 V $< V < -0.86$ V and 0.86 V $< V <$ 0.96 V. (**c**,**d**). The corresponding Lyapunov exponent spectrums.

In Figure 14a, when $V = -0.937$ V, the system exhibits complex dynamics from periodic behavior to chaotic behavior. When $V > -0.884$ V, the chaotic oscillation disappears and gradually evolves into periodic oscillation through the inverse period-doubling bifurcation. The chaotic behaviors of the circuit appear for ranges of -0.937 V $< V < -0.884$ V and 0.884 V $< V < 0.937$ V. Observably, the third-order circuit has symmetrical bifurcation behaviors with bias voltage V.

Figure 15a–i show attractors on the $x - v_C$ plane with different bias voltages. The system generates a single cycle with bias voltages of ±0.95 V and -0.87 V. When the bias voltages are ±0.94 V, ±0.938 V, and ±0.93 V, the system will generate double cycle, quadruple cycle, and chaos, respectively. These dynamic behaviors are consistent with those analyzed in Figure 14.

The dynamic behaviors with the capacitance C are visualized by Figure 16a,c, where $V = \pm0.92$ V, $L = 7.2$ mH, and initial state is $(0, 0, 0)$. The effect of inductance L is visualized by Figure 16b,d, where $V = \pm0.92$ V, $C = 10$ μF, and initial state is $(0, 0, 0)$. Figure 15a,b show the bifurcation diagram of the capacitor voltage v_C when C ranges from 1 μF to 300 μF and L varies from 6 mH to 9 mH, respectively, where the voltage of the red part is 0.92 V and the blue part is -0.92 V. Figure 15c,d show the Lyapunov exponent spectrums with $V = 0.92$ V, corresponding to Figure 15a,b, respectively. The minimum Lyapunov exponent value LE_3 is too small and omitted. They are observed that the capacitor voltage v_C has the consistent bifurcation behavior when $V = \pm0.92$ V, which is caused by the symmetry of the system.

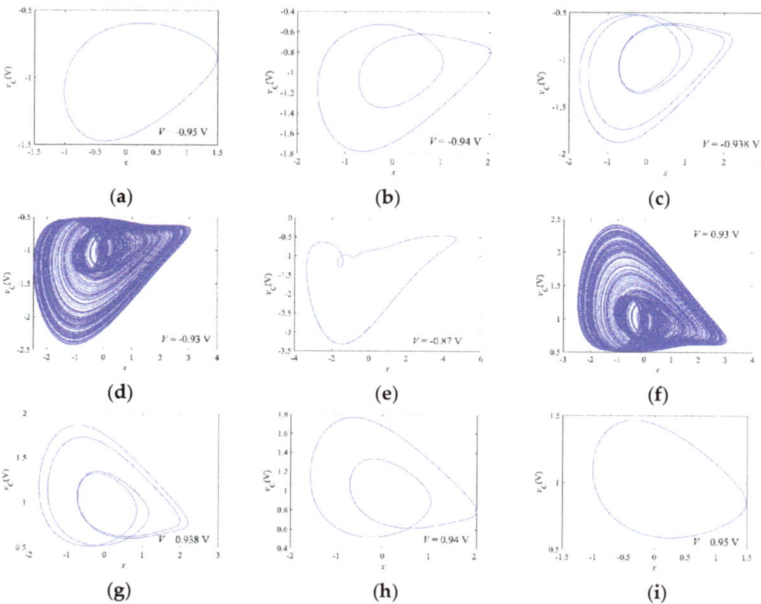

Figure 15. The oscillation attractors with different bias voltage V of (**a**) -0.95 V, (**b**) -0.94 V, (**c**) -0.938 V, (**d**) -0.93 V, (**e**) -0.87 V, (**f**) 0.93 V, (**g**) 0.938 V, (**h**) 0.94 V, and (**i**) 0.95 V.

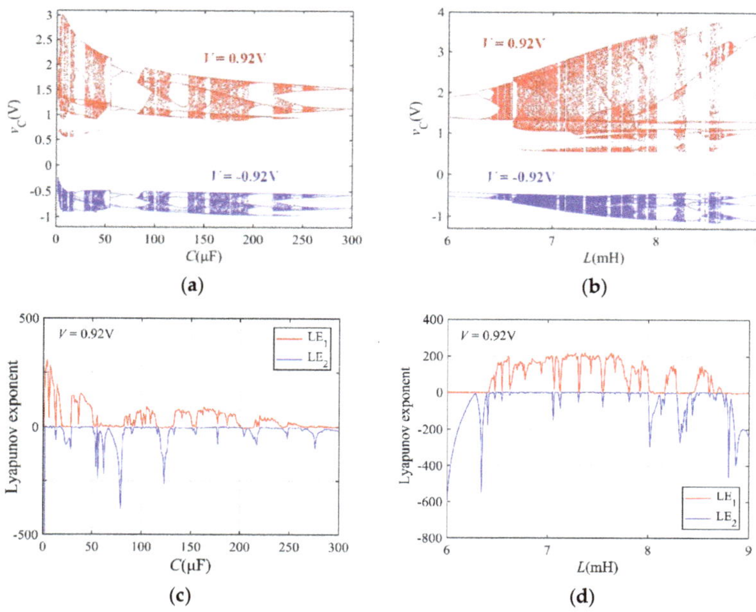

Figure 16. The dynamics behaviors over the capacitance range 0.1 μF < C < 300 μF: (**a**) the bifurcation diagram with bias voltage $\pm$ 0.92 V, and (**b**) the Lyapunov exponent spectrum with bias voltage 0.92 V. The dynamics behaviors over the inductance range 6 mH < C < 9 mH: (**c**) the bifurcation diagram with bias voltage $\pm$ 0.92 V, and (**d**) the Lyapunov exponent spectrum with bias voltage 0.92 V.

To observe the dynamic behaviors varying with the bias voltage V and the capacitance C, we plot the dynamic maps in Figure 17 with the inductance $L = 7.2$ mH. In Figure 17, the blue, green, and yellow areas represent the chaos, periodic oscillation, and stable point, respectively. In addition, the boundary between the blue area and the green area represents that the dynamic behavior changes from periodic oscillation to chaos or from chaos to periodic oscillation. To observe the chaotic behavior of the third-order circuit, we zoomed in on the region where chaotic oscillations appear.

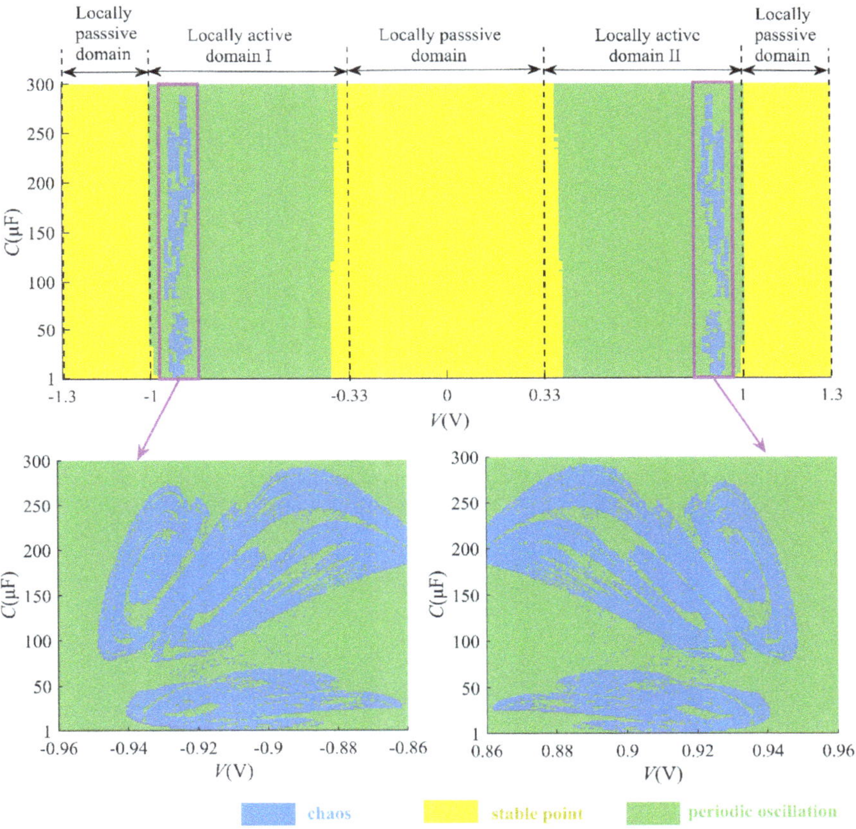

Figure 17. The dynamic map over the ranges 1 μF < C < 300 μF and -1.3 V < V < 1.3 V.

The locally active domains are -1 V < V < -0.33 V and 0.33 V < V < 1 V. Observed from Figure 17 that the third-order circuit has consistent symmetrical dynamics, and the chaos and periodic oscillation only occur in the locally active domains.

With the bias voltage $V = -0.92$ V, the dynamic map depending on both the inductance L and capacitance C is shown in Figure 18, where C ranges from 1 μF to 300 μF and L varies from 6 mH to 9 mH. In Figure 18, the blue and green areas represent the chaos and periodic oscillation, respectively. The dynamic map shows the complex behaviors of the periodic oscillation and chaotic oscillation of the third-order system.

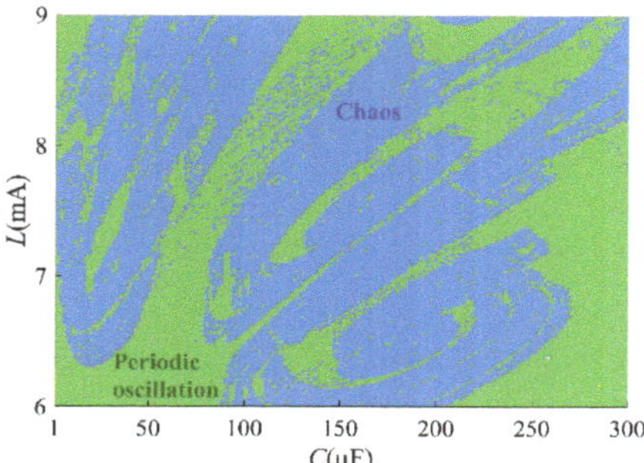

Figure 18. The dynamic map over the ranges 1 μF < C < 300 μF and 6 mH < L < 9 mH.

6. Conclusions

In this paper, a modified Chua Corsage Memristor is proposed, which has symmetrical locally active domains. The local activity and the edge of chaos of the SCCM are explored by analyzing the small-signal equivalent circuit. By analyzing the small-signal admittance function, the SCCM is determined to exhibit the capacitive characteristic in its locally active domain, so it can be connected with the inductor to form a second-order circuit. The edge of chaos domain is determined using the conjugate poles of the admittance function and local activity. It is proven that the second-order circuit generates complex symmetrical oscillation either on or near the edge of chaos domains. Furthermore, the third-order circuit is built by paralleling a capacitor with the SCCM in the second-order circuit. Based on the theory of local activity and edge of chaos, the oscillation mechanism of the chaos and periodic oscillation in the third-order circuit has been expounded.

The action potential may emerge either on or near the edge of chaos, so the CCM could be used to neuronal circuits to simulate neuromorphic behavior. The research to explore the mechanism of neuromorphic dynamics is of great significance for neurons and even neural networks.

Author Contributions: Conceptualization, J.Y., Y.L. and G.W.; methodology, J.Y., Y.S. and G.W.; formal analysis J.Y. and F.L.; writing—original draft preparation, J.Y., F.L. and G.W.; writing—review and editing, J.Y., Y.S. and G.W.; supervision, Y.L. and G.W.; funding acquisition, Y.L. and G.W. All authors have read and agreed to the published version of the manuscript.

Funding: This research was funded in part by the National Natural Science Foundation of China under Grant number 62171173, Grant number 61771176, and Grant number 61801154, and in part by the Zhejiang Provincial Natural Science Foundation of China under Grant number LY20F010008.

Conflicts of Interest: The authors declare no conflict of interest.

References

1. Chua, L.O. Local activity is the origin of complexity. *Int. J. Bifurc. Chaos* **2011**, *15*, 3435–3456. [CrossRef]
2. Ascoli, A.; Slesazeck, S.; Mähne, H.; Tetzlaff, R.; Mikolajick, T. Nonlinear dynamics of a locally-active memristor. *IEEE Trans. Circuits Syst. I-Regul. Pap.* **2015**, *62*, 1165–1174. [CrossRef]
3. Liang, Y.; Wang, G.; Chen, G.; Dong, Y.; Yu, D.; Iu, H.H.C. S-Type Locally active memristor-based periodic and chaotic oscillators. *IEEE Trans. Circuits Syst. I-Regul. Pap.* **2020**, *67*, 5139–5152. [CrossRef]
4. Chua, L.O.; Sbitnev, V.; Kim, H. Neurons are poised near the edge of chaos. *Int. J. Bifurc. Chaos* **2012**, *22*, 1250098. [CrossRef]
5. Chua, L.O. Everything you wish to know about memristors but are afraid to ask. *Radioengineering* **2015**, *24*, 319–368. [CrossRef]
6. Itoh, M.; Chua, L.O. Chaotic oscillation via edge of chaos criteria. *Int. J. Bifurc. Chaos* **2017**, *27*, 1730035. [CrossRef]
7. Chua, L.O. Memristor, Hodgkin–Huxley, and edge of chaos. *Nanotechnology* **2013**, *24*, 67–94. [CrossRef] [PubMed]
8. Pickett, M.D.; Williams, R.S. Sub-100 fJ and sub-nanosecond thermally driven threshold switching in niobium oxide crosspoint nanodevices. *Nanotechnology* **2012**, *23*, 215202. [CrossRef] [PubMed]
9. Li, S.; Liu, X.; Nandi, S.K.; Nath, S.K.; Elliman, R.G. Origin of current-controlled negative differential resistance modes and the emergence of composite characteristics with high complexity. *Adv. Funct. Mater.* **2019**, *29*, 1905060. [CrossRef]

10. Kumar, S.; Williams, R.S. Separation of current density and electric field domains caused by nonlinear electronic instabilities. *Nat. Commun.* **2018**, *9*, 2030. [CrossRef] [PubMed]

11. Goodwill, J.M.; Ramer, G.; Li, D.; Hoskins, B.D.; Pavlidis, G.; McClelland, J.J.; Centrone, A.; Bain, J.A.; Skowronski, M. Spontaneous current constriction in threshold switching devices. *Nat. Commun.* **2019**, *10*, 1628. [CrossRef] [PubMed]

12. Dong, Y.; Wang, G.; Chen, G.; Shen, Y.R.; Ying, J.J. A bistable nonvolatile locally-active memristor and its complex dynamics. *Commun. Nonlinear Sci. Numer. Simul.* **2020**, *84*, 105203. [CrossRef]

13. Ying, J.; Liang, Y.; Wang, J.; Dong, Y.; Wang, G.; Gu, M. A tristable locally-active memristor and its complex dynamics. *Chaos Solitons Fract.* **2021**, *148*, 111038. [CrossRef]

14. Sharma, A.A.; Bain, J.A.; Weldon, J.A. Phase coupling and control of oxide-based oscillators for neuromorphic computing. *IEEE J. Explor. Solid-State Comput. Devices Circuits* **2015**, *1*, 2329–9231. [CrossRef]

15. Zhang, X.; Zhuo, Y.; Luo, Q.; Wu, Z.; Midya, R.; Wang, Z.; Song, W.; Wang, R.; Upadhyay, N.K.; Fang, Y.; et al. An artificial spiking afferent nerve based on Mott memristors for neurorobotics. *Nat. Commun.* **2020**, *11*, 51. [CrossRef]

16. Kumar, S.; Strachan, J.P.; Williams, R.S. Chaotic dynamics in nanoscale NbO_2 mott memristor for analogue computing. *Nature* **2017**, *548*, 318–321. [CrossRef]

17. Ying, J.; Liang, Y.; Wang, G.; Iu, H.H.C.; Zhang, J.; Jin, P. Locally active memristor based oscillators: The dynamic route from period to chaos and hyperchaos. *Chaos* **2021**, *31*, 063114. [CrossRef]

18. Dong, Y.; Wang, G.; Liang, Y.; Chen, G. Complex dynamics of a bi-directional N-type locally-active memristor. *Commun. Nonlinear Sci. Numer. Simul.* **2022**, *105*, 106086. [CrossRef]

19. Mannan, Z.I.; Choi, H.; Kim, H. Chua corsage memristor oscillator via Hopf bifurcation. *Int. J. Bifurc. Chaos* **2016**, *26*, 1630009. [CrossRef]

20. Mannan, Z.I.; Yang, C.; Kim, H. Oscillation with 4-lobe Chua corsage memristor. *IEEE Circuits Syst. Mag.* **2018**, *18*, 14–27. [CrossRef]

21. Mannan, Z.I.; Yang, C.; Adhikari, S.P.; Kim, H. Exact analysis and physical realization of the 6-lobe Chua corsage memristor. *Complexity* **2018**, *2018*, 8405978. [CrossRef]

22. Mannan, Z.I.; Choi, H.; Rajamani, V.; Kim, H.; Chua, L.O. Chua corsage memristor: Phase portraits, basin of attraction, and coexisting pinched hysteresis loops. *Int. J. Bifurc. Chaos* **2017**, *27*, 1730011. [CrossRef]

23. Etemadi, M.; Ghobaei-Arani, M.; Shahidinejad, A. Resource provisioning for IoT services in the fog computing environment: An autonomic approach. *Comput. Commun.* **2020**, *161*, 109–131. [CrossRef]

24. Aslanpour, M.S.; Dashti, S.E.; Ghobaei-Arani, M.; Rahmanian, A.A. Resource provisioning for cloud applications: A 3-D, provident and flexible approach. *J. Supercomput.* **2018**, *74*, 6470–6501. [CrossRef]

25. Zhang, W.; Li, C.D.; Huang, T.W.; He, X. Synchronization of memristor-based coupling recurrent neural networks with time-varying delays and impulses. *IEEE Trans. Neural Netw. Learn. Syst.* **2015**, *26*, 3308–3313. [CrossRef]

26. Vadivel, R.; Ali, M.S.; Joo, Y.H. Robust H∞ performance for discrete time T-S fuzzy switched memristive stochastic neural networks with mixed time-varying delays. *J. Exp. Theor. Artif. Intell.* **2021**, *33*, 79–107. [CrossRef]

27. Kumar, S.; Williams, R.S.; Wang, Z. Third-order nanocircuit elements for neuromorphic engineering. *Nature* **2020**, *585*, 518–523. [CrossRef]

28. Yi, W.; Tsang, K.K.; Lam, S.K.; Bai, X.; Crowell, J.A.; Flores, E.A. Biological plausibility and stochasticity in scalable VO_2 active memristor neurons. *Nat. Commun.* **2018**, *9*, 4661. [CrossRef]

29. Dong, Y.J.; Liang, Y.; Wang, G.Y.; Iu, H.H.C. Chua corsage memristor based neuron models. *Electron. Lett.* **2021**, *57*, 903–905. [CrossRef]

30. Jin, P.P.; Wang, G.Y.; Liang, Y.; Iu, H.H.C.; Chua, L.O. Neuromorphic dynamics of Chua corsage memristor. *IEEE Trans. Circuits Syst. I-Regul. Pap.* **2021**, *68*, 4419–4432. [CrossRef]

31. Dogaru, R.; Chua, L.O. Edge of chaos and local activity domain of FitzHugh–Nagumo equation. *Int. J. Bifurc. Chaos* **1998**, *8*, 211–257. [CrossRef]

32. Li, C.; Min, F.; Li, C. Multiple coexisting attractors of the serial–parallel memristor-based chaotic system and its adaptive generalized synchronization. *Nonlinear Dyn.* **2018**, *94*, 2785–2806. [CrossRef]

33. Peng, G.; Min, F. Multistability analysis, circuit implementations and application in image encryption of a novel memristive chaotic circuit. *Nonlinear Dyn.* **2017**, *90*, 1607–1625. [CrossRef]

34. Chua, L.O.; Kang, S.M. Memristive devices and systems. *Proc. IEEE* **1976**, *64*, 209–223. [CrossRef]

35. Corinto, F.; Ascoli, A. Memristive diode bridge with LCR filter. *Electron. Lett.* **2012**, *48*, 824–825. [CrossRef]

36. Yuan, F.; Li, Y. A chaotic circuit constructed by a memristor, a memcapacitor and a meminductor. *Chaos* **2019**, *29*, 101101. [CrossRef] [PubMed]

Article

Characteristic Analysis of Fractional-Order Memristor-Based Hypogenetic Jerk System and Its DSP Implementation

Chuan Qin, Kehui Sun * and Shaobo He

School of Physics and Electronics, Central South University, Changsha 410083, China; Qinchuan@csu.edu.cn (C.Q.); heshaobo@csu.edu.cn (S.H.)
* Correspondence: kehui@csu.edu.cn

Abstract: In this paper, a fractional-order memristive model with infinite coexisting attractors is investigated. The numerical solution of the system is derived based on the Adomian decomposition method (ADM), and its dynamic behaviors are analyzed by means of phase diagrams, bifurcation diagrams, Lyapunov exponent spectrum (LEs), dynamic map based on SE complexity and maximum Lyapunov exponent (MLE). Simulation results show that it has rich dynamic characteristics, including asymmetric coexisting attractors with different structures and offset boosting. Finally, the digital signal processor (DSP) implementation verifies the correctness of the solution algorithm and the physical feasibility of the system.

Keywords: chaos; fractional-order calculus; memristor model; coexisting attractors; Adomian decomposition method

Citation: Qin, C.; Sun, K.; He, S. Characteristic Analysis of Fractional-Order Memristor-Based Hypogenetic Jerk System and Its DSP Implementation. *Electronics* **2021**, *10*, 841. https://doi.org/10.3390/electronics10070841

Academic Editors: Chun Sing Lai, Zhekang Dong and Donglian Qi

Received: 7 March 2021
Accepted: 29 March 2021
Published: 1 April 2021

Publisher's Note: MDPI stays neutral with regard to jurisdictional claims in published maps and institutional affiliations.

1. Introduction

Chaotic systems have initial sensitivity, long-term unpredictability and other excellent characteristics; therefore, they can be cross-combined with other scientific fields such as biology, information science, security, and engineering [1–5]. For this reason, more and more scholars are focusing on establishing new chaotic systems with better chaotic characteristics. Among them, building a memristive chaotic system is an effective method [6,7]. A memristor is a bridge connecting magnetic flux and electric charge [8], although it took a long time from the concept of the memristor to the advent of the real memristor [9]. However, in recent years, memristors have been widely studied due to their special properties, which have promoted memristors in electrical and electronics [10], communication [11], neural networks [12], biological simulation [13], and security [14] and other fields of application. In the field of chaos, memristors have become a research focus due to their rich nonlinear characteristics. For example, by introducing memristors with different properties into the existing dynamic systems, some chaotic or hyperchaotic systems with rich characteristics have been studied [15,16]. A new memristive chaotic circuit was obtained by replacing the non-linear resistor with a memristor in a chaotic circuit [17–20]. Most of these studies are based on integer-order calculus systems. Fractional-order calculus can more accurately describe physical models, so it has attracted the attention of researchers and became a focus of nonlinear research.

By combining fractional calculus and memristors, people pay more attention to the behavior of the system with its control parameters. Study results show that many fractional-order memristive systems have rich dynamic characteristics. For example, Mou et al. [21] analyzed the dynamic behavior of a 4D hyperchaotic memristive circuit with different parameters. Li et al. [22] reported a 4D system with an infinite equilibrium point of order memristor, but with further research, some scholars found that the system parameters are not the only factors that affect the dynamic characteristics of the system. Recently, Bao's team found that some memristive systems have extreme multistability [23,24], which is reflected in the complete bifurcation path of the system with changes in initial values.

For example, the memristor-based system proposed in Ref. [24] has unlimited coexistence attractors and the transition behavior is completely different from the transient chaos. Moreover, as a special dynamic characteristic, extreme multistability does not only appear in memristive systems. For example, Wan et al. [25] reported super multistability in discrete neural networks. Chen et al. [26] made a more in-depth study of extreme multistability. In Ref. [26], Chen et al. constructed a 3rd-order dimensionality-reducing flux capable of maintaining the original dynamics of the original 5th-order memristive Chua circuit. The charge model confirms that sensitive extreme multistability phenomena can be detected in the magnetic flux domain.

Offset boosting is a method of chaotic control. It is usually achieved by adding a constant term after a certain parameter of the system. By changing the constant, the attractor of the system can be copied and panned. In 2016, Li et al. presented many systems by applying offset boosting and summarized the rules [27]. Offset boosting can effectively produce multistability phenomena. Therefore, many scholars have used it in their own research [28–32], but most of them focused on integer-order systems. In contrast, offset boosting has fewer applications in fractional-order systems [33,34].

In this paper, we constructed a 4D fractional-order hypogenetic jerk system based on a memristor and implemented digital circuit implementation. The introduction of the memristor led the system to show extreme multistability phenomena. In addition, offset boosting is realized by introducing constants. In Section 2, the fractional-order hypogenetic Jerk system model based on a memristor is presented and the solution of this system is derived based on the ADM algorithm. In Section 3, the dynamic characteristics of the system are analyzed from three aspects: order change, control parameter change and system initial value change. In Section 4, coexistence of multiple attractors is shown and the existence of these coexistence attractors is verified with DSP technology. Offset boosting as a chaos control method is successfully implemented in this system. Finally, the research results are summarized and the future research directions are pointed out.

2. Solution of the Fractional-Order Memristor-Based Hypogenetic Jerk System

2.1. Description of Adomian Decomposition Method

The Adomian decomposition algorithm is an analytical algorithm. The main idea is to decompose the differential equation into three parts: linear, nonlinear and constant terms. The nonlinear term needs to be transformed into an equivalent special polynomial, and then the inverse operator method is used for step by step derivation, and finally, the sum of the deduced components is the high-precision approximate solution of the differential equation. The Adomian algorithm has been widely used to solve fractional chaotic systems [35–39] due to its fast calculation speed and high solution accuracy.

For the fractional order system $D_{t_0}^q x(t) = f(x(t)) + g(t)$, where $x(t) = [x_1(t), x_2(t), \cdots x_n(t)]^T$ are the system state variables, $g(t) = [g_1, g_2 \cdots g_n]^T$ is the constant of the system, and f represents a functional formula containing linear and nonlinear parts. The system can be expressed as the following form

$$\begin{cases} D_{t_0}^q x(t) = Lx + Nx + g(t) \\ x^{(k)}(t_0^+) = b_k, k = 0, 1, \cdots, m-1 \ , \\ m \in N, m-1 < q < m \end{cases} \tag{1}$$

where L and N are the linear and nonlinear terms of this equations, respectively, b_k is the initial condition. After multiplying both sides of the equation by the integral operator $J_{t_0}^q$, we can obtain

$$x = J_{t_0}^q Lx + J_{t_0}^q Nx + \phi, \tag{2}$$

where $\phi = \sum_{k=0}^{m-1} b_k \frac{(t-t_0)^k}{k!}$ is the initial value. According to the principle of adomian decomposition algorithm [35], the solution of the system is expressed by

$$x(t) = \sum_{i=0}^{\infty} x^i = F(x(t_0)) ,$$ (3)

Decompose the nonlinear term

$$\begin{cases} A_j^i = \frac{1}{i!} [\frac{d^i}{d\lambda^i} N(v_j^i(\lambda))]_{\lambda=0} \\ v_j^i(\lambda) = \sum_{k=0}^{i} (\lambda)^k x_j^k \end{cases} ,$$ (4)

where $i = 0, 1, 2, \cdots, \infty, j = 0, 1, 2, \cdots, n$, then the nonlinear term is expressed by

$$Nx = \sum_{i=0}^{\infty} A^i(x^0, x^1, \cdots, x^i) .$$ (5)

Thus, the following equation is obtained

$$x = \sum_{i=0}^{\infty} x^i = J_{t_0}^q L \sum_{i=0}^{\infty} x^i + J_{t_0}^q N \sum_{i=0}^{\infty} x^i + J_{t_0}^q g + \phi .$$ (6)

By applying the following recursive relation, we have

$$\begin{cases} x^0 = \phi \\ x^1 = J_{t_0}^q Lx^0 + J_{t_0}^q A^0(x^0) \\ x^2 = J_{t_0}^q Lx^1 + J_{t_0}^q A^1(x^0, x^1) \\ \vdots \\ x^i = J_{t_0}^q Lx^{i-1} + J_{t_0}^q A^{i-1}(x^0, x^1, \cdots, x^{i-1}) \\ \vdots \end{cases} .$$ (7)

2.2. Solution of the Fractional-Order Memristor-Based Hypogenetic Jerk System Based on ADM

In recent years, many memristive chaotic systems have been proposed. A memristor-based hypogenetic chaotic jerk system is reported in Ref. [24]. Through replacing the newly proposed memristor featured by $W(\phi) = \alpha + 3\beta\varphi^2$ and introducing fractional calculus into the hypogenetic chaotic jerk system, the new system is established by

$$\begin{cases} D_{t_0}^q x = |y| - b \\ D_{t_0}^q y = (\alpha + 3\beta w^2)z \\ D_{t_0}^q z = |x| - y - az - c \\ D_{t_0}^q w = z \end{cases} ,$$ (8)

where x, y, z are state variables, and w is the state variable of the memristor. a, b, c are the control parameters. α and β are the control parameters of the memristor. q is the order number of the system. According to the Ref. [35], the solution of this system is expressed by

$$X(t) = \begin{bmatrix} x(t) \\ y(t) \\ z(t) \\ w(t) \end{bmatrix} = \begin{bmatrix} x(t_0) \\ y(t_0) \\ z(t_0) \\ w(t_0) \end{bmatrix} + J_{t_0}^q \begin{bmatrix} |y| - b \\ \alpha z \\ |x| - y - az - c \\ z \end{bmatrix} + J_{t_0}^q \begin{bmatrix} 0 \\ 3\beta zw^2 \\ 0 \\ 0 \end{bmatrix} .$$ (9)

Decomposing the non-linear terms $3\beta zw^2$, we obtain

$$\begin{cases} A_0 = 3\beta z_0 w_0^2 \\ A_1 = 3\beta \cdot (2z_0 w_0 w_1 + z_1 w_0^2) \\ A_2 = 3\beta \cdot (z_0 w_1^2 + 2z_0 w_0 w_2 + z_1 w_0 w_1 + z_1 w_0 w_2 + z_2 w_0^2) \\ A_3 = 3\beta \cdot (4z_0 w_0 w_3 + 4z_0 w_1 w_2 + 2z_1 w_0 w_3 + \frac{8}{3}z_1 w_0 w_2 + \frac{2}{3}z_1 w_1 w_2 \\ \quad + \frac{8}{3}z_2 w_0 w_1 + \frac{4}{3}z_2 w_0 w_2 + 2z_3 w_0^2) \end{cases}$$

According to the following initial conditions

$$\begin{cases} x^0 = x(t_0) \\ y^0 = y(t_0) \\ z^0 = z(t_0) \\ w^0 = w(t_0) \end{cases} , \tag{10}$$

make $c_1^0 = x^0, c_2^0 = y^0, c_3^0 = z^0, c_4^0 = w^0$ and according to Formula (9) and fractional calculus properties, we can obtain

$$\begin{cases} x^1 = (|c_2^0| - b)\frac{(t-t_0)^q}{\Gamma(q+1)} \\ y^1 = (\alpha c_3^0 + 3\beta c_3^0(c_4^0)^2)\frac{(t-t_0)^q}{\Gamma(q+1)} \\ z^1 = (|c_1^0| - c_2^0 - ac_3^0 - c)\frac{(t-t_0)^q}{\Gamma(q+1)} \\ w^1 = c_3^0\frac{(t-t_0)^q}{\Gamma(q+1)} \end{cases} , \tag{11}$$

then assign the coefficient value of the above formula to the corresponding variable. That is, assign the first coefficient to c_1^1, the second coefficient to c_2^1, and so on. After three iterations, the other three coefficients of the equation are derived as

$$\begin{cases} c_1^2 = (|c_2^1| - b) \\ c_2^2 = (\alpha c_3^1 + 3\beta[2c_3^0 c_4^0 c_4^1 + c_3^1(c_4^0)^2)] \\ c_3^2 = (|c_1^1| - c_2^1 - ac_3^1 - c) \\ c_4^2 = c_3^1 \end{cases} , \tag{12}$$

$$\begin{cases} c_1^3 = (|c_2^2| - b) \\ c_2^3 = (\alpha c_3^2 + 3\beta[c_3^0(c_4^1)^2) + 2c_3^0 c_4^0 c_4^2 + c_3^1 c_4^0 c_4^1 + c_3^1 c_4^0 c_4^2 + c_3^2(c_4^0)^2)] \\ c_3^3 = (|c_1^2| - c_2^2 - ac_3^2 - c) \\ c_4^3 = c_3^2 \end{cases} , \tag{13}$$

$$\begin{cases} c_1^4 = |c_2^3| - b \\ c_2^4 = \alpha \cdot c_3^3 + 3\beta \cdot [4c_3^0 c_4^0 c_4^3 + 4c_3^0 c_4^1 c_4^2 + 2c_3^1 c_4^0 c_4^3 + \frac{8}{3}c_3^1 c_4^0 c_4^2 + \frac{2}{3}c_3^1 c_4^1 c_4^2 \\ \quad + \frac{8}{3}c_3^2 c_4^0 c_4^1 + \frac{4}{3}c_3^2 c_4^0 c_4^2 + 2c_3^3 \cdot (c_4^0)^2] \\ c_3^4 = |c_1^3| - c_2^3 - a \cdot c_3^3 - c \\ c_4^4 = c_3^3 \end{cases}$$

Finally, the fractional order approximate solution of the system is expressed by

$$\tilde{x}_j(t) = c_j^0 + c_j^1\frac{(t-t_0)^q}{\Gamma(q+1)} + c_j^2\frac{(t-t_0)^{2q}}{\Gamma(2q+1)} + c_j^3\frac{(t-t_0)^{3q}}{\Gamma(3q+1)} + c_j^4\frac{(t-t_0)^{4q}}{\Gamma(4q+1)} . \tag{14}$$

Based on this approximate solution, we let the system order $q = 0.95$, the system parameters $a = 0.4$, $b = 2.1$, $c = 2.6$, $\alpha = 1.8$, $\beta = 0.01$, and the initial value $(0.1, 0.1, 0.1, 0.1)$. Under this parameter condition, the system was simulated by MATLAB. Figure 1 displays the phase trajectories of two different planes under this condition. It shows a strange attractor symbolizing chaos.

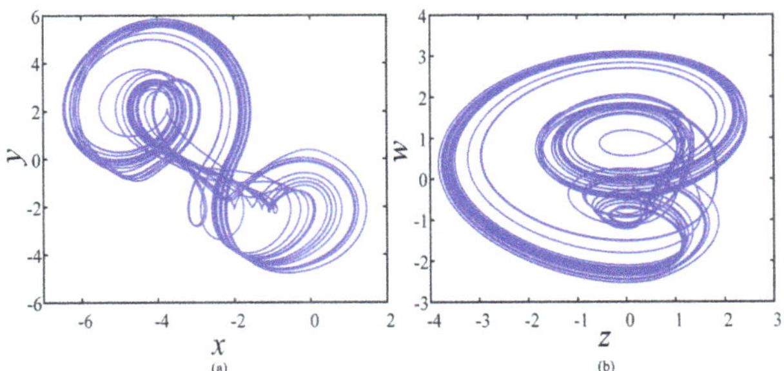

Figure 1. Phase diagrams of the system: (**a**) $x - y$ plane; (**b**) $z - w$ plane.

The QR decomposition method [40] is an effective method used to calculate the Lyapunov exponent. By this method, the Jacobian matrix of the system is decomposed into the product of the orthogonal matrix Q and the upper triangular matrix R. Then, the Lyapunov exponent of the system can be calculated:

$$\lambda_j = \frac{1}{Mh} \sum_{i=1}^{M} \ln(|R_i(j,j)|) , \tag{15}$$

where j is the dimensionality of the system, M is the number of iterations, and h is the iteration step size. In this case, the system Lyapunov exponents are calculated by the QR method as $LE_1 = 0.1917$, $LE_2 = 0$, $LE_3 = -0.0272$, $LE_4 = -0.6812$. The Lyapunov exponent distribution is $[+\ 0\ -\ -]$, so it is a chaotic system.

3. Dynamical Analysis of the System

In this section, the phase portraits, bifurcation diagrams, Lyapunov exponent spectra and dynamic map are utilized to analyze the system dynamics.

3.1. Dynamical Analysis with the Order q

The control parameters are set as $a = 0.4$, $b = 2.6$, $c = 2.1$, $\alpha = 1.8$, $\beta = 0.01$, and the initial value $(1, 1, 1, 2)$. The phase portraits with different q are shown in Figure 2. This figure shows that the attractor structure of the system is also different for different q. Figure 2a,b show two densely structured strange attractors. A single scroll attractor is shown in Figure 2c, and Figure 2d is periodic. The attractor shown in Figure 2e is interesting, and it looks like a combination of the attractors in Figure 2c,f.

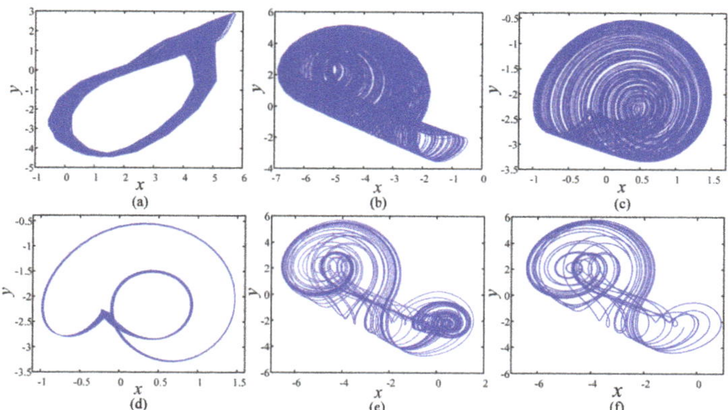

Figure 2. Phase diagrams with different q: (**a**) $q = 0.21$; (**b**) $q = 0.35$; (**c**) $q = 0.5$; (**d**) $q = 0.52$; (**e**) $q = 0.6$; (**f**) $q = 0.9$.

To identify the transition from periods to chaos in the system, we set the control parameters and initial values remain the same as above and q varies at the range [0.2, 1] with the step size of 0.004. The bifurcation diagram and the LEs of system are obtained in Figure 3. The lowest order that makes the system chaotic is $q = 0.21$ in this case. It was found that there is an obvious periodic window at the interval range $q \in [0.507, 0.524]$, and from the LEs, it can be seen that other regions except for this interval and some narrower periodic windows are chaotic. Some special properties are displayed in Figure 3a. First, unlike most bifurcation diagrams, it has a no period-doubling bifurcation path and is not a continuous whole. In some regions, it changes abruptly, and the bifurcation area jumps without portent from one area to another. Then, the chaotic system stays in the state of chaos at a large range of order, except for several windows. Finally, the system evolves into a periodic state through reverse-period-doubling bifurcation. By observing Figures 2 and 3, we can find that different bifurcation behaviors correspond to different attractor structures.

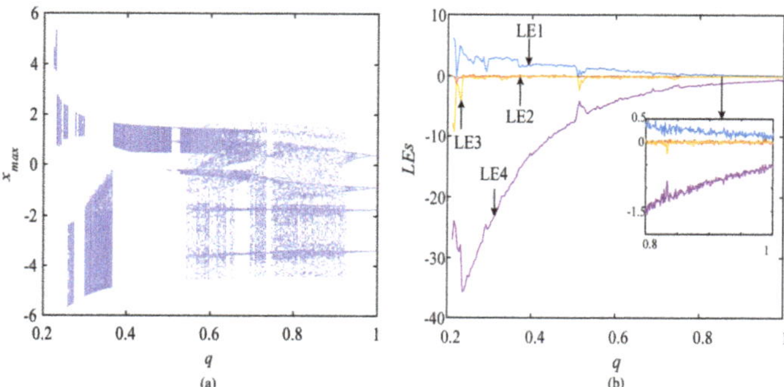

Figure 3. Dynamics with q change: (**a**) bifurcation diagram, (**b**) Lyapunov exponents.

3.2. Dynamical Analysis with the Parameters

Set a as the bifurcation parameter, and set the remaining parameters as $b = 2.1, c = 2.6$, $\alpha = 1.8$, $\beta = 0.01$, and the order $q = 0.95$. When a is changed at the range [0.35, 0.7], the bifurcation diagram and its LEs are shown in Figure 4. When the control parameter a gradually increases, the system starts from the chaotic state, and several period windows appear as a increases. When $a = 0.445$, there is a jump in the bifurcation diagram. After the system returns to the original bifurcation path, it goes to the periodic state through the reverse period-doubling bifurcation. Figure 4b also proves the existence of these periodic windows, which verifies the above analysis.

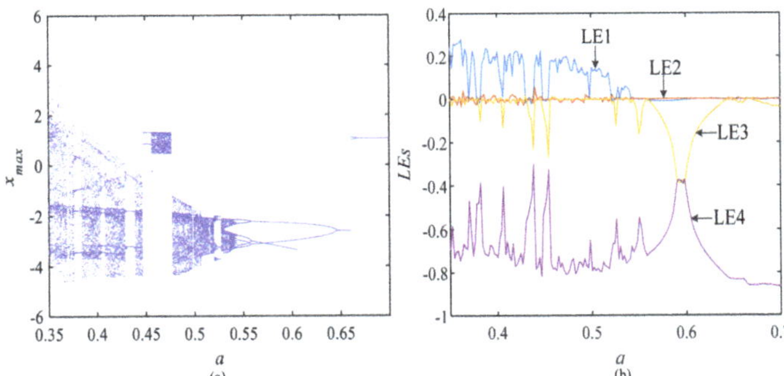

Figure 4. Dynamics with a change: (**a**) bifurcation diagram; (**b**) Lyapunov exponents.

Set $a = 0.4$, while q and other parameters remain unchanged, We studied the influence of parameter c on system behavior. When c is changed at the range [0, 3], the bifurcation diagram and its LEs are shown in Figure 5. It can be seen that the system stays in the state of chaos at a large range of parameter c, except for three small period windows $c \in [0.66, 0.77]$, $[1.27, 1.38]$ and $[2.3, 2.43]$.

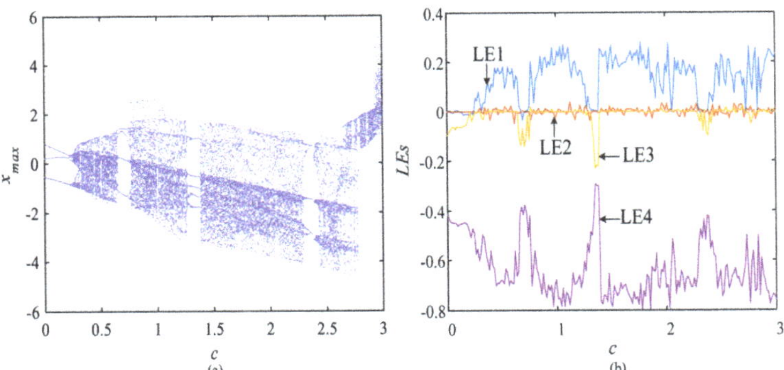

Figure 5. Dynamics with c change: (**a**) bifurcation diagram; (**b**) Lyapunov exponents.

3.3. Dynamical Analysis with the Initial Values

Generally, chaotic systems are sensitive to initial values, but the structure of the attractor remains stable. Even if some systems are capable of coexisting attractors due to the existence of multiple stable states, the number of coexisting attractors is usually limited. Ref. [24] reported that the four-line balanced deformed Jerk system has extreme multistability. The bifurcation diagram and Lyapunov exponent spectrum of the system (8) with the initial value are plotted to analyze the behavior of the system.

Set the control parameters as $a = 0.4$, $b = 2.6$, $c = 2.1$, $\alpha = 1.8$, $\beta = 0.01$, and the order $q = 0.95$, and the remaining three initial values are all set to 1. Figure 6 shows the bifurcation diagram of the system changing x_0 and z_0, where x_0 varies at the range $[-7, 5]$ and z_0 varies at the range $[-3, 6]$. The bifurcation behavior of the system remains unchanged when x_0 and z_0 change. So, we mainly analyze the dynamic characteristics of the system with y_0 and w_0.

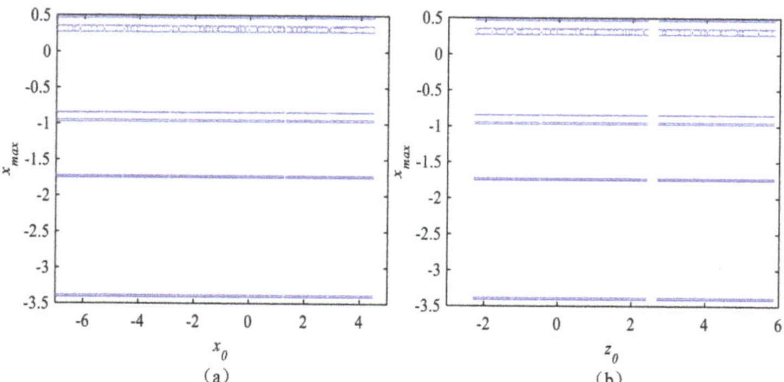

Figure 6. Bifurcation diagrams with x_0 and z_0 change: (**a**) x_0 change; (**b**) z_0 change.

Set the same system parameters and order as before, the initial conditions are assigned as $x_0 = 1$, $z_0 = 1$, $w_0 = 4$. y_0 varies at the range $[-7, 9]$. The bifurcation diagram and the LEs of the system (8) are shown in Figure 7. In the interval $[-7, -4]$, the bifurcation behavior of the system is special. When the initial condition y_0 increases from -7, the system breaks into chaos at first through a period-doubling bifurcation. The bifurcation paths have many narrow periodic windows, and the bifurcation points corresponding to these windows form another bifurcation path with a breakpoint. Then, the system suddenly jumps to another chaotic state. As y_0 continues to increase, the bifurcation becomes normal. There are three obvious periodic windows, and the system quickly evolves into chaotic state again through the period-doubling bifurcation. The LEs shown in Figure 7b verify the accuracy of bifurcation diagrams.

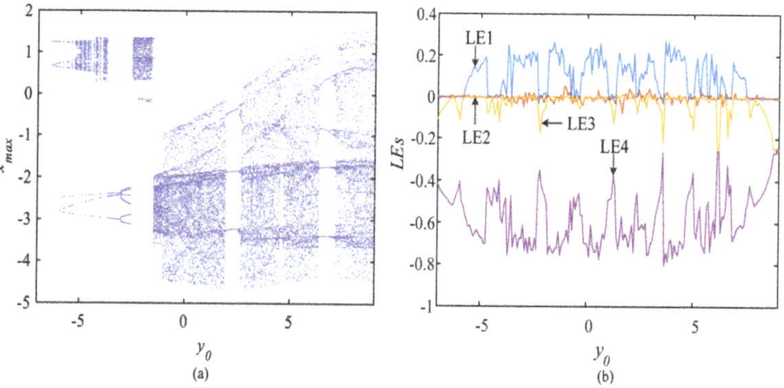

Figure 7. Dynamics with y_0 change: (**a**) bifurcation diagram; (**b**) Lyapunov exponents.

The control parameters and order of the system (8) remain unchanged, and change the initial value to $x_0 = 1$, $y_0 = 1$, $z_0 = 1$; w_0 varies at the range $[-7, 8]$. The bifurcation diagram and the LEs of the system (8) are shown in Figure 8. As w_0 increases, it is obvious that the bifurcation diagram can be divided into five intervals of $[-7, -5]$, $(-5, -1.9]$, $(-1.9, 5]$, $(5, 6.7]$, $(6.7, 8]$ numbered 1–5. These five intervals have a certain degree of symmetry. In interval 1 and 5, the system evolves into a chaotic state through forward (reverse) period-doubling bifurcation. Then, the system entered interval 2 and 4 and re-evolved. In interval 3, the system is chaotic, except for a few periodic windows. LEs have more severe oscillations than Figure 7b. This is because the system state switches rapidly between periodic and chaotic. This phenomenon can be seen from interval 4 of the bifurcation diagram.

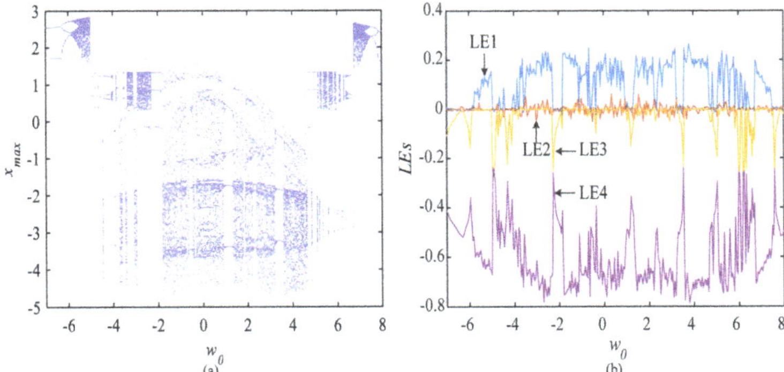

Figure 8. Dynamics with w_0 change: (**a**) bifurcation diagram; (**b**) Lyapunov exponents.

Dynamical maps based on the SE complexity [41] and the maximum Lyapunove exponents in the $y_0 - w_0$ plane with $a = 0.4$, $b = 2.6$, $c = 2.1$, $\alpha = 1.8$, $\beta = 0.01$ are shown in Figure 9. In Figure 9a, dark colors indicate a system is chaotic, and light colors indicate that the system may be periodic states, and white indicates divergence. The dynamic map based on the maximum Lyapunove exponent is more precise. Orange indicates the system is chaotic ($LEmax > 0.03$), and yellow indicates stable resting behavior($0 < LEmax < 0.03$), and cyan-blue indicates periodic states and blue indicates divergence. Dynamic map based on the maximum Lyapunov exponent can distinguish the state of the system at the critical region.

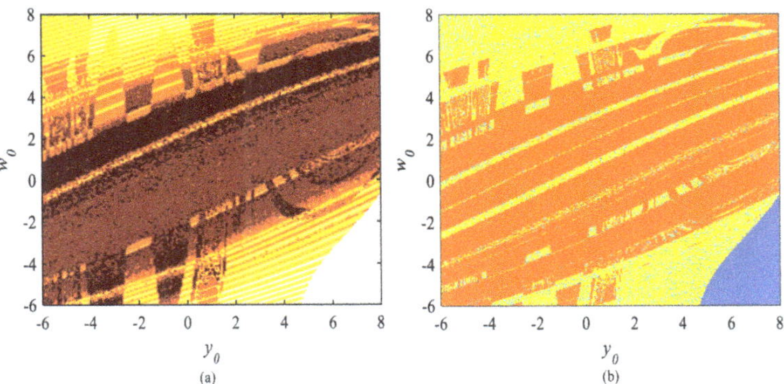

Figure 9. Dynamic maps in $y_0 - w_0$: (**a**) based on SE complexity; (**b**) based on maximum Lyapunov exponent.

4. Multiple Coexisting Attractors of System

4.1. Multiple Coexisting Attractors and Its Digital Circuit Implementation

The control parameters and the order of system (8) remain unchanged, and we set different initial values to plot the phase diagrams. Figure 10 shows nine asymmetric coexisting attractors. In order to observe more clearly, the coexistence attractors at each initial value are separately plotted. There are four chaotic attractors with different structures and five periodic attractors. It illustrates the multiple stability of the system, and it is just a dynamic characteristic exhibited by a few sample points in the initial value space.

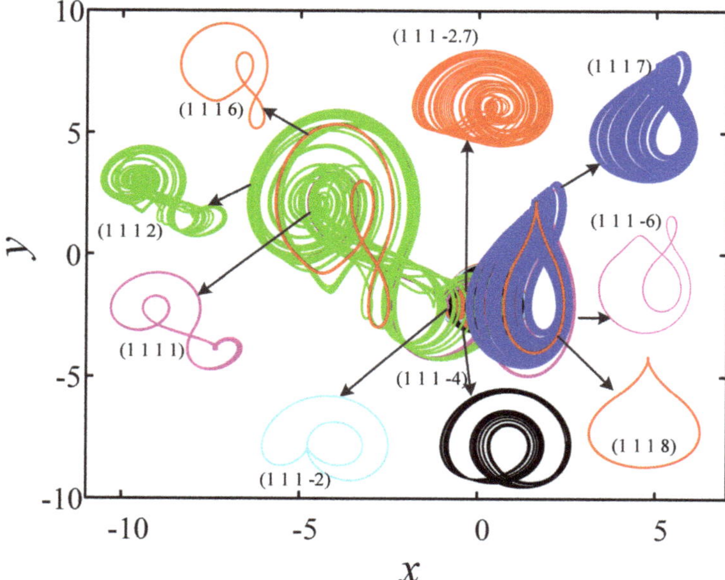

Figure 10. Multiple coexisting attractors with different initial values.

The hardware implementation of a chaotic system is an important method for verifying the feasibility of the system. Due to the tolerance of electronic component parameters, this increases the difficulty of using analog circuits to implement chaotic systems. However, the digital circuit implementation scheme based on the DSP platform used in this article does not have this problem. Figure 11 shows the DSP hardware connection schematic diagram. In the experiment, the IDE (Integrated Development Environment) of the DSP platform uses CCS (Code Composer Studio). We can use it to set various parameters such as system parameters and iteration step length. The initialized data are transmitted to the DSP through the communication interface for calculation, and the result is transmitted to the oscilloscope (Tektronix MDO 3104, Tektronix, Hong Kong, China) through the D/A converter (DAC8552, Texas Instruments, Dallas, TX, USA) for display.

Figure 12 shows the program flowchart. After the DSP is initialized, the various parameters of the system are set, and then iterative calculations are started. Push the result into the stack to facilitate the next calculation to call the result. The result after data processing is output through D/A. In the experiment, we set the same initial conditions as when the system has coexistence attractors. The DSP implementation hardware connection diagram is shown in Figure 13. After debugging, the system experimental phase diagram is obtained. Comparing Figures 10 and 14, it can be concluded that we have successfully completed the DSP implementation.

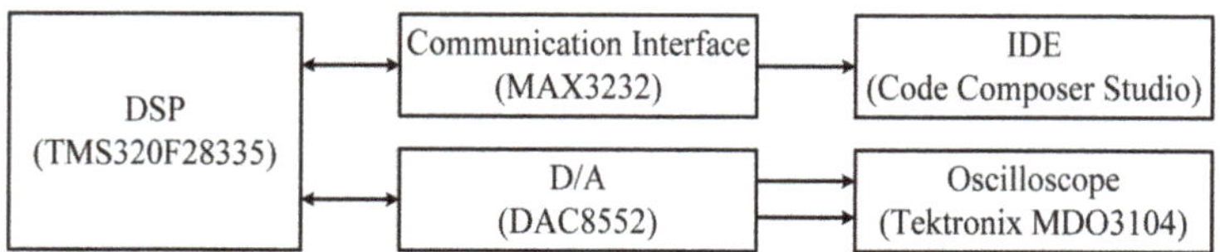

Figure 11. The digital signal processor (DSP) hardware connection schematic diagram of fractional-order memristor-based hypogenetic jerk system.

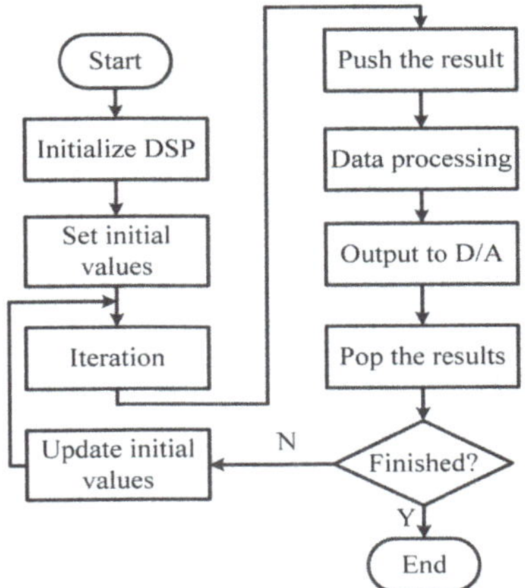

Figure 12. Flowchart for DSP implementation program.

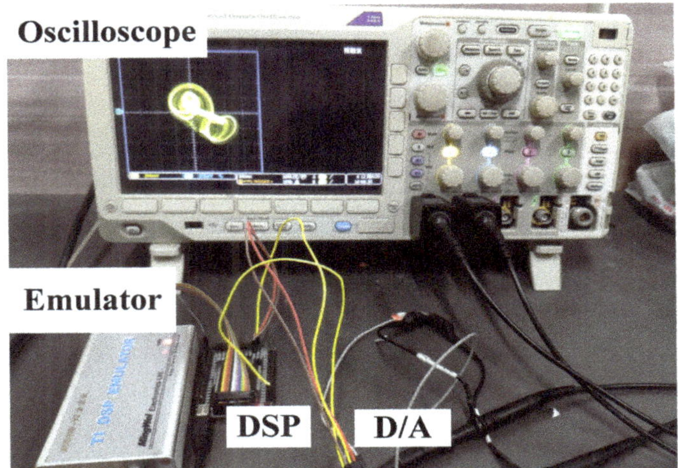

Figure 13. The DSP implementation hardware connection diagram.

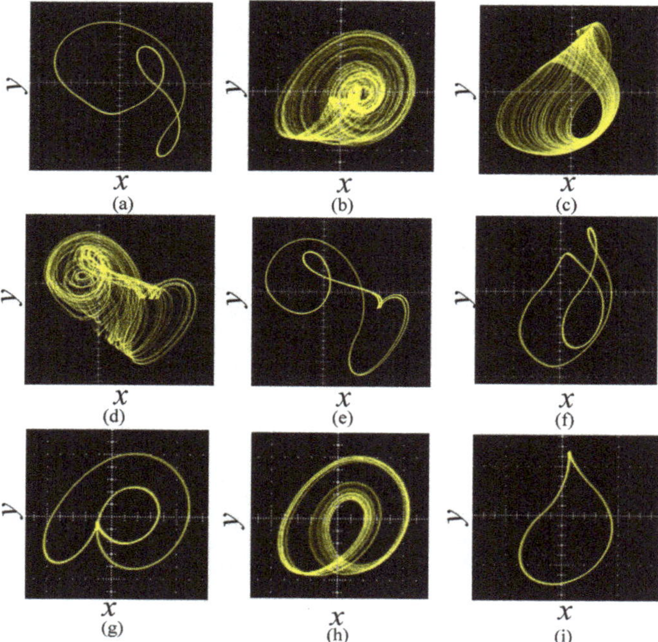

Figure 14. Multiple coexisting attractor with different initial values: (**a**) (1 1 1 6); (**b**) (1 1 1 −2.7); (**c**) (1 1 1 7); (**d**) (1 1 1 2); (**e**) (1 1 1 1); (**f**) (1 1 1 −6); (**g**) (1 1 1 −2); (**h**) (1 1 1 −4); (**i**) (1 1 1 8).

4.2. Offset Boosting

Offset boosting control is discussed in this section. According to Ref. [27], we can generate the offset by adding a constant term after the variable that has only appeared once in the system. By observing Formula (8), we can find that the variable x satisfies the conditions for constructing offset boosting. The constant term p is added to the third dimension, so we can obtain

$$
\begin{cases}
D_{t_0}^q x = |y| - b \\
D_{t_0}^q y = (\alpha + 3\beta w^2)z \\
D_{t_0}^q z = |x + p| - y - az - c \\
D_{t_0}^q w = z
\end{cases}
\tag{16}
$$

Set the system parameters and order to remain the same as during characteristic analysis, the initial conditions [1, 1, 1, 4], and the offset parameter p are set to $-3, 0, 3$. The offset boosting phenomenon is illustrated in Figure 15. After the offset boosting control is applied, the system has richer dynamic behavior under certain initial values. Only change the initial conditions to [1, 1, 1, 5.5] without changing other conditions. Figure 16 shows the offset boosting phenomenon under this condition. It shows the boosting phenomenon of three different states under the same initial value. When $p = 3, 2, 1, 0$, the system remains in a chaotic state for boosting. When $p = -1, -1.2, -1.4, -1.6$, the system remains a double-periodic orbit for boosting. When $p = 3, -4, -5, -6$, the system remains a single-periodic orbit for boosting.

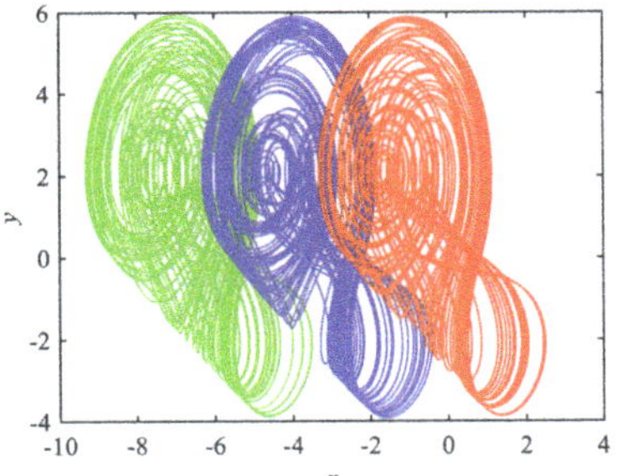

Figure 15. Offset boosting with control parameter p. $p = 3$ (green), $p = 0$ (blue), $p = -3$ (red).

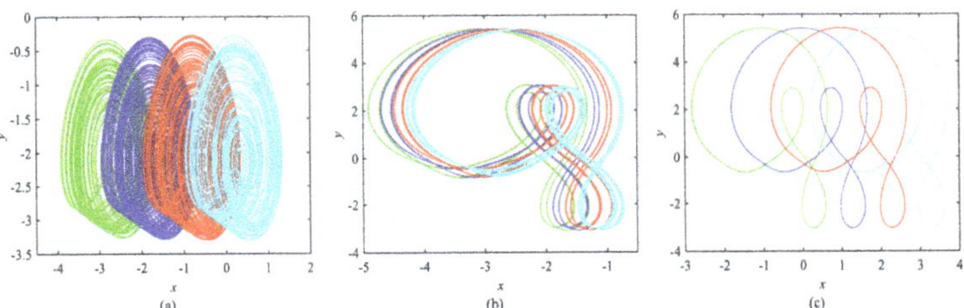

Figure 16. Offset boosting with control parameter p. (**a**) $p = 3, 2, 1, 0$; (**b**) $p = -1, -1.2, -1.4, -1.6$; (**c**) $p = -3, -4, -5, -6$.

5. Conclusions

In this paper, many analysis methods are used to analyze the dynamic characteristics of this fractional-order menristor-based hypogenetic jerk system, such as a phase diagram, bifurcation diagram, and Lyapunov exponent spectrum. DSP technology is used to successfully verify the feasibility of the system. It is found that the system not only has rich dynamic characteristics with the change of the order and system parameters, but also has a complete period-doubling bifurcation path from single-cycle to multi-cycle with the change in initial values. A change in the bifurcation path implies a change in the structure of the attractor. Through phase diagram analysis, at least nine coexisting attractors were found. The control and application of this fractional-order menristor-based hypogenetic jerk system will be studied next.

Author Contributions: Numerical simulation, theoretical calculation and manuscript writing, C.Q., supervision, K.S., direction, S.H. All authors have read and agreed to the published version of the manuscript.

Funding: This work was supported by the Natural Science Foundation of China (No. 61901530, 62071496), the Research and Innovation Project of Graduate of Central South University (No. 2020zzts381), the Natural Science Foundation of Hunan Province (No.2020JJ5767).

Data Availability Statement: The data used to support the findings of this study are included within the article.

Conflicts of Interest: The authors declare no conflict of interest.

References

1. Luo, J.; Xu, X.; Ding, Y.; Yuan, Y.; Yang, B.; Sun, K.; Yin, L. Application of a memristor-based oscillator to weak signal detecbtion. *Eur. Phys. J. Plus* **2018**, *133*, 239. [CrossRef]
2. Hua, Z.; Zhou, Y.; Huang, H. Cosine-transform-based chaotic system for image encryption. *Inf. Sci.* **2019**, *480*, 403–419. [CrossRef]

3. Ye, X.; Mou, J.; Luo, C.; Wang, Z. Dynamics analysis of Wienbridge hyperchaotic memristive circuit system. *Nonlinear Dyn.* **2018**, *92*, 923–933. [CrossRef]
4. Wang, X.; Wang, S.; Zhang, Y.; Luo, C. A one-time pad color image cryptosystem based on SHA-3 and multiple chaotic systems.*Opt. Lasers Eng.* **2018**, *103*, 1–8. [CrossRef]
5. Mou, J.; Sun, K.; Ruan, J.; He, S. A nonlinear circuit with two memcapacitors.*Nonlinear Dyn.* **2016**, *86*, 1–10. [CrossRef]
6. Wang, G.; Zang, S.; Wang, X.; Yuan, F.; Iu, H.H.-C. Memcapacitor model and its application in chaotic oscillator with memristor. *Chaos* **2017**, *27*, 013110. [CrossRef]
7. Bao, H.; Wang, N.; Wu, H.; Song, Z.; Bao, B. Bi-stability in an improved memristor-based third-order Wien-bridge oscillator. *IETE Tech. Rev.* **2019**, *36*, 109–116. [CrossRef]
8. Chua, L. Memristor-the missing circuit element. *IEEE Trans. Circuit Theory* **1971**, *18*, 507–509. [CrossRef]
9. Strukov, D.B.; Snider, G.S.; Stewart, D.R.; Williams, R.S. The missing memristor found. *Nature* **2008**, *453*, 80–83. [CrossRef]
10. Yalagala, B.; Khandelwal, S.; Deepika, J.; Badhulika, S. Wirelessly destructible MgO-PVP-Graphene composite based flexible transient memristor for security applications. *Mat. Sci. Semicon. Proc.* **2019**, *104*, 104673. [CrossRef]
11. Wang, W.; Jia, X.; Luo, X.; Kurths, J.; Yuan, M. Fixed-time synchronization control of memristive MAM neural networks with mixed delays and application in chaotic secure communication. *Chaos Soliton Fract.* **2019**, *126*, 85–96. [CrossRef]
12. Zhang, W.; Cao, J.; Wu, R.; Chen, D.; Alsaadi, F.E. Novel results on projective synchronization of fractional-order neural networks with multiple time delays. *Chaos Soliton Fract.* **2018**, *117*, 76–83. [CrossRef]
13. Lunelli, L.; Collini, C.; Jimenez-Garduño, A.; Roncador, A.; Giusti, G.; Verucchi, R.; Pasquardini, L.; Iannotta, S.; Macchi, P.; Lorenzelli, L.; et al. Prototyping a memristive-based device to analyze neuronal excitability. *Biophys. Chem.* **2019**, *253*, 106212. [CrossRef]
14. Sun, J.; Yang, Q.; Wang, Y. Dynamical analysis of novel memristor chaotic system and DNA encryption application. *IJST-T Electr. Eng.* **2020**, *44*, 449–460. [CrossRef]
15. Rajagopal, K.; Guessas, L.; Karthikeyan, A.; Srinivasan, A.; Adam, G. Fractional-order memristor no equilibrium chaotic system with its adaptive sliding mode synchronization and genetically optimized fractional order PID synchronization. *Complexity* **2017**, *2017*, 1–19. [CrossRef]
16. Chen, M.; Feng, Y.; Bao, H.; Bao, B.; Wu, H.; Xu, Q. Hybrid state variable incremental integral for reconstructing extreme multistability in memristive jerk system with cubic nonlinearity. *Complexity* **2019**, *2019*, 1–16. [CrossRef]
17. Xu, B.; Wang, G.; Iu, H.H.-C.; Yu, S.; Yuan, F. A memristor-meminductor-based chaotic system with abundant dynamical behaviors. *Nonlinear Dyn.* **2019**, *96*, 765–788. [CrossRef]
18. Sadecki, J.; Marszalek, W. Complex oscillations and two-parameter bifurcations of a memristive circuit with diode bridge rectifier. *Microelectron. J.* **2019**, *93*, 104636. [CrossRef]
19. Rajagopal, K.; Li, C.; Nazarimehr, F.; Karthikeyan, A.; Duraisamy, P.; Jafari, S. Chaotic dynamics of modified wien bridge oscillator with fractional order memristor. *Radioengineering* **2019**, *27*, 165–174. [CrossRef]
20. Ruan, J.; Sun, K.; Mou, J.; He, S.; Zhang, L. Fractional-order simplest memristor-based chaotic circuit with new derivative. *Eur. Phys. J. Plus* **2018**, *133*, 3. [CrossRef]
21. Mou, J.; Sun, K.; Wang, H.; Ruan, J. Characteristic analysis of fractional-order 4D hyperchaotic memristive circuit. *Math. Probl. Eng.* **2017**, *2017*, 2313768. [CrossRef]
22. Li, R.; Huang, D. Stability analysis and synchronization application for a 4D fractional-order system with infinite equilibria. *Phys. Scripta* **2020**, *95*, 015202. [CrossRef]
23. Chen, C.; Chen, J.; Bao, H.; Chen, M.; Bao, B. Coexisting multi-stable patterns in memristor synapse-coupled Hopfield neural network with two neurons. *Nonlinear Dyn.* **2019**, *95*, 3385–3399. [CrossRef]
24. Bao, H.; Wang, N.; Bao, B.; Chen, M.; Jin, P.; Wang, G. Initial condition-dependent dynamics and transient period in memristor-based hypogenetic jerk system with four line equilibria. *Commun. Nonlinear Sci.* **2018**, *57*, 264–275. [CrossRef]
25. Wan, P.; Sun, D.; Zhao, M.; Wan, L.; Jin, S. Multistability and attraction basins of discrete-time neural networks with nonmonotonic piecewise linear activation functions. *Neural Netw.* **2020**, *122*, 231–238. [CrossRef]
26. Chen, M.; Sun, M.; Bao, H.; Hu, Y.; Bao, B. Flux-charge analysis of two-memristor-based chua's circuit: Dimensionality decreasing model for detecting extreme multistability. *IEEE T Ind. Electron.* **2020**, *67*, 2197–2206. [CrossRef]
27. Li, C.; Sprott, J.C. Variable-boostable chaotic flows. *Optik* **2016**, *127*, 10389–10398. [CrossRef]
28. Bayani, A.; Rajagopal, K.; Khalaf, A.J.M.; Jafari, S.; Leutcho, G.; Kengne, J. Dynamical analysis of a new multistable chaotic system with hidden attractor: Antimonotonicity, coexisting multiple attractors, and offset boosting. *Phys. Lett. A* **2019**, *383*, 1450–1456. [CrossRef]
29. Li, C.; Lei, T.; Wang, X.; Chen, G. Dynamics editing based on offset boosting. *Chaos* **2020**, *30*, 063124. [CrossRef] [PubMed]
30. Li, H.; Yang, Y.; Li, W.; He, S.; Li, C. Extremely rich dynamics in a memristor-based chaotic system. *Eur. Phys. J. Plus* **2020**, *135*, 579. [CrossRef]
31. Zhang, S.; Zheng, J.; Wang, X.; Zeng, Z.; He, S. Initial offset boosting coexisting attractors in memristive multi-double-scroll hopfield neural network. *Nonlinear Dyn.* **2020**, *102*, 2821–2841. [CrossRef]
32. Yu, F.; Liu, L.; Shen, H.; Zhang, Z.; Huang, Y.; Cai, S.; Deng, Z.; Wan, Q. Multistability analysis, coexisting multiple attractors, and FPGA implementation of Yu-Wang four-wing chaotic system. *Math. Probl. Eng.* **2020**, *2020*, 1–16. [CrossRef]

33. Ding, D.; Shan, X.; Jun, L.; Hu, Y.; Yang, Z.; Ding, L. Initial boosting phenomenon of a fractional-order hyperchaotic system based on dual memristors. *Mod. Phys. Lett. B* **2020**, *34*, 2050191. [CrossRef]
34. Tamba, V.K.; Kom, G.H.; Kingni, S.T.; Mboupda Pone, J.R.; Fotsin, H.B. Analysis and electronic circuit implementation of an integer- and fractional-order four-dimensional chaotic system with offset boosting and hidden attractors. *Eur. Phys. J.Spec. Top.* **2020**, *229*, 1211–1230. [CrossRef]
35. He, S.; Sun, K.; Wang, H. Solution of the fractional-order chaotic system based on Adomian decomposition algorithm and its complexity analysis. *Acta Phys. Sin. Ed.* **2014**, *63*, 030502.
36. Ye, X.; Wang, X.; Mou, J.; Yan, X.; Xian, Y. Characteristic analysis of the fractional-order hyperchaotic memristive circuit based on the Wien bridge oscillator. *Eur. Phys. J. Plus* **2018**, *133*, 516. [CrossRef]
37. Yang, F.; Li, P. Characteristics analysis of the fractional-order chaotic memristive circuit based on Chua's circuit. *Mob. Netw. Appl.* **2019**, *5*, 1–9. [CrossRef]
38. Yang, F.; Mou, J.; Ma, C.; Cao, Y. Dynamic analysis of an improper fractional-order laser chaotic system and its image encryption application. *Opt. Laser Eng.* **2020**, *129*, 106031. [CrossRef]
39. He, S.; Sun, K.; Wang, H. Complexity analysis and DSP implementation of the fractional-order Lorenz hyperchaotic system. *Entropy* **2015**, *17*, 8299–8311. [CrossRef]
40. He, S.; Sun, K.; Wang, H. Solution and dynamics analysis of a fractional-order hyperchaotic system. *Math. Methods Appl. Sci.* **2016**, *39*, 2965–2973. [CrossRef]
41. Sun, K.H.; He, S.B.; He, Y.; Yin, L.Z. Complexity analysis of chaotic pseudo-random sequences based on spectral entropy algorithm. *Acta. Phys. Sin. Ed.* **2013**, *62*, 010501.

 electronics

MDPI

Article

VO$_2$ Carbon Nanotube Composite Memristor-Based Cellular Neural Network Pattern Formation

Yiran Shen and Guangyi Wang *

Institute of Modern Circuits and Intelligent Information, Hangzhou Dianzi University, Hangzhou 310018, China; yrshen@hdu.edu.cn
* Correspondence: wanggyi@163.com

Abstract: A cellular neural network (CNN) based on a VO$_2$ carbon nanotube memristor is proposed in this paper. The device is modeled by SPICE at first, and then the cell dynamic characteristics based on the device are analyzed. It is pointed out that only when the cell is at the sharp edge of chaos can the cell be successfully awakened after the CNN is formed. In this paper, we give the example of a 5×5 CNN, set specific initial conditions and observe the formed pattern. Because the generated patterns are affected by the initial conditions, the cell power supply can be pre-programmed to obtain specific patterns, which can be applied to the future information processing system based on complex space–time patterns, especially in the field of computer vision.

Keywords: VO$_2$ carbon nanotube composite memristor; cellular neural network (CNN); von Neumann structure; local activity; edge of chaos

Citation: Shen, Y.; Wang, G. VO$_2$ Carbon Nanotube Composite Memristor-Based Cellular Neural Network Pattern Formation. *Electronics* **2021**, *10*, 1198. https://doi.org/10.3390/electronics10101198

Academic Editor: Kris Campbell

Received: 10 April 2021
Accepted: 11 May 2021
Published: 18 May 2021

Publisher's Note: MDPI stays neutral with regard to jurisdictional claims in published maps and institutional affiliations.

1. Introduction

The traditional processor uses the von Neumann structure, in which the storage unit and the processing unit are separated and connected by bus. In recent years, with the development of semiconductor technology, the speed of processing units has been greatly improved. However, due to the limitation of bus bandwidth, the operation speed of the whole processor is limited, which is called the "von Neumann bottleneck problem" [1,2]. In order to solve this problem, inspired by the biological way of processing information, many pieces of literature have proposed various non-von Neumann processor solutions [3–5]. Typical bio-inspired computing relies on nonlinear networks that contain the same cells; each cell has a relatively simple structure and interacts with the surrounding cells. This structure, called a CNN, a cellular neural network, has been applied in fields such as computer vision [6–8].

The concept of a CNN can be traced back to two articles by Chua in 1988 [9,10], which respectively give the theory and application of CNNs. Recently, Itoh [11] summarized some characteristics of CNNs in a long paper. Weiher et al. [12] discussed the pattern formation of CNNs based on the NbO$_2$ memristor.

Recently, brain-like computing has become a hot topic in research [13–15], and the key is to find devices that can produce spike signals like neurons and that have very low power consumption. The VO$_2$ carbon nanotube (CNT) composite device is a Mott memristor recently proposed [16]. It can generate periodic peak pulses with a pulse width of less than 20 ns, it is driven by a DC current or voltage, and it does not need additional capacitance. It uses metal–carbon nanotubes as heaters, and, compared with the pure Mott VO$_2$ proposed earlier, adding CNTs can greatly reduce the transient duration and pulse energy and increase the frequency of a peak pulse by three orders of magnitude.

The VO$_2$ nano crossbar device does not need the process of electric forming and has low device size dispersion. For devices with a critical size between 50 and 600 nm, the change coefficient of the switch threshold voltage is less than 13%, the switch durability is more than 26.6 million cycles, and the IV characteristics of the device are not changed significantly. The VO$_2$ device technology in the process of non-electric formation accelerates the development of an active memristor neuron circuit. It can simulate the most known neuron dynamics and clear the way for the realization of a large-scale integrated circuit (IC). In addition, the VO$_2$ memristor is superior to its NbO$_2$ counterpart in both switch speed and switching energy. The simulated Mott transition in the VO$_2$ is 100 times faster than in the NbO$_2$ and consumes only about one-sixth (16%) of the energy [17].

In this paper, a cellular neural network based on a VO$_2$ CNT is proposed. As a cell itself, a VO$_2$ CNT is set to be stable and static, that is, in "sleep". If the memristor is on the "edge of chaos", two or more cells are connected by the RC coupling, which will make it in the state of "wake-up", namely, in the dynamic oscillation mode. The second part of this paper is the modeling of VO$_2$ CNT composite devices, providing the SPICE model. The third part is the cell circuit, which gives the analysis process of the decoupled circuit. The response of the memristor is expanded near the operating point and the small-signal equivalent circuit is given. Based on this, the influence of coupling R and C device parameters on the input impedance of one port is analyzed. The fourth part is the CNN simulation, which describes the pattern formation characteristics of a CNN composed of memristor-based cells.

Compared with the traditional CMOS realization of CNN, the memristor counterpart can largely save power and chip area, and provide ultra-high processing speeds.

2. VO$_2$ Carbon Nanotube Composite Device Modeling

The structure of the VO$_2$ carbon nanotube composite device is shown in Figure 1. It contains a transverse active region, which is defined by a VO$_2$ thin metal strip with a thickness of about 5 nm (as shown in the red region of the figure), and its two ends are connected with Pd electrodes (as shown in the blue region of the figure). This is a planar Mott metal-insulator transition device. The aligned carbon nanotubes (as shown in the black line) were first grown on quartz substrate, and then transferred to the surface of a VO$_2$ thin metal strip before the whole device was formed.

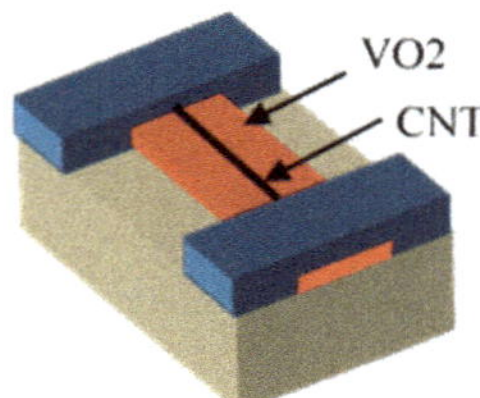

Figure 1. A VO$_2$ carbon nanotube composite device structure. The red part is the VO$_2$ and the black line over it represents the carbon nanotube.

As a Mott metal-insulator transition device, it satisfies the following equation [16]:

$$i_m = AT^2 e^{\left(\frac{\beta\sqrt{v_m/d}-\phi}{kT}\right)}$$
$$\frac{dT}{dt} = \frac{i_m v_m + v_m^2/R_{CNT}}{C_{th}} - \frac{T-T_{amb}}{C_{th}R_{th}} \tag{1}$$

The first formula of Equation (1) is the current emission equation of the Schottky diode, and the second formula is Newton's cooling law, where, i_m and v_m are the current and voltage through the memristor NDR device, respectively; T is the absolute temperature of the device; T_{amb} is the ambient temperature; A and β are scaling constants; d is the effective device length; k is Boltzmann constant; ϕ is the energy barrier; C_{th} is the heat capacity (effective thermal mass); R_{th} is the thermal resistance; R_{CNT} is the resistance value of the CNT (carbon nanotube), taking 600 kΩ. The second expression of Equation (1) represents the electrical and thermal coupling between R_{CNT} and VO$_2$. The parameters of the device are shown in Table 1.

Table 1. Device parameters. The data is cited directly from the literature [16].

Parameter	VO$_2$ CNT Device
T_{amb}	296 K
β	3.3×10^{-4} eV $\cdot$ m$^{0.5}$ $\cdot$ V$^{-0.5}$
A	1.7×10^{-9} A $\cdot$ K^{-2}
d	5×10^{-6} m
R_{th}	2.5×10^{8} K $\cdot$ W^{-1}
C_{th}	5×10^{-17} J $\cdot$ K^{-1}
R_S	5500 Ω
V_S	10 V
R_{CNT}	600,000 Ω
ϕ	0.58 eV
k	8.62×10^{-5} eV $\cdot$ K^{-1}

According to Equation (1), the SPICE model of the device can be established, as shown in Figure 2

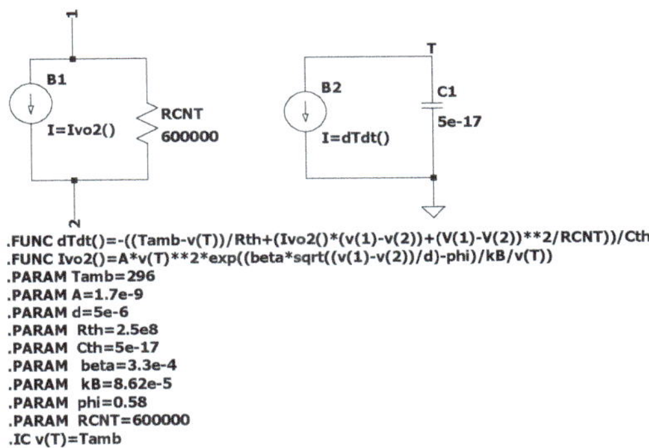

Figure 2. SPICE model of the device.

The quasi-static volt-ampere characteristics of the device can be obtained by using the SPICE model, as shown in Figure 3.

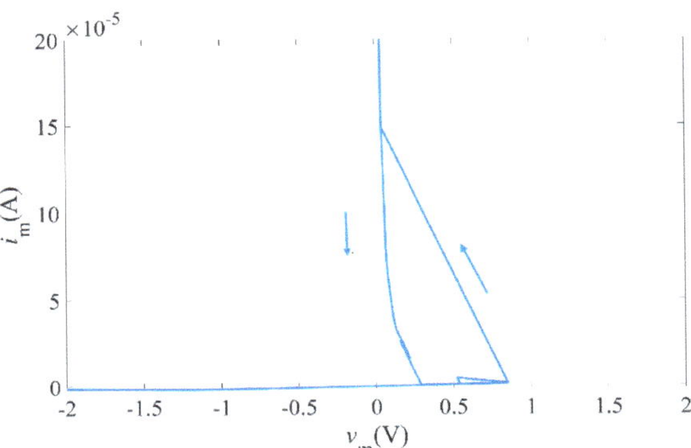

Figure 3. Quasi-static volt-ampere characteristics of the VO$_2$ carbon nanotube composite device.

The simulation measurement method of the quasi-static volt-ampere characteristics of the device is to use a sinusoidal voltage source with a frequency of 1 Hz and amplitude of 10 V to excite the device through a small series resistor, and then draw the V–I relationship curve by measuring the current flowing through the device. When you look at Figure 3, you can clearly see a hysteresis region, which is a typical feature of the local active memristor.

3. Cell Circuit

Figure 4 is the designed circuit of a memristor cell, in which R_S is the bias resistor, M is the VO$_2$ memristor and R_{CNT} is the carbon nanotube resistance. The parallel resistor and capacitor combination is the circuit coupled with adjacent cells, and it is also a part of the cell.

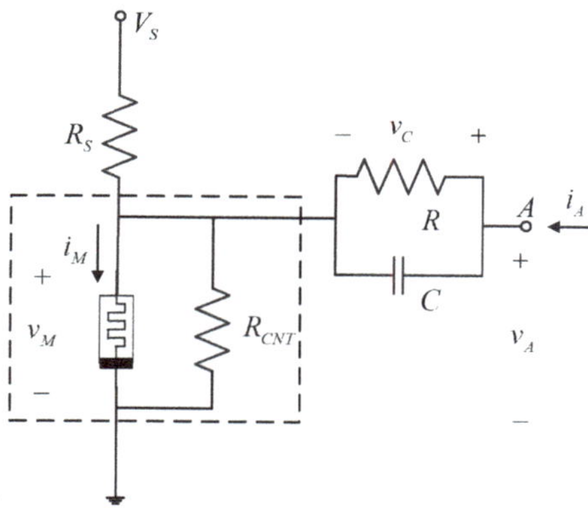

Figure 4. Cell circuit. The part shown in the dotted box is the VO$_2$ carbon nanotube composite device, R_s is the bias resistor and R and C are the coupling resistor and capacitor, respectively.

According to Figure 4, the circuit equation can be obtained by using the Kirchhoff current law, the Kirchhoff voltage law and Equation (1), as follows:

$$\begin{aligned}
\frac{dT}{dt} &= \frac{v_m i_m + v_m^2/R_{CNT}}{C_{th}} - \frac{T - T_{amb}}{C_{th} R_{th}} \\
C\frac{dv_C}{dt} &= -\frac{v_C}{R} + i_A \\
0 &= -i_m + AT^2 e^{\frac{\beta\sqrt{v_m/d} - \phi}{kT}} \\
0 &= -V_S + R_S i_m + (1 + R_S/R_{CNT})v_m - R_S i_A
\end{aligned} \tag{2}$$

$$v_A = v_C + v_m \tag{3}$$

In order to make it easy to write a MATLAB simulation program, Equations (2) and (3) can be further arranged into a matrix form, as follows:

$$\begin{aligned}
\mathbf{E\dot{x}} &= \mathbf{f}(\mathbf{x}, i_A) \\
v_A &= \mathbf{Cx}
\end{aligned} \tag{4}$$

where

$$\mathbf{x} = \begin{pmatrix} x_1 & x_2 & x_3 & x_4 \end{pmatrix}^{\mathrm{T}} = \begin{pmatrix} T & v_C & v_m & i_m \end{pmatrix}^{\mathrm{T}} \tag{5}$$

$$\mathbf{E} = \begin{bmatrix} 1 & 0 & 0 & 0 \\ 0 & C & 0 & 0 \\ 0 & 0 & 0 & 0 \\ 0 & 0 & 0 & 0 \end{bmatrix} \tag{6}$$

$$\mathbf{f}(\mathbf{x}, i_A) = \begin{bmatrix} \frac{v_m i_m + v_m^2/R_{CNT}}{C_{th}} - \frac{T - T_{amb}}{C_{th} R_{th}} \\ -\frac{v_C}{R} + i_A \\ -i_m + AT^2 e^{\frac{\beta\sqrt{v_m/d} - \phi}{kT}} \\ -V_S + R_S i_m + (1 + R_S/R_{CNT})v_m - R_S i_A \end{bmatrix} \tag{7}$$

$$\mathbf{C} = [0\ 1\ 1\ 0] \tag{8}$$

The necessary conditions for the local activity of resistive-coupled reaction–diffusion CNNs (RD-CNNs) are given in [18]. In order to apply the theory to the designed network, the input impedance Z of port A must be calculated. Therefore, it is necessary to linearize Equation (4) near the operating point without coupling, that is to say, the small-signal equivalent circuit analysis condition is satisfied under the condition of a zero-port current. The memristor equation is expanded near the operating point.

$$
\begin{aligned}
T &= T_0 + \delta T \\
v_m &= V_M + \delta v_m \\
i_m &= I_M + \delta i_m
\end{aligned}
\tag{9}
$$

Here, T_0, V_M and I_M are the state, voltage and current values of the DC operating point, respectively.

We can expand the expression of the memristor current near the operating point:

$$
\begin{aligned}
i_m &= I_M + \delta i_m \\
&= a'_{00}(Q) + a'_{11}(Q)\delta T + a'_{12}(Q)\delta v_m + h.o.t
\end{aligned}
\tag{10}
$$

where

$$
\begin{aligned}
I_M &= a'_{00}(Q) = i_m = AT_0^2 e^{\left(\frac{\beta\sqrt{V_M/d}-\phi}{kT_0}\right)} \\
a'_{11}(Q) &= \frac{\partial i_m}{\partial T}\Big|_Q = \left(2T_0 - \frac{\beta\sqrt{V_M/d}-\phi}{k}\right) Ae^{\frac{\beta\sqrt{V_M/d}-\phi}{kT_0}} \\
a'_{12}(Q) &= \frac{\partial i_m}{\partial v_m}\Big|_Q = \frac{A\beta T_0}{2k\sqrt{d}}\frac{1}{\sqrt{V_M}}e^{\frac{\beta\sqrt{V_M/d}-\phi}{kT_0}}
\end{aligned}
\tag{11}
$$

h.o.t. is the higher-order term relative to δT and δv_m. If $|\delta T| \ll 1$ and $|\delta v_m| \ll 1$, we can get the following approximate linear relationship:

$$\delta i_m = a'_{11}(Q)\delta T + a'_{12}(Q)\delta v_m \tag{12}$$

Next, we will expand the memristor equation of state with a Taylor series near the operating point $(V_M, T_0)_Q$:

$$\tfrac{dT}{dt}\Big|_Q = f(T, v_m) = f(T_0 + \delta T, V_M + \delta v_m) = f(T_0, V_M) + b'_{11}(Q)\delta T + b'_{12}(Q)\delta v_m + h.o.t \tag{13}$$

where

$$
\begin{aligned}
b'_{11}(Q) &= \frac{\partial f(T,v_m)}{\partial T}\Big|_Q = -\frac{1}{C_{th}R_{th}} + \frac{V_M}{C_{th}}\left(2T_0 - \frac{\beta\sqrt{V_M/d}-\phi}{k}\right) Ae^{\frac{\beta\sqrt{V_M/d}-\phi}{kT_0}} \\
b'_{12}(Q) &= \frac{\partial f(T,v_m)}{\partial v_m}\Big|_Q = \frac{1}{C_{th}}\left(\frac{A\beta T_0}{2\sqrt{d}}\sqrt{V_M}e^{\frac{\sqrt{V_M/d}-\phi}{kT_0}} + AT_0 e^{\frac{\sqrt{V_M/d}-\phi}{kT_0}} + \frac{2V_M}{R_{CNT}}\right)
\end{aligned}
\tag{14}
$$

Notice that $f(T_0, V_M) = 0$, because (T_0, V_M) is a point on the DC V–I curve. Ignoring the high-order small term, let us linearize the memristor state Equation (13) near the operating point:

$$\frac{d(\delta T)}{dt} = b'_{11}(Q)\delta T + b'_{12}(Q)\delta v_m \tag{15}$$

Taking Laplace transform for Equations (12) and (15), we obtain

$$
\begin{aligned}
\hat{i}_m(s) &= a'_{11}(Q)\hat{T}(s) + a'_{12}(Q)\hat{v}_m(s) \\
s\hat{T}(s) &= b'_{11}(Q)\hat{T}(s) + b'_{12}(Q)\hat{v}_m(s)
\end{aligned}
\tag{16}
$$

The Laplace transforms of δT, δi_m and δv_m are $\hat{T}(s)$, $\hat{i}_m(s)$ and $\hat{v}_m(s)$. According to the second formula of (16), we obtain:

$$\hat{T}(s) = \frac{b'_{12}(Q)\hat{v}_m(s)}{s - b'_{11}(Q)} \tag{17}$$

According to the first formula of (16) and Equation (17), the admittance function of the memristor can be obtained as follows:

$$
\begin{aligned}
Y(s,Q) &\triangleq \frac{\hat{i}_m(s)}{\hat{v}_m(s)} = \frac{a'_{11}(Q)b'_{12}(Q)}{s - b'_{11}(Q)} + a'_{12}(Q) \\
&= \frac{1}{s\frac{1}{a'_{11}(Q)b'_{12}(Q)} + \frac{(-b'_{11}(Q))}{a'_{11}(Q)b'_{12}(Q)}} + a'_{12}(Q)
\end{aligned} \tag{18}
$$

Change Equation (18) into the following form:

$$Y(s,Q) = \frac{1}{sL_s + R_s} + \frac{1}{R_p} \tag{19}$$

$Y(s,Q)$ is the admittance of the memristor. The small signal equivalent circuit of the memristor is shown in Figure 5. L_s, R_s and R_p are defined as follows:

$$
\begin{aligned}
L_s &= \frac{1}{a'_{11}(Q)b'_{12}(Q)} \\
R_s &= \frac{(-b'_{11}(Q))}{a'_{11}(Q)b'_{12}(Q)} \\
R_p &= \frac{1}{a'_{12}(Q)}
\end{aligned} \tag{20}
$$

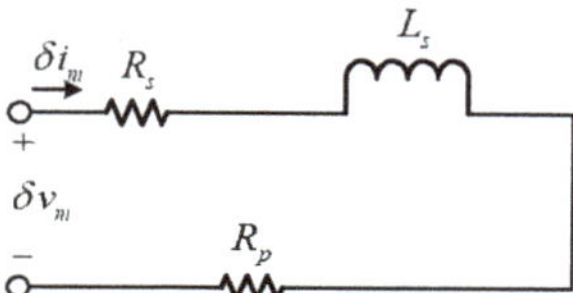

Figure 5. Small signal equivalent circuit of the memristor.

Considering the DC bias of the cell circuit, according to Equation (2), and that the R and C parameter values do not affect the DC operating point of the system, let $i_A = 0$, $f(\mathbf{x}, i_A) = f(\mathbf{x}, 0) = 0$, respectively. Then, we use MATLAB to solve the DC operating point Q, $(T_0, V_M,)_Q = (3.049042473697802e + 03, 0.006060366246039)$, and substitute the operating point value into Equation (20), we can evaluate L_s, R_s and R_p respectively.

The equivalent impedance of port A in the frequency domain is obtained.

$$Z(j\omega) = Z(s)|_{s=j\omega} = \frac{1}{Y(s,Q)} + \left. \frac{1}{\frac{1}{R} + \frac{1}{sC}} \right|_{s=j\omega} \tag{21}$$

In [18], Professor Chua derived the local active condition of resistive coupled RD-CNNs based on the reaction–diffusion equation. It is said that the cell is in the local active region if the input impedance of the one port cell satisfies at least one of the following conditions:

(a) There is a pole in the right half plane;
(b) There are higher-order poles on the imaginary axis;
(c) There is a pole $s = j\omega_p$ of order one on the imaginary axis and $\lim\limits_{s \to j\omega_p} (s - j\omega_p)Z(s)$ is a negative real number or a complex number with non-zero imaginary part;
(d) There is at least one angular frequency value ω, such that the real part of the impedance $Z(j\omega)$ is less than zero.

Through calculation, the system impedance $Z(s, Q)$ has the following form:

$$Z(s, Q) = \frac{A(s - z_1)(s - z_2)}{(s - p_1)(s - p_2)} \tag{22}$$

where A is the real coefficient. According to Chua's theory [18], the system described by conditions (a)–(c) is unstable, while the system is a stable local active system when condition (d) is satisfied along with the poles being in the left half plane, that is, the system is located in the edge of chaos. In addition, according to the theory of signals and systems, $Y(s, Q)$ is the transfer function of the system. When the poles of $Y(s, Q)$, corresponding to the zeroes, is located in the right half plane of the complex plane, the system will be unstable, and the system is said to be on the sharp edge of chaos. Specifically, because the numerator of the input impedance is a quadratic polynomial, its zeroes correspond to two cases: the system has two positive zeroes (two zeroes whose real part is greater than zero; they are conjugate zeroes, i.e., $z_1 \neq z_2$, $Re(z_1) > 0$ and $Re(z_2) > 0$) or one positive real zero (zero is a positive real number as multiple roots, $(z_1 = z_2) > 0$), corresponding to dynamic pattern and static pattern, respectively.

In Figure 6, yellow indicates that the system is in the local passive area and the input impedance does not meet Condition (d); the other areas are in the local active area and the input impedance meets Condition (d). Blue indicates that the system is at the sharp edge of chaos. At this time, the impedance function of the system has two positive zeroes (two zeroes whose real parts are greater than zero; they are conjugate zeroes) or one positive real zero (zero is a positive real number as multiple roots), which correspond to the dark blue and light blue regions, respectively. Other regions in the figure, that is, when green corresponds to other cases of zeroes, indicate that the system is at the edge of chaos. In Figure 6, the edge region of chaos, that is, the light green Region II, contains coupling parameters that may not destabilize the system because the resulting local input impedance of the cell does not have the zero point of the positive real part. On the other hand, the union of the dark blue Region III and the light blue Region IV represents the sharp edge of chaos. In the dark blue Region III, the local input impedance of the cell allows a pair of complex conjugate zeroes with positive real parts, which makes the system unstable and leads to the formation of a dynamic pattern in a steady state. In the light blue Region IV, a zero point of the local input impedance of a single cell is positive and real, which triggers the instability of the system and gradually leads to a static mode. It has been proved by literature [12] that when the cell is in the sleeping mode, the "cell" equation has only a steady-state homogeneous solution; only when the cell is in the sharp edge of chaos can it be successfully "awakened" when it is connected to the CNN; that is to say, the system equation has a non-homogeneous solution, and the light blue region is in so-called "static wake-up". When a cell is connected to a network, it has a non-homogeneous static stable solution, which is different from when the cell is isolated; the dark blue region is called a dynamic wake-up, which causes oscillation when the cell is connected to the network, so it has a dynamic oscillation solution, which is different from the static solution when the cell is isolated. Because the capacitance value of static wake-up is too large, it is not practical for an integrated system. Only a dynamic wake-up is considered.

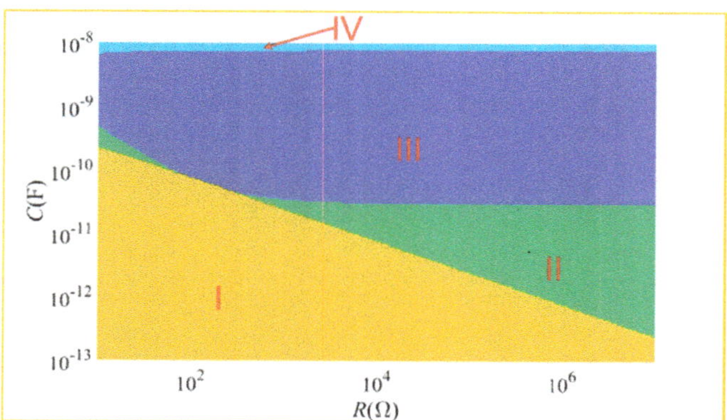

Figure 6. Input impedance characteristic diagram depending on *R* and *C* parameters. I—passive region; II—edge of chaos region; III—sharp edge of chaos region (dynamic); IV—sharp edge of chaos region (static).

4. CNN and Simulation

Connect two cells, and the common resistance and capacitance can be equivalent to two identical series connections [12] by the circuit principle, so as to distribute them to each cell, as shown in Figure 7.

$$\widetilde{R} = 2R$$
$$\widetilde{C} = C/2 \tag{23}$$

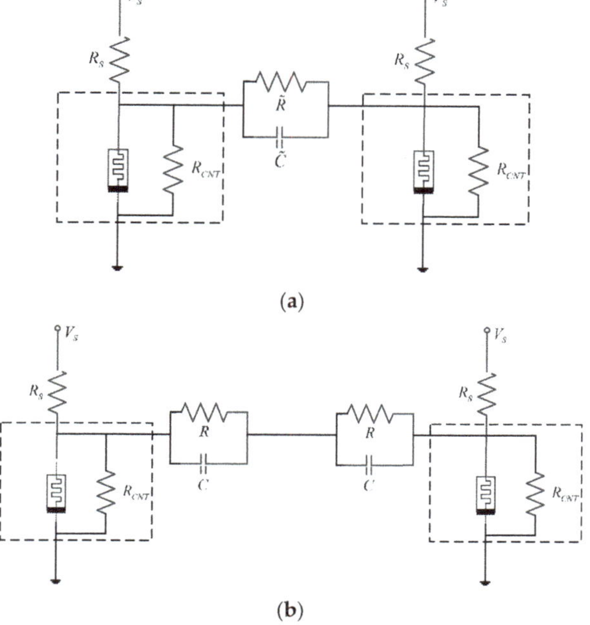

(a)

(b)

Figure 7. The common resistor and capacitor are equivalent to two parts in the series. (**a**) is the original configuration while (**b**) is its equivalent circuit.

Select $R = 1$ MΩ, $C = 720$ p, corresponding to $\widetilde{R} = 2$ MΩ, $\widetilde{C} = 360$ p; the two cells are in sleeping mode before coupling, as shown in Figure 8. In Figure 8b, the voltage decreases from 10 V to 0 V, and the cell does not oscillate.

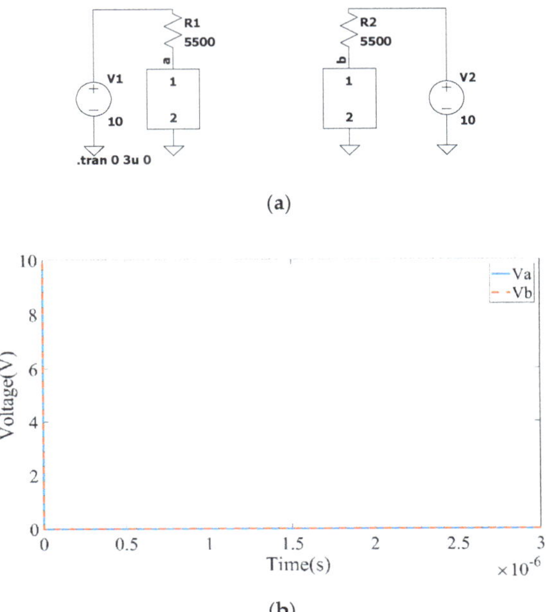

(a)

(b)

Figure 8. The two cells are uncoupled and in sleeping mode; (**a**) is the configuration of two uncoupled cells (**b**) shows there is no oscillation.

After coupling, it is in wake-up mode, as shown in Figure 9. Figure 9b shows that the voltage drops from the initial 10 V to oscillate near 0 V.

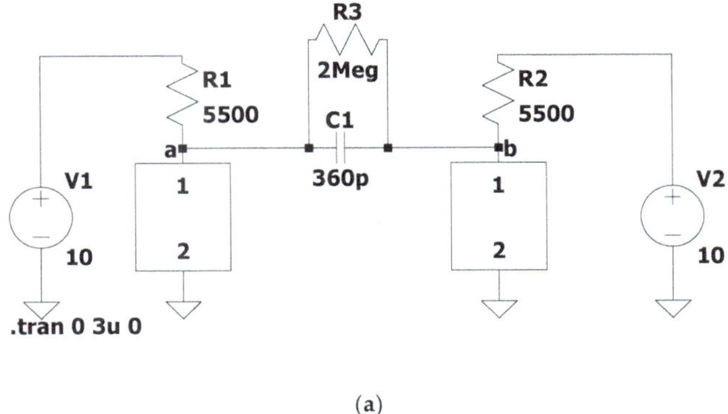

(a)

Figure 9. *Cont.*

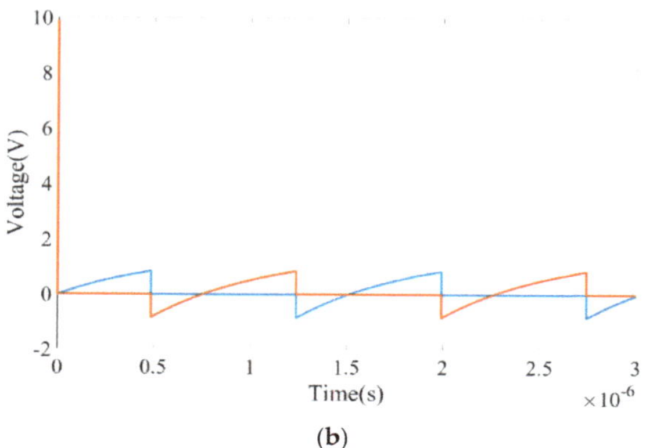

(b)

Figure 9. The cells are in wake-up mode after coupling. (**a**) is configuration of two coupled cells. In (**b**), we see there is oscillation in the circuit.

It can be seen from Figures 8 and 9 that when the cell is in the sleeping mode, both cells do not oscillate, and the differential equation of the system has only a trivial solution, that is, a zero solution. When the cell is in wake-up mode, the differential equation of the cell system has oscillatory solutions, and different cells get different oscillatory solutions.

In order to observe the characteristics of a CNN composed of cells, 5×5 cells are arrayed, as shown in Figure 10, and the DC excitation of each cell rises to 10 V in 1 u seconds. The cells in the central position (3,3) are deliberately raised to 10 V in 0.9 u seconds. The voltage response of 25 cells is observed, as shown in Figure 11.

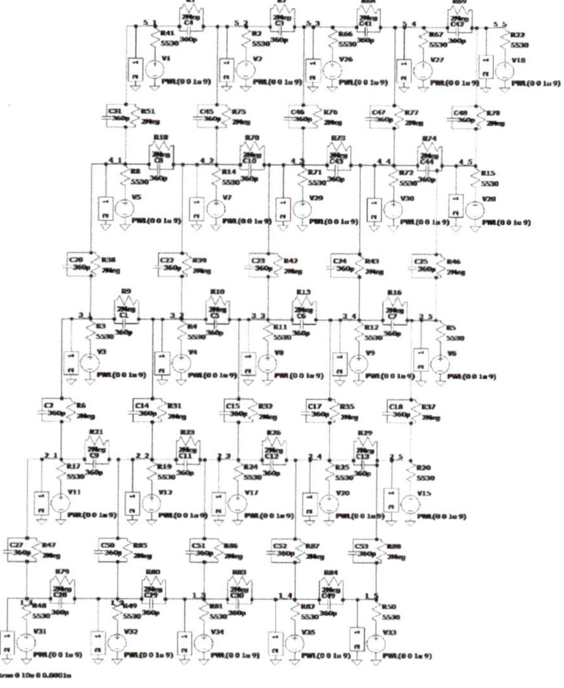

Figure 10. CNN composed of 5×5 cells.

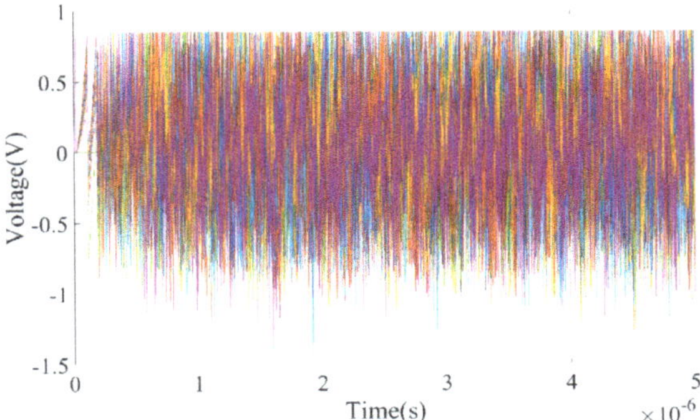

Figure 11. Terminal voltage response of each memristor.

It can be seen from Figure 11 that after a short transient response, each cell reaches steady-state oscillation. According to Figure 11, the network pattern diagram at different times can be made, as shown in Figure 12, in which different colors represent different terminal voltages of the cellular memristor. Figure 12 is just a reproduction of Figure 11 from another view of the spatial distribution of the sampled voltage at a specific time, where, in the vertical axis, "1" represents the highest voltage of the cell, and other values are just the normalized voltage relating to that cell respectively. This is called the pattern of the network. From the figure, it can be seen that the initial conditions can affect the network pattern formation, so that the power source voltage of each cell can be programmed to obtain a specific pattern, so as to complete the information processing.

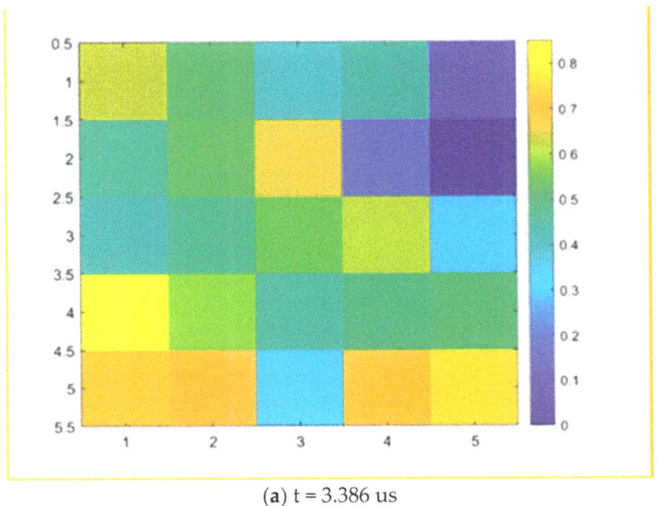

(**a**) t = 3.386 us

Figure 12. *Cont.*

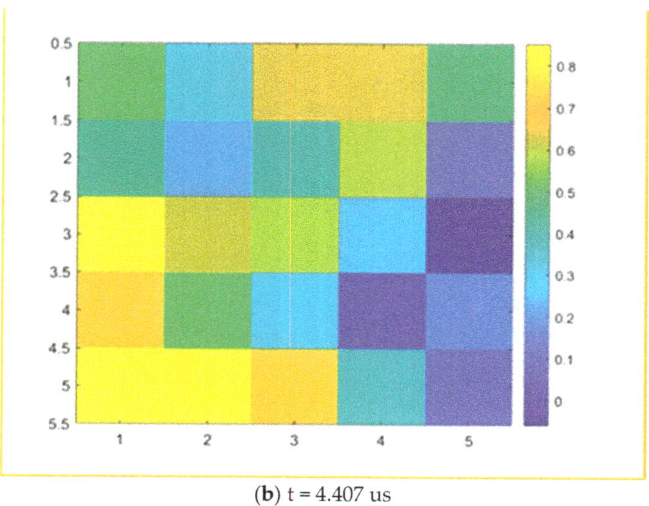

(**b**) t = 4.407 us

Figure 12. Pattern formation diagrams of the network. They are directly derived from Figure 11. (**a**) is at t = 3.386 us and (**b**) is at t = 4.407 us.

5. Conclusions

In this paper, a CNN based on a VO_2 carbon nanotube composite memristor was proposed. The analysis shows that the coupling *RC* parameters will affect the network pattern formation. According to the local active and edge of chaos theory proposed by Professor Chua, it is pointed out that only when the cell is in the sharp edge of chaos region can the cell be successfully awakened after forming the network, that is, the differential equation of the system has a non-homogeneous solution. If the system equation has a non-homogeneous static solution, it is said that the system is in static mode; if the system equation has an oscillatory solution, it is said that the system is in dynamic mode. When the cell is biased in other regions, the differential equation of the system has only a trivial solution, that is, a zero solution, and the cell is in the sleeping mode. At the same time, the formation of the mode is also affected by the initial conditions. In our experiment, we deliberately made the cells located in the middle of the 5×5 CNN network earlier than other cells, so that we could program the cell power supply voltage to get a specific mode and complete the information processing. In summary, the architecture proposed in this paper can be applied to future computing occasions based on complex space–time patterns, especially in the field of computer vision.

Author Contributions: Conceptualization, Y.S.; Formal analysis, Y.S.; Funding acquisition, G.W.; Methodology, Y.S.; Supervision, G.W. All authors have read and agreed to the published version of the manuscript.

Funding: This project was supported by the National Natural Science Foundation of China (Grant No. 61771176).

Conflicts of Interest: The authors declare no conflict of interest.

References

1. Keshav, P. The von Neumann Bottleneck Revisited. *Computer Architecture News*, 26 July 2018.
2. Lu, C.H.; Lin, C.S.; Chao, H.L.; Shen, J.S.; Hsiung, P.A. Reconfigurable multi-core architecture—A plausible solution to the von Neumann performance bottleneck. *IJAIS* **2015**, *2*, 217. [CrossRef]
3. Shin, D.; Yoo, H.J. The Heterogeneous Deep Neural Network Processor with a Non-von Neumann Architecture. *Proc. IEEE* **2019**, *108*, 1245–1260. [CrossRef]

4. Inoue, K.; Pham, C.-K. The Memorism Processor: Towards a Memory-Based Artificially Intelligence Complementing the von Neumann Architecture. *SICE J. Control. Meas. Syst. Integr.* **2017**, *10*, 544–550. [CrossRef]

5. Yao, P.; Wu, H.; Gao, B.; Tang, J.; Zhang, Q.; Zhang, W.; Yang, J.J.; Qian, H. Fully hardware-implemented memristor convolutional neural network. *Nature* **2020**, *577*, 641–646. [CrossRef] [PubMed]

6. Dellaert, F.; Vandewalle, J. Automatic Design of Cellular Neural Networks by means of Genetic Algorithms: Finding a Feature Detector. In Proceedings of the Third IEEE International Workshop on Cellular Neural Networks and their Applications (CNNA-94), Rome, Italy, 18–21 December 1994.

7. Lee, S.; Kim, M.; Kim, K.; Kim, J.Y.; Yoo, H.J. 24-GOPS 4.5-mm2 Digital Cellular Neural Network for Rapid Visual Attention in an Object-Recognition SoC. *IEEE Trans. Neural Netw.* **2011**, *22*, 64–73. [PubMed]

8. Cuevas, E.; Diaz-Cortes, M.-A.; Mezura-Montes, E. Corner Detection of Intensity Images with Cellular Neural Networks (CNN) and Evolutionary Techniques. *Neurocomputing* **2019**, *347*, 82–93. [CrossRef]

9. Chua, L.O.; Yang, L. Cellular neural networks: Theory. Circuits and Systems. *IEEE Trans. Circuits Syst.* **1988**, *35*, 1257–1272. [CrossRef]

10. Chua, L.O.; Yang, L. Cellular neural networks Applications. *IEEE Trans. Circuits Syst.* **1988**, *35*, 1273–1290. [CrossRef]

11. Itoh, M. Some Interesting Features of Memristor CNN. 2019. Available online: https://arxiv.org/abs/1902.05167 (accessed on 2 March 2012).

12. Weiher, M.; Herzig, M.; Tetzlaff, R.; Ascoli, A.; Mikolajick, T.; Slesazeck, S. Pattern Formation with Locally Active S-Type NbO_x Memristors. Circuits and Systems I: Regular Papers. *IEEE Trans. Circuits Syst.* **2019**, *99*, 1–12.

13. Kuzum, D.; Jeyasingh, R.G.D.; Lee, B.; Wong, H.-S.P. Nanoelectronic Programmable Synapses Based on Phase Change Materials for Brain-Inspired Computing. *Nano Lett.* **2012**, *12*, 2179–2186. [CrossRef] [PubMed]

14. Fan, D.; Sharad, M.; Sengupta, A.; Roy, K. Hierarchical Temporal Memory Based on Spin-Neurons and Resistive Memory for Energy-Efficient Brain-Inspired Computing. *IEEE Trans. Neural Netw. Learn. Syst.* **2016**, *27*, 1907–1919. [CrossRef]

15. Wu, X.; Saxena, V.; Zhu, K. A CMOS Spiking Neuron for Dense Memristor-Synapse Connectivity for Brain-Inspired Computing. In Proceedings of the International Joint Conference on Neural Networks (IJCNN, 2015), Killarney, Ireland, 12–17 July 2015.

16. Bohaichuk, S.M.; Kumar, S.; Pitner, G.; McClellan, C.J.; Jeong, J.; Samant, M.G.; Wong, H.-S.P.; Parkin, S.S.P.; Williams, R.S.; Pop, E. Fast Spiking of a Mott VO_2-Carbon Nanotube Composite Device. *Nano Lett.* **2019**, *19*, 6751–6755. [CrossRef] [PubMed]

17. Yi, W.; Tsang, K.K.; Lam, S.K.; Bai, X.; Crowell, J.A.; Flores, E.A. Biological plausibility and stochasticity in scalable VO_2 active memristor neurons. *Nat. Commun.* **2018**, *9*, 4661. [CrossRef] [PubMed]

18. Chua, L.O. Local activity is the origin of complexity. *Int. J. Bifurc. Chaos* **2011**, *15*, 3435–3456. [CrossRef]

 *electronics*

Article

History Erase Effect of Real Memristors

Yiran Shen and Guangyi Wang *

Institute of Modern Circuits and Intelligent Information, Hangzhou Dianzi University, Hangzhou 310018, China; yrshen@hdu.edu.cn
* Correspondence: wanggyi@hdu.edu.cn

Abstract: Different from the static (power-off) nonvolatile property of a memristor, the history erase effect of a memristor is a dynamic characteristic, which means that under the excitation of switching or different signals, the memristor can forget its initial value and reach a unique stable state. The stable state is determined only by the excitation signal and has nothing to do with its initial state. The history erase effect is a desired effect in memristor applications such as memory. It can simplify the complexity of the writing circuit and improve the storage speed. If the memristor's response depends on the initial state, a state reset operation is required before each writing operation. Therefore, it is of great theoretical and practical significance to judge whether the memristor has a history erase effect. Based on the study of the history erase effect of real memristors, this paper focuses on the history erase effect of a Hewlett-Packard (HP) TiO_2 memristor and the Self-Directed Channel (SDC) memristor of Knowm Company. The DC and AC responses of the HP TiO_2 memristor are given, and it is pointed out that there is no AC history erase effect. However, considering the parasitic memcapacitance effect, it is found that it has the effect. Based on the theoretical model of the SDC memristor, its history erase properties with and without considering parasitic effects are studied. It should be noted that this study method can be useful for other materials such as Al_2O_3 and MoS_2.

Keywords: memristor; history erase effect; dynamic route; power-off plot

Citation: Shen, Y.; Wang, G. History Erase Effect of Real Memristors. *Electronics* **2021**, *10*, 303. https://doi.org/10.3390/electronics10030303

Academic Editor: George Angelos Papadopoulos
Received: 31 December 2020
Accepted: 23 January 2021
Published: 27 January 2021

Publisher's Note: MDPI stays neutral with regard to jurisdictional claims in published maps and institutional affiliations.

1. Introduction

A memristor has many unique properties, such as a simple physical structure, easy high-density integration, nonvolatility, historical behavior, low power consumption, and good scaling, which have attracted the extensive attention of scientists and researchers [1–3]. However, unlike other important inventions and discoveries throughout history, the memristor was first proposed as a theoretical concept. In 1971, Chua proposed the memristor as the fourth basic circuit element in his groundbreaking paper "Memristor: The Missing Circuit Element" [4]. In his paper, starting from the basic theory of circuits, professor Chua pointed out that just as resistance links voltage and current, capacitance links charge and voltage, and inductance links magnetic flux and current, there must be a fourth basic circuit element that links charge and magnetic flux. He considered this kind of element a memory resistor and coined the term memristor. He proved that this kind of element is not equivalent to any circuit consisting of the simple connection of the other three basic circuit elements; thus, it is a new basic circuit element. Since then, for more than 30 years, research on memristors has been limited to a few circuit theorists. It was not until 2008 that Dmitri [5] and others of Hewlett-Packard (HP) laboratories announced that nanomemristor devices had been manufactured artificially. Memristors returned to the public's attention and thus ushered in the upsurge of memristor research. Since 2008, a number of papers have been published in peer-reviewed journals involving memristor manufacturing and memristor applications in different scientific and technological fields [6–8].

Since memristors are extensively used as memory devices, the study of their switching dynamics is timely and interesting. The motivation of this paper is, by studying one of a memristor's properties, the history erase effect, to provide useful information both theoretically and practically to guide scientists and engineers toward improving device performance.

Boyd introduced the concept of history erase in 1985 [9], but his strict mathematical definition was abstract and difficult to apply. Until 2016, Ascoli studied a tantalum dioxide memristor at HP laboratories using the concept of history erase and described it in a popular way [10]. After that, Tezlaff et al. [11,12] studied the local history erase effect of bistable memristors. In 2017, Menzel et al. [13] analyzed the origin of the history erase effect in ReRAM. In the same year, Ascoli et al. [14] analyzed its historical erasure effect by means of a closed analytical solution of the dynamic characteristics of the TaO memristor switch. In 2018, Ascoli, Tetzlaff, and Menzel published a summary [15]. Based on the circuit theory model, the history erase effect of a variety of practical memristors was studied.

In fact, the history erase effect describes the dynamic characteristics of memristors, which should be distinguished from their static nonvolatile memory characteristics. The so-called nonvolatility of the memristor refers to the state before power off is memorized when the power supply is cut off. The behavior of the memristor can be described by the power-off plot (POP). The history erase effect refers to the property that the memristor can forget its initial value under the excitation of the switching signal and reach its unique stable state after a period of time. This state is only determined by the excitation signal and is independent of the initial state of the memristor. Therefore, the history erase effect is a desired effect that can simplify the complexity of the writing circuit of the memory and improve the storage speed. If the memristor has no history erase effect and its response depends on the initial state, there is still an operation of state clearing before each writing operation. It is undoubtedly ideal to avoid such additional overhead.

Based on the above understanding, this paper studies the history erase effect of two practical memristor devices, namely the HP TiO_2 memristor and the Knowm self-aligning channel memristor. The authors found that the HP TiO_2 memory device has no history erase effect under AC signal excitation, but because of its parasitic memcapacitance, it leads to the history erase phenomenon in the actual device. As a supplement to this study, the authors also studied the latest commercial memristor device, the discrete Self-Directed Channel (SDC) memristor of Knowm Company [16–19]. There is no suitable model description for this kind of memristor. Therefore, the author first used the generic VTEAM model [20,21] to fit the characteristic curve of the memristor measured from the experiment, followed by the optimization method of simulated annealing to determine the model parameters. The history erase property of the SDC memristor is then studied by using the VTEAM model, and it is found that it has this effect.

Section 2 describes the concept of the history erase effect and points out that the ideal general memristor device does not have the history erase property. Section 3 studies the historical erase characteristics of the HP TiO_2 memristor and discusses the influence of the parasitic effect on the performance of the device. Section 4 studies the modeling and history erase effect of the discrete SDC memristor of Knowm Company. Finally, the conclusion of this paper is given in Section 5.

2. History Erase Effect of the Ideal Generic Memristor

It should be noted that not all memristors have a history erase effect. For example, it can be proved that the ideal generic memristor does not have a history erase effect. The lack of history erase effects in ideal generic memristors was already revealed in [10]; however, we rederive the results here by another method.

For the ideal generic memristor, the definition equation is as follows:

$$\frac{dx}{dt} = g(x)v_m \tag{1}$$

$$i_m = G(x)\,v_m \tag{2}$$

where x is the state variable, and v_m and i_m are the voltage and current across both ends of the memristor, respectively. $G(x)$ is the memconductance, and $g(x)$ is a function of state x.

Note that in (1), if $g(x) = 0$, then $dx/dt = 0$, $x = x_0$. Then, the response of the memristor depends on the initial state (we assume $G(x_0) \neq$ constant). If $g(x) \neq 0$, for (1), the variables are separated and the two sides are integrated, then we obtain:

$$\varphi_m = \varphi_{m0} + \int_0^x \frac{1}{g(x)} dx \tag{3}$$

Obviously, the memristor flux response φ_m is related to the initial state φ_{m0}, hence the memristor has no history erase effect.

3. HP TiO$_2$ Memristor

The HP TiO$_2$ memristor was invented by Strukov et al. of HP laboratories in 2008 [5,22]. Its linear ion drift model is as follows:

$$\begin{cases} v_m(t) = (R_{ON}\frac{w(t)}{D} + R_{OFF}(1 - \frac{w(t)}{D}))i_m(t) \\ \frac{dw(t)}{dt} = \mu_V \frac{R_{ON}}{D} i_m(t) \end{cases} \tag{4}$$

where $w(t)$ is the width of the doping region, D is the length of the device, R_{OFF} is the resistance value when the device is completely off, R_{ON} is the resistance value when the device is fully on, μ_V is the ion mobility, and v_m and i_m are the voltage and current of the device, respectively. Considering the dimensional limitation of the actual device, Equation (4) should be modified as follows:

$$\frac{dw(t)}{dt} = \begin{cases} \mu_V \frac{R_{ON}}{D} i_m(t) & w(t) \in (0, D) \text{ or } (w(t) = D, v_m(t) < 0) \text{ or } (w(t) = 0, v_m(t) > 0) \\ 0 & (w(t) = D, v_m(t) \geq 0) \text{ or } (w(t) = 0, v_m(t) \leq 0) \end{cases} \tag{5}$$

By substituting the current in Equation (5) with voltage, the following results can be obtained:

$$\frac{dw(t)}{dt} = \begin{cases} \mu_V \frac{R_{ON}}{D} \frac{v_m(t)}{(R_{ON}\frac{w(t)}{D} + R_{OFF}(1 - \frac{w(t)}{D}))}, & w(t) \in (0, D) \text{ or } (w(t) = D, v_m(t) < 0) \\ & \text{or } (w(t) = 0, v_m(t) > 0); \\ 0 & (w(t) = D, v_m(t) \geq 0) \\ & \text{or } (w(t) = 0, v_m(t) \leq 0) \end{cases} \tag{6}$$

In Equation (6), if the parameters of the equation are assigned according to the data in Strukov's original paper [5], $D = 10$ nm, $R_{OFF} = 16$ kΩ, $R_{ON} = 100$ Ω, and $\mu_V = 10^{-14}$ m^{2}s^{-1}V^{-1}, according to which the dynamic route of the memristor can be drawn, as shown in Figure 1.

The dynamic route is a diagram of dw/dt vs. $w(t)$ with the voltage v_m as a parameter, which is based on the state equation $dw/dt = g(w(t), v_m)$, as shown in Equation (6). Each dynamic route in the graph corresponds to a voltage of the memristor. The dynamic route of voltage v equal to zero is the power-off plot (POP). From the power-off plot $dw/dt = 0$, the corresponding continuous $w(t)$ value is the equilibrium point of the memristor, which shows that the memristor has infinite stable states and is distributed between $(0, D)$. The physical meaning is that the memristor can store any state between $(0, D)$ after power off. The arrow on the graph indicates the direction of change of the memristor state w, $dw/dt > 0$. When $v_m > 0$, state w moves to the right and converges to $w = D$ and $dw/dt < 0$. When $v_m < 0$, state w moves to the left and converges to $w = 0$. Note that the circle on the curve indicates that the value at that point is discontinuous. The circle on the left indicates that the value of all curves should be $(0, 0)$, and the circle on the right indicates that the value of all curves should be $(D, 0)$.

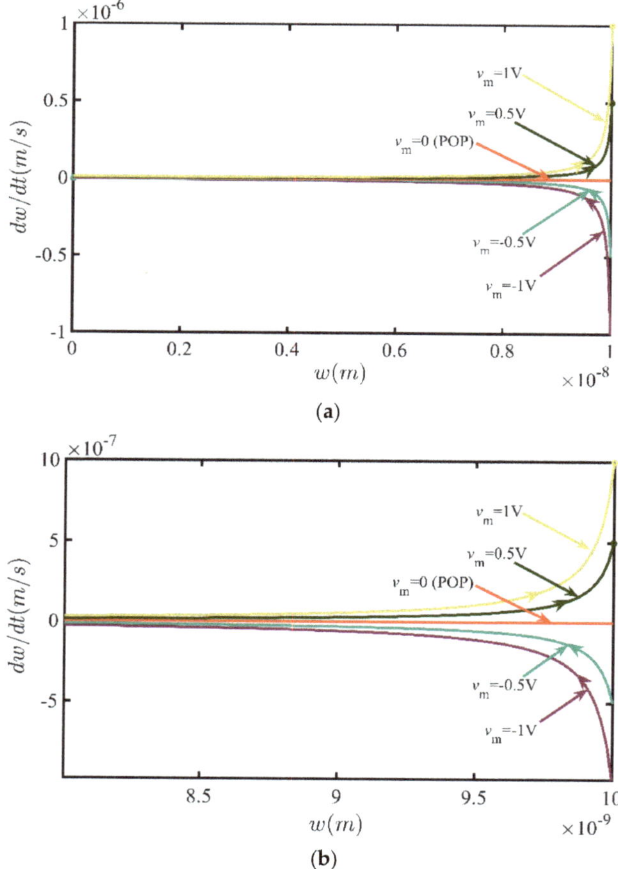

Figure 1. The dynamic route of the Hewlett-Packard (HP) TiO$_2$ memristor, in which the red line is the power-off plot (POP); that is, the relationship between $\frac{dw}{dt}$ and w when the voltage is 0. (**b**) Local zoom-in of (**a**). Note that the small circles in the curve indicate that the curve cannot obtain the value of that point.

3.1. DC Response

As for the DC response of the memristor, a large number of papers have been published stating that the memristor is excited by a low-frequency AC signal, which is called quasistatic excitation. However, this method is strictly incorrect. The correct method is to make the state equation of the memristor zero. For each V_K value of DC input, the stable equilibrium point X_K satisfying the equation is solved from the state equation, and then the corresponding I_K is obtained by substituting it into Ohm's law equation of the memristor. Then. the points (V_K, I_K) are drawn in the V–I plane and connected with arcs to obtain the relationship curve between I and V, i.e., the DC V–I diagram.

For the HP TiO$_2$ memristor, we have found that the system has two stable equilibrium points: $w = 0$ and $w = D$. Let $v_m = 0.5$ V and $v_m = -0.2$ V. According to Equation (6), the transient response of the memristor is obtained by solving the differential equation numerically, as shown in Figure 2. Figure 2a corresponds to the excitation of the $v_m = 0.5$ V DC voltage, and the memristor responds with different initial values that converge to $w = D$, i.e., to show the ON resistance R_{ON}. Figure 2b corresponds to $v_m = -0.2$ V. The response of the memristor with different initial values converges to $w = 0$ under the excitation of -0.2 V, i.e., to show high resistance R_{OFF}. It can be seen from the figure that there are two stable equilibrium points in the memristor.

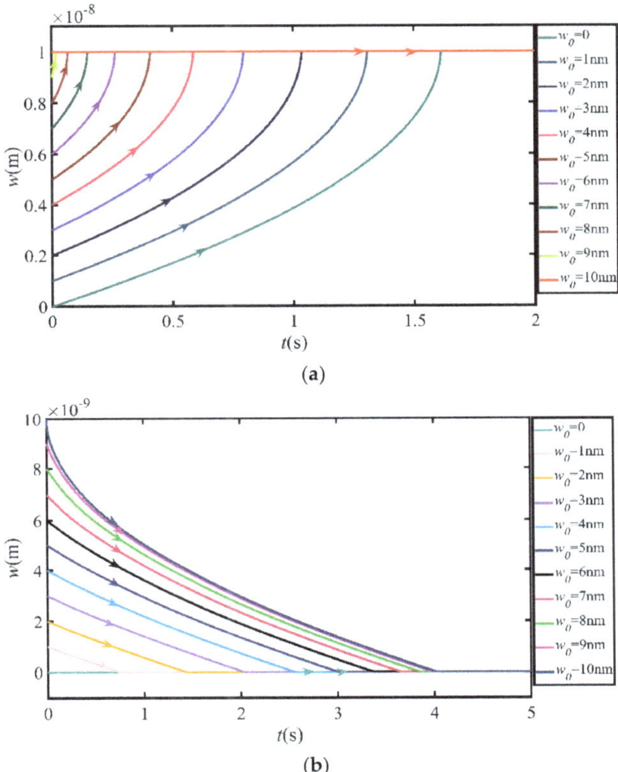

Figure 2. Transient response of the memristor under DC. (**a**) Corresponding voltage $v_m = 0.5$ V; (**b**) corresponding voltage $v_m = -0.2$ V. The initial value is $w_0 \in \{0, 1$ nm, 2 nm, 3 nm, 4 nm, 5 nm, 6 nm, 7 nm, 8 nm, 9 nm, 10 nm$\}$.

By using the method mentioned above and substituting the two equilibrium points $w = 0$ and $w = D$ of the HP memristor into the first formula in (4), and noting the condition of the existence of the equilibrium point in (6) ($v_m \leq 0$, $w = 0$, and $v_m \geq 0$, $w = D$), the DC response curve can be obtained, as shown in Figure 3.

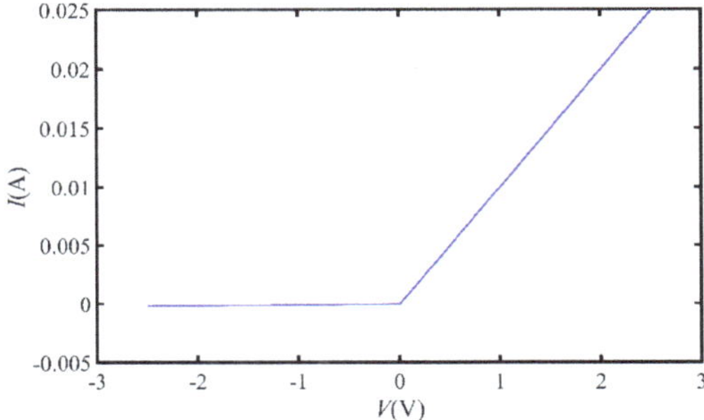

Figure 3. DC response of the HP TiO$_2$ memristor.

It can be seen from Figure 3 that the voltage and current correspond one-to-one. When $v_m < 0$, there is a high resistance R_{OFF}, and when $v_m > 0$, there is a low resistance R_{ON}. The response is independent of the initial value, and there is a history erase phenomenon. The method to derive Figure 3 is the same as [10], where the DC characteristic of a TaO memristor from HP labs was obtained similarly from the Strachan model, presented in [23] after opportune boundary conditions were imposed on the allowable memristor state existence domain.

3.2. AC Response

The AC response of the HP TiO$_2$ memristor is studied below. In other words, $v_m(t) = 0.5 sin(2\pi f t)$ is substituted into Equation (6) to solve the differential equation with different initial values by a numerical method. The response of the solution is shown in Figure 4. Figure 4a corresponds to f = 1 Hz, which is equivalent to the excitation under quasistatic conditions, and Figure 4c corresponds to f = 10 Hz. The response in both cases shows that there is no history erase effect. In fact, this is predictable. If we look at Figure 1 carefully, we find that the curve family is symmetrical about the horizontal axis, which indicates that the rate of change dw/dt is symmetrical about the positive and negative swings of the AC signal. Thus, the net contribution of dw/dt to the state w in a signal cycle is 0. Therefore, periodic oscillation should be made around the initial value, and the curve distinguishes with different initial values. It should be noted that the amplitude of the response decreases when frequency increases. This is because the $v_m - i_m$ pinched loop shrinks when the frequency increases (see Figure 4b,d).

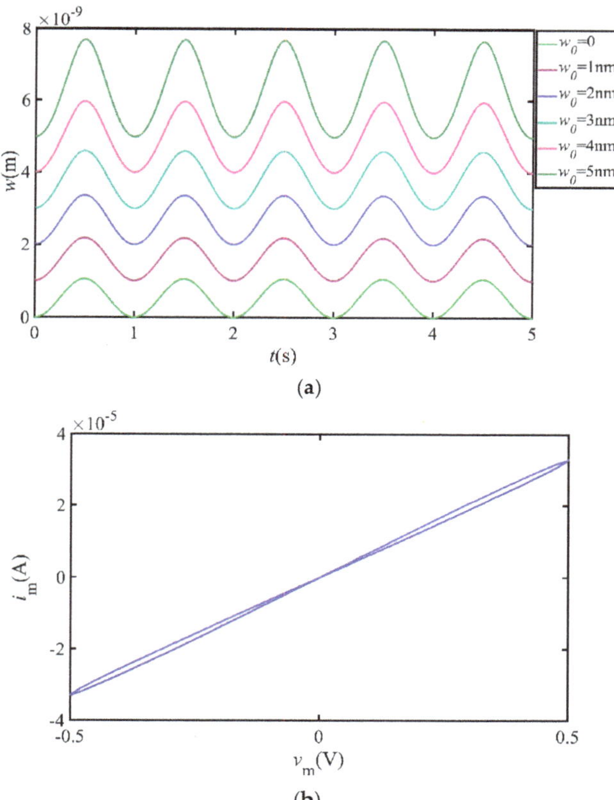

(a)

(b)

Figure 4. *Cont.*

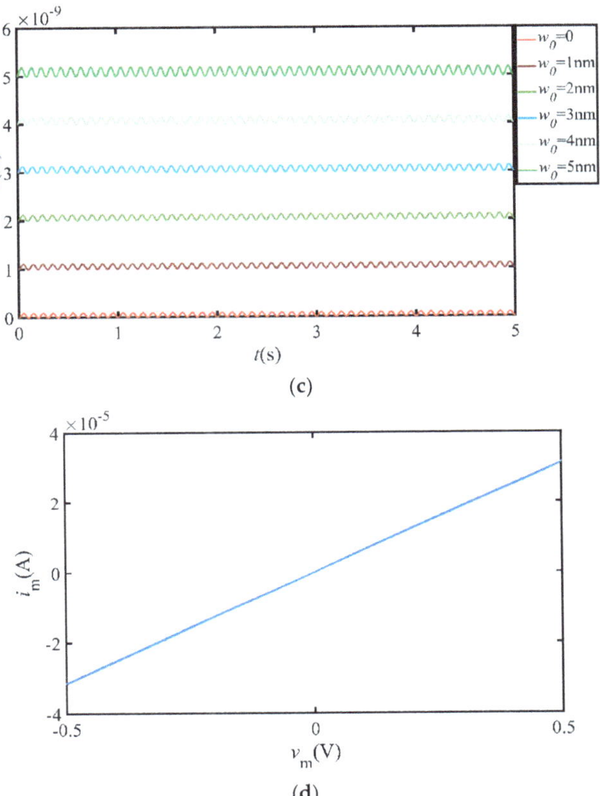

Figure 4. $v_m(t) = V_0 sin(2\pi f t)$ is used as the excitation signal, where $V_0 = 0.5$ V. (**a**) corresponds to $f = 1$ Hz, equivalent to the excitation under quasistatic conditions; (**b**) corresponds to the locus of $v_m - i_m$ at $f = 1$ Hz; (**c**,**d**) correspond to $f = 10$ Hz.

3.3. Closed-Form Solution of th eDynamic Equation of the HP TiO$_2$ Memriston

It is observed that the HP TiO$_2$ memristor does not have a history erase effect according to the numerical method above. Next, the response properties of the HP TiO$_2$ memristor are given by a strict mathematical method.

Substituting the first equation of (4) into the second, the following relationship is obtained:

$$\frac{dw(t)}{dt} = \mu_v \frac{R_{ON}}{D} \frac{v_m(t)}{R_{ON}\frac{w(t)}{D} + R_{OFF}(1 - \frac{w(t)}{D})} \tag{7}$$

By separating the variables and integrating the two sides, the following results are obtained:

$$\int_{w(0)}^{w(t)} \left(R_{ON}\frac{w(t)}{D} + R_{OFF}(1 - \frac{w(t)}{D}) \right) dw(t) = \int_0^t \mu_v \frac{R_{ON}}{D} v_m(t)dt \tag{8}$$

Equation (8) can be further expressed as:

$$\tfrac{1}{2D}(R_{OFF} - R_{ON})w^2(t) - R_{OFF}w(t) + \mu_v \tfrac{R_{ON}}{D}\varphi_m(t) - \mu_v \tfrac{R_{ON}}{D}\varphi_m(0) - [\tfrac{1}{2D}(R_{OFF} - R_{ON})w^2(0) - R_{OFF}w(0)] = 0 \tag{9}$$

The expression of $w(t)$ is obtained by solving the equation.

$$w(t) = \frac{R_{OFF} \pm \sqrt{R_{OFF}^2 - \frac{2}{D}(R_{OFF} - R_{ON})[\mu_v \frac{R_{ON}}{D}\varphi_m(t) - \mu_v \frac{R_{ON}}{D}\varphi_m(0) - (\frac{1}{2D}R_{OFF}w^2(0) - R_{OFF}w(0))]}}{\frac{R_{OFF} - R_{ON}}{D}} \tag{10}$$

Note that because $0 \leq w(t) \leq D$ holds, we should drop the plus sign in Equation (10). Note also that $R_{OFF} \gg R_{ON}$, and the above formula is simplified as:

$$w(t) = (D - D\sqrt{1 - \frac{2}{D}[\mu_v \frac{R_{ON}}{DR_{OFF}}\varphi_m(t) - \mu_v \frac{R_{ON}}{DR_{OFF}}\varphi_m(0) - (\frac{1}{2D}w^2(0) - w(0))]}) \tag{11}$$

Under DC,

$$\frac{d\varphi_m(t)}{dt} = V \tag{12}$$

By integrating the two sides, we obtain:

$$\varphi_m(t) = \varphi_m(0) + Vt \tag{13}$$

Substituting Equation (13) into Equation (11), we obtain:

$$w(t) = D - D\sqrt{1 - T} \tag{14}$$

where

$$T = \frac{2}{D}[\mu_v \frac{R_{ON}}{DR_{OFF}}Vt - (\frac{1}{2D}w^2(0) - w(0))]$$

In Formula (14), $w(t)$ is a real number and $0 \leq w(t) \leq D$; therefore, $0 \leq \sqrt{1-T} \leq 1$ and $0 \leq T \leq 1$. If a positive DC voltage $V > 0$ is applied to the memristor, then T is a monotonic increasing function and $\sqrt{1-T}$ a monotonic decreasing function. When the memristor reaches $T = 1$, $w(t)$ reaches the upper limit D; if a negative DC voltage $V < 0$ is applied to the memristor, T is a monotonic decreasing function and $\sqrt{1-T}$ a monotonic increasing function. When $T = 0$ and $\sqrt{1-T} = 1$, with t increasing, $w(t)$ reaches the lower limit of 0.

Although the state variable w in Equation (11) is a function of the initial value $w(0)$, under the excitation of positive and negative DC voltages, the memristor rapidly reaches two states, namely low resistance R_{ON} and high resistance R_{OFF}, which are independent of the initial value, indicating that the memristor has a history erase effect. If the memristor is excited by positive and negative pulse voltages, as long as its amplitude and pulse width meet the condition of the state transition, the memristor will become a nonvolatile binary switch device, which can be used in binary memory.

Figure 5 shows the DC response diagram of the memristor state variables when $V = 0.5$ V and $V = -0.2$ V, showing the two limit states under different initial values.

It will be proved that the AC response of the HP memristor has no history erase effect under any AC voltage excitation. When AC voltage is v, we have

$$\frac{d\varphi_m(t)}{dt} = v(t) \tag{15}$$

by integrating the two sides, we obtain:

$$\varphi_m(t) = \varphi_m(0) + \int_0^t v_m(t)dt \tag{16}$$

substituting Equation (16) into Equation (11) yields:

$$w(t) = (D - D\sqrt{1 - \frac{2}{D}[\mu_v \frac{R_{ON}}{DR_{OFF}}\int_0^t v_m(t)dt - (\frac{1}{2D}w^2(0) - w(0))]}) \tag{17}$$

Because the AC signal $v_m(t)$ is arbitrarily chosen, the root term in Equation (17) is not a monotone increasing or decreasing function and is closely related to the initial value $w(0)$ of the memristor. The state variable $w(t)$ is related to the initial value, and there is no history erase effect. Taking the cosine excitation signal as an example, the excitation voltage is assumed to be $v_m = cos(t)$, then $\int_0^t v_m(t)dt = \sin t$. Substituting it into Equation (17) will obtain:

$$w(t) = (D - D\sqrt{1 - \frac{2}{D}[\mu_v \frac{R_{ON}}{DR_{OFF}} \sin t - (\frac{1}{2D}w^2(0) - w(0))]}) \tag{18}$$

It is obvious that the state variable w varies with time t and initial value $w(0)$. Especially when $t = k\pi$, $k \in$ N Equation (18) becomes

$$w(k\pi) = (D - D\sqrt{1 + \frac{2}{D}(\frac{1}{2D}w^2(0) - w(0))}) \tag{19}$$

$w(k\pi)$ depends directly on the initial value $w(0)$, which shows that the AC response of the memristor is related to the initial value, and there is no history erase effect.

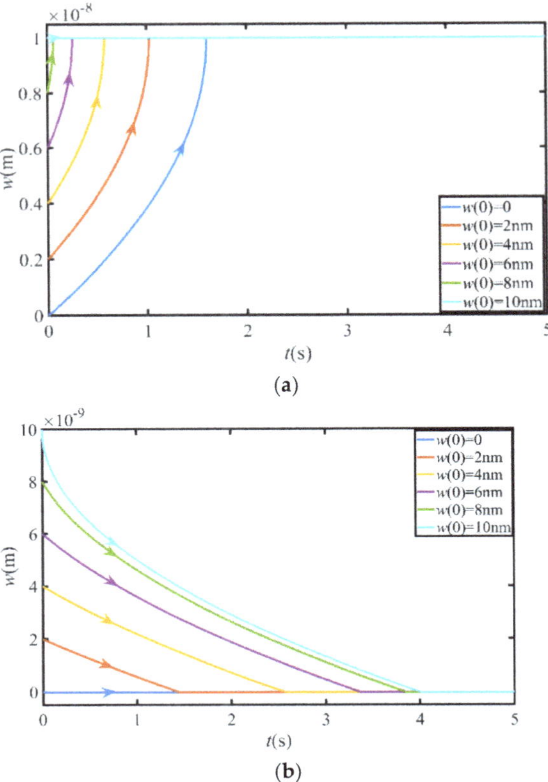

Figure 5. DC response diagram of the memristor, according to Equation (15). (**a**) Under a positive voltage $v_m = 0.5$ V, the responses of initial values $w(0) \in \{0,\ 2\,\text{nm}, 4\,\text{nm}, 6\,\text{nm}, 8\,\text{nm}, 10\,\text{nm}\}$ are obtained. (**b**) Under a negative voltage $v_m = -0.2$ V, the responses of initial values $w(0) \in \{0,\ 2\,\text{nm}, 4\,\text{nm}, 6\,\text{nm}, 8\,\text{nm}, 10\,\text{nm}\}$ are obtained.

3.4. HP TiO$_2$ Memristor Model with a Window Function

The HP memristor model represented by Equation (4) is an ideal model, in which the equation of state is linear, which means that the ions in the doped region move linearly under the external electric field. In practice, however, the drift motion is nonlinear. The boundary effect at the two ends of $w = 0$ and $w = D$ especially is strongly nonlinear. In order to describe this nonlinear effect, Strukov added a window function $w(D - w)/D^2$ to his memristor model in his original reference [5], and modified the memristor (Equation (4)) into the following form:

$$\begin{cases} v_m(t) = (R_{ON}\frac{w(t)}{D} + R_{OFF}(1 - \frac{w(t)}{D}))i_m(t) \\ \frac{dw(t)}{dt} = \mu_V \frac{R_{ON}}{D} i_m(t)\frac{w(t)(D-w(t))}{D^2} \end{cases} \tag{20}$$

By inserting the first formula of Equation (20) with the second, separating variables, and integrating both sides, we obtain:

$$\int_{w(0)}^{w(t)} D\left(\frac{R_{ON}}{D - w(t)} + \frac{R_{OFF}}{w(t)}\right)dw(t) = \int_0^t \mu_v \frac{R_{ON}}{D}v_m(t)dt \tag{21}$$

In (21), the integral is solved and simplified to

$$\frac{w(t)^{R_{OFF}}}{(D - w(t))^{R_{ON}}} = \frac{w(0)^{R_{OFF}}}{(D - w(0))^{R_{ON}}}e^{-\mu_v \frac{R_{ON}}{D^2}\varphi_m(0)}e^{\mu_v \frac{R_{ON}}{D^2}\varphi_m(t)} \tag{22}$$

when DC excitation is applied:

$$\frac{d\varphi_m(t)}{dt} = V \tag{23}$$

By integrating the two sides, we obtain:

$$\varphi_m(t) = \varphi_m(0) + Vt \tag{24}$$

Substituting Equation (24) into Equation (22), we obtain the following results:

$$\frac{w(t)^{R_{OFF}}}{(D - w(t))^{R_{ON}}} = \frac{w(0)^{R_{OFF}}}{(D - w(0))^{R_{ON}}}e^{\mu_v \frac{R_{ON}}{D^2}Vt} \tag{25}$$

If a positive voltage is applied and the right side of the equation is an increasing function of t, which tends to infinity after a period of time, then there must be $w(t) \to D$ on the left side of the equation; if a negative voltage is applied and the right side of the equation is a decreasing function of t and tends to 0 after a period of time, then there must be $w(t) \to 0$ on the left side of the equation. It can be seen that under the excitation of positive and negative DC voltages, the memristor state variables w tend to D and 0, respectively; that is, the memristor is in R_{ON} and R_{OFF} states, respectively, and its response is independent of the initial value.

When an AC cosine voltage is applied, Equation (24) becomes $\varphi_m(t) = \varphi_m(0) + \sin(t)$. Let the responses of the state variable under two different initial values, respectively, be $w_1(t)$ and $w_2(t)$, which can be substituted into Equation (22) and divided:

$$\frac{\frac{w_2(t)^{R_{OFF}}}{(D-w_2(t))^{R_{ON}}}}{\frac{w_1(t)^{R_{OFF}}}{(D-w_1(t))^{R_{ON}}}} = \frac{\frac{w_2(0)^{R_{OFF}}}{(D-w_2(0))^{R_{ON}}}}{\frac{w_1(0)^{R_{OFF}}}{(D-w_1(0))^{R_{ON}}}} \tag{26}$$

when $t \to \infty$, if $w_1(t)$ tends to $w_2(t)$, the left side of Equation (26) is 1, while the right side is equal to a constant, which is not 1:

$$R.H.S. = \frac{\frac{w_2(0)^{R_{OFF}}}{(D-w_2(0))^{R_{ON}}}}{\frac{w_1(0)^{R_{OFF}}}{(D-w_1(0))^{R_{ON}}}} \tag{27}$$

In this case, Equation (26) would not hold, i.e., $w_1(\infty) \neq w_2(\infty)$. Therefore, there is no history erase effect in the memristor.

3.5. Parasitic Memcapacitance Effect of the HP TiO$_2$ Emristor

If we further observe the structure of the HP TiO$_2$ memristor, we will find that there is a parasitic parallel plate capacitor in the memristor; in fact, as shown next, it is a memcapacitor, because the capacitance is dependent on state variable w, and the dielectric between the two plates is undoped TiO$_2$. This is because undoped TiO$_2$ has a higher resistivity, equivalent to an insulator medium, while doped TiO$_2$ has a lower resistivity, equivalent to a conductor, as shown in Figure 6.

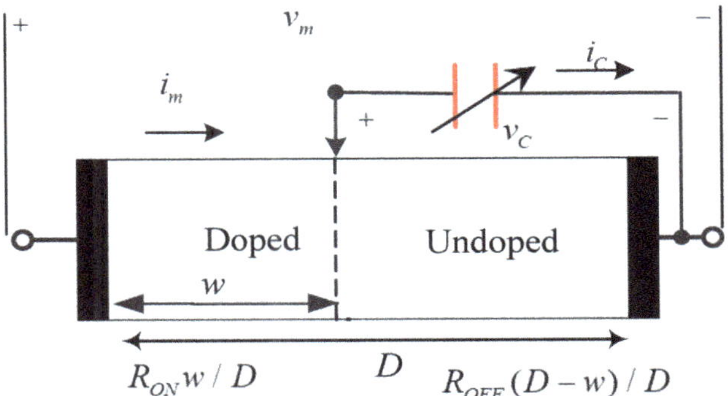

Figure 6. Structure of the HP TiO$_2$ memristor.

The parasitic capacitance is equivalent to a flat-plate capacitor, and because the capacitance is dependent on state variable $w(t)$, it is indeed a memcapacitor:

$$C = \frac{\varepsilon_{TiO_2}S}{D - w} \tag{28}$$

According to [24], we know $\varepsilon_{TiO_2} = 5\varepsilon_0 F/m$, $S = 1 \times 10^4$ nm^2. When the parasitic memcapacitance is considered, by KCl and KVL, we have:

$$\begin{cases} v_m(t) = R_{ON}\frac{w(t)}{D}i_m(t) + R_{OFF}(1 - \frac{w(t)}{D})(i_m(t) - i_C(t)) \\ i_C(t) = \frac{dC}{dt}v_C(t) + C\frac{dv_C(t)}{dt} \\ v_C(t) = v_m(t) - R_{ON}\frac{w(t)}{D}i_m(t) \\ \frac{dw(t)}{dt} = \mu_V\frac{R_{ON}}{D}i_m(t) \end{cases} \tag{29}$$

where $v_C(t)$ is the voltage across both ends of the capacitor, and dC/dt is the derivative of capacitance to time. Expression (28) of capacitance is substituted into the second equation of (29), and then the second equation is substituted into the first equation and the third equation into the forth, and Equation (30) is obtained:

$$\begin{cases} \dfrac{dw(t)}{dt} = \dfrac{\mu_V}{w(t)}(v_m(t) - v_C(t)) \\[2mm] \dfrac{dv_C(t)}{dt} = \dfrac{D(1-\frac{w(t)}{D})}{\varepsilon_{TiO_2}S}\left[\dfrac{v_m(t)-v_C(t)}{R_{ON}\frac{w(t)}{D}} - \dfrac{v_C(t)}{R_{OFF}(1-\frac{w(t)}{D})} - \dfrac{\varepsilon_{TiO_2}S}{D^3(1-\frac{w(t)}{D})^2}\dfrac{\mu_V}{\frac{w(t)}{D}}(v_m(t) - v_C(t))v_C(t)\right] \end{cases} \tag{30}$$

According to Equation (30), the AC response of state variable $w(t)$ to t is solved by MATLAB and shown as Figure 7. The excitation signal in the diagram is $v_m(t) = V_0 sin(2\pi ft)$, V_0 is selected as 0.5 V, and f is set to 10 Hz.

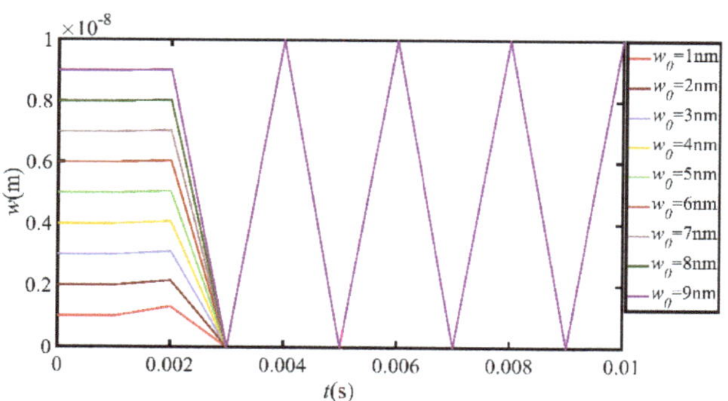

Figure 7. AC response of state variable $w(t)$ to t considering parasitic memcapacitance, in which the excitation signal is $v_m(t) = V_0 sin(2\pi ft)$, V_0 is selected as 0.5 V, and f is set to 10 Hz.

It can be seen from the figure that under different initial values, the AC response converges to a stable value after a period of time, so there is a history erase effect.

4. The Self-Directed Channel (SDC) Memristor

The Self-Directed Channel (SDC) memristor is a product developed by Knowm Company. This is an ion conduction device (also known as an electrochemical metallization (ECM) device), which changes the resistance of the device by relying on Ag+ entering the channel of the active layer of the device. The hysteretic curve can be measured by software and hardware tools provided by Knowm Company. However, there is no mathematical model. In order to analyze it theoretically and design a memristor circuit with it, it is necessary to establish a mathematical model for it.

4.1. Establishment of the SDC Memristor Model

In his 2015 paper [20], Kvatinsky proposed a generic voltage-controlled memristor model VTEAM. The model's equation is as follows:

$$i_m(t) = v_m(t)/\left(R_{ON} + \frac{w - w_{ON}}{w_{OFF} - w_{ON}}(R_{OFF} - R_{ON})\right) \tag{31}$$

$$\frac{dw(t)}{dt} = \begin{cases} k_{OFF}\left(\frac{v_m(t)}{v_{OFF}} - 1\right)^{\alpha_{OFF}}, & 0 < v_{OFF} < v, \\ 0 & v_{ON} < v_m < v_{OFF}, \\ k_{ON}\left(\frac{v_m(t)}{v_{ON}} - 1\right)^{\alpha_{ON}}, & v_m < v_{ON} < 0. \end{cases} \tag{32}$$

The equation has ten parameters, where R_{OFF}, R_{ON}, v_{OFF}, v_{ON}, w_{OFF}, w_{ON}, k_{OFF} and k_{ON} are real numbers, and α_{OFF}, α_{ON} are natural numbers. w_{OFF}, w_{ON} are the boundaries of the internal variable w, and R_{OFF}, R_{ON} are resistance values when the values of state variable are w_{OFF}, w_{ON}, respectively. k_{OFF}, k_{ON}, α_{OFF} and α_{ON} are constants, and v_{OFF}, v_{ON} are the threshold voltages.

VTEAM model is a generic model. Because of its inherent universality and robustness, it can be applied to a large number of memristor models and experimental data. Considering the current voltage characteristics of a specific memristor, a set of parameters are selected to make the VTEAM model conform to the reference I–V relationship of the SDC memristor. In order to determine the I–V curve, simulated annealing algorithms can be used to minimize the relative root mean square error. The relative root mean square error is

$$f(\mathsf{x}) = \sqrt{\frac{\left(\dfrac{\sum\limits_{i=1}^{N} (v(\mathsf{x}) - v_{ref})^2}{\sum\limits_{i=1}^{N} v_{ref}^2} + \dfrac{\sum\limits_{i=1}^{N} (i(\mathsf{x}) - i_{ref})^2}{\sum\limits_{i=1}^{N} i_{ref}^2} \right)}{N}} \tag{33}$$

where N is the number of samples, $v(\mathsf{x})$ and $i(\mathsf{x})$ are the sampling voltage and current values of the VTEAM model, respectively, and v_{ref} and i_{ref} are the actual measured sampling voltage and current values, respectively.

The fitting process is to make the program iterate over k_{OFF}, k_{ON} to minimize the error function given in (33). In order to avoid convergence to the local minimum instead of the best global fit, other parameters ($R_{OFF}, R_{ON}, v_{OFF}, v_{ON}, w_{OFF}, w_{ON}, \alpha_{OFF}, \alpha_{ON}$) are selected manually to show as much similarity as possible to the reference I–V relationship (relative root mean square error is less than 1.5%). In addition, the ideal window function can be used to constrain the state variables in the process of fitting.

In the experiment, the memristor is connected with a 1KΩ resistor in series and an applied AC signal as $v_m(t) = V_0 sin(2\pi ft)$. Considering that only part of the voltage falls on the memristor, V_0 can be taken as 1 V, and the frequency f can be set to 5 Hz. After the volt–ampere data are obtained, the data are imported into MATLAB, and the simulated annealing algorithm is used to set the objective function to (34).

The VTEAM model was used to fit the experimental data, and Figure 8 was obtained. Looking at Figure 8, it can be found that the model fitting is very good in the first quadrant, while there is a little difference between the fitting data and the measured data in the third quadrant. This is mainly due to the insufficient sampling accuracy of the measurement software and hardware provided by Knowm Company (200 time points in total).

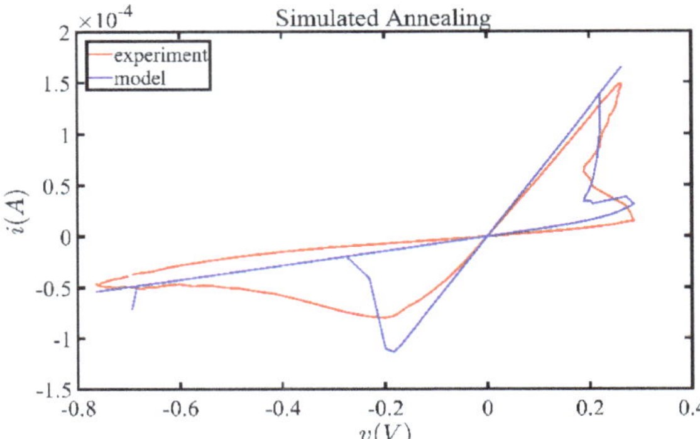

Figure 8. Volt–ampere characteristic curve of memristor. Red is the experimental data, blue is the model fitting data under the optimization algorithm.

According to the computer simulation, the optimal parameters of the model can be determined as shown in Table 1.

Table 1. Model parameters obtained by the simulation of the optimization algorithm.

Parameters	Value
R_{OFF}	14.277 kΩ
R_{ON}	1.5936 kΩ
v_{OFF}	0.02 V
v_{ON}	-0.13 V
k_{OFF}	538,530.50 nm
k_{ON}	-2.6213 m
α_{OFF}	2
α_{ON}	8
w_{ON}	0
w_{OFF}	0.001 m

4.2. Dynamic Routes and AC Response of the SDC Memristor

According to the model parameters, the dynamic routes of the SDC memristor can be drawn in a similar way to that of the HP TiO$_2$ memristor and shown as Figure 9. It is noted that in Equation (32), dw/dt does not depend on w but only on v_m. Therefore, dw/dt vs. w is a set of horizontal lines.

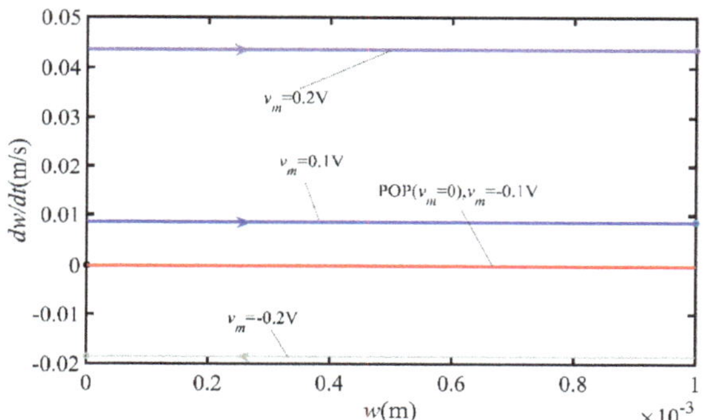

Figure 9. Dynamic routes of the Self-Directed Channel (SDC) memristor. Tote that POP (v_m = 0) and the line corresponding to v_m = 0.1 V are in fact two different lines.

It can be seen from the diagram that the dynamic routes are asymmetric with respect to the horizontal axis, which indicates that the net contribution of dw/dt to w is not zero when the signal amplitude is positive and negative for one cycle under AC excitation. Therefore, it is speculated that under AC excitation, $w(t)$ may gradually tend to a stable value after a period of time under the state of net increase or decrease of $w(t)$, thus it has a history erase effect. We use a sinusoidal signal $v_m(t) = V_0 sin(2\pi f t)$, where f is 5 Hz and with $V_0 = 0.5$ V as the AC excitation and applied to the memristor. Thus, Figure 10 can be obtained. It can be seen from Figure 10 that the SDC memristor has a history erase effect under AC conditions.

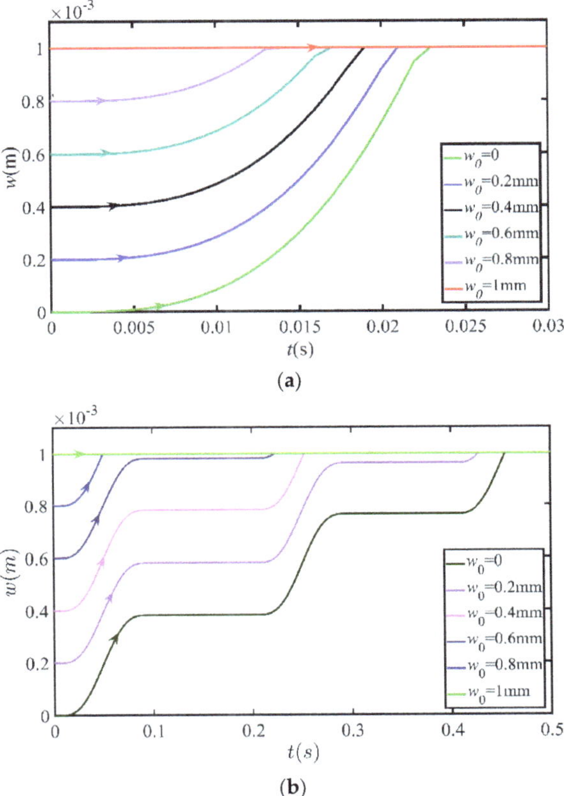

(a)

(b)

Figure 10. Sinusoidal signal $v_m(t) = V_0 sin(2\pi ft)$, with f = 5 Hz and V_0 = 0.5 V as the AC excitation and applied to the memristor (**a**) can be obtained, (**b**) corresponds to V_0 = 0.1 V.

4.3. Considering the Parasitic Capacitance Effect

Similar to the case of the HP TiO_2 memristor (refer to Figure 6), considering that the undoped part is also equivalent to a moving parallel plate capacitor, it can be considered to have parallel a small parasitic capacitor on it. Assuming 1nF, the circuit equation becomes:

$$\begin{cases} \dfrac{dv_C(t)}{dt} = \dfrac{1}{C}\left[\dfrac{v_m(t)-v_C(t)}{(1-\frac{w-w_{ON}}{w_{OFF}-w_{ON}})R_{ON}} - \dfrac{v_C(t)}{\frac{w-w_{ON}}{w_{OFF}-w_{ON}}R_{OFF}}\right] \\ \dfrac{dw(t)}{dt} = \begin{cases} k_{OFF}\left(\dfrac{v_m(t)}{v_{OFF}}-1\right)^{\alpha_{OFF}}, & 0 < v_{OFF} < v, \\ 0 & v_{ON} < v_m < v_{OFF}, \\ k_{ON}\left(\dfrac{v_m(t)}{v_{ON}}-1\right)^{\alpha_{ON}}, & v_m < v_{ON} < 0. \end{cases} \end{cases} \tag{34}$$

The solution of the state variable $w(t)$ can be obtained by numerically solving the above differential equation with MATLAB, as shown in Figure 11.

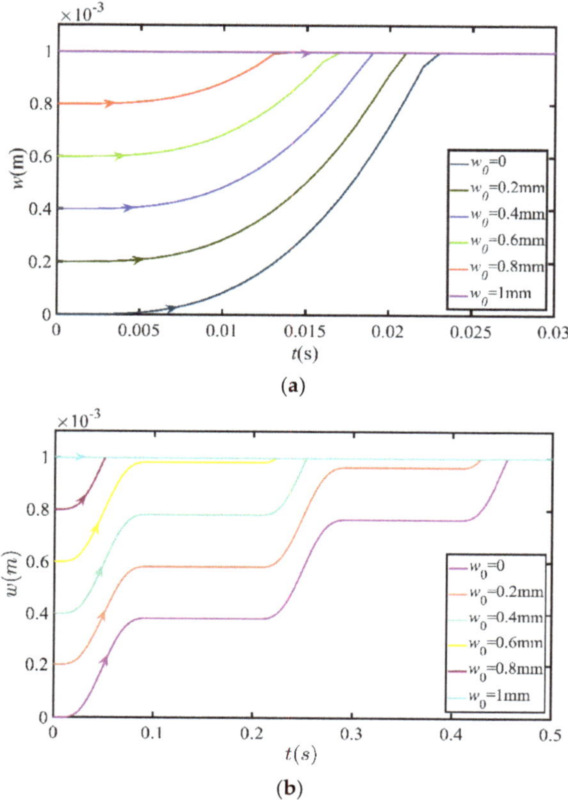

Figure 11. AC response with parasitic capacitance taken into account. Using sinusoidal signal $v_m(t) = V_0 sin\,(2\pi f t)$, with f = 5 Hz and V_0 = 0.5 V as the AC excitation, (**a**) is obtained. (**b**) corresponds to V_0 = 0.1 V.

Compared with Figures 10 and 11, the parasitic capacitance has little effect on the SDC memristor.

5. Conclusions

In this paper, the history erase effect of the HP TiO$_2$ memristor is studied. Firstly, the existence of the history erase effect under DC and AC conditions is predicted from the dynamic routes. The historical erase properties of the HP TiO$_2$ memristor under AC and DC are then verified by a MATLAB numerical simulation. It can be seen that the HP TiO$_2$ memristor has a DC history erase effect but no AC history erase effect. After the numerical simulation, the closed-form solution of the dynamic equation of the TiO$_2$ memristor is given. Thus, the history erase properties of the TiO$_2$ memristor under DC and AC conditions are explained theoretically. Furthermore, the parasitic capacitance effect of the HP TiO$_2$ memristor is considered, and it is pointed out that the parasitic capacitance effect can cause the HP TiO$_2$ memristor to have an AC history erase effect. It is worth noting that Menzel et al. pointed out that the resistance of the doping region in the memristor is the origin of history erase effect [13]. From the work of this paper, it seems that the resistance of the doped region and the parasitic memcapacitance of the undoped region work together to form a discharge path, which leads to the history erase effect. As a supplement to the research work on the history erase effect of the HP TiO$_2$ memristor, we studied the history erase effect of the latest discrete SDC memristor made by Knowm

Company. The DC and AC voltages and parasitic capacitance are considered, respectively. It is worth mentioning that the method of modeling the SDC memristor with the generic voltage-controlled memristor model VTEAM is also given in this paper. This method uses a simulated annealing algorithm to fit the actual measured data, which solves the problem that Knowm Company's discrete SDC memristor still lacks an accurate mathematical model. It can be seen that no matter with and without the parasitic capacitance, the SDC memristor has an AC history erase effect.

Author Contributions: Numerical simulation, theoretical calculation and manuscript writing, Y.S.; supervision, G.W. All authors have read and agreed to the published version of the manuscript.

Funding: This research was funded by The National Natural Science Foundation of China (NNSFC), grant number 61771176.

References

1. Maikap, S.; Banerjee, W. In Quest of Nonfilamentary Switching: A Synergistic Approach of Dual Nanostructure Engineering to Improve the Variability and Reliability of Resistive Random-Access-Memory Devices. *Adv. Electron. Mater.* **2020**. [CrossRef]
2. Ginnaram, S.; Qiu, J.T.; Maikap, S. Role of the Hf/Si Interfacial Layer on the High Performance of MoS2-Based Conductive Bridge RAM for Artificial Synapse Application. *IEEE Electron. Device Lett.* **2020**, *41*, 709–712. [CrossRef]
3. Li, H.; Hu, M. Compact model of memristors and its application in computing systems. In Proceedings of the 2010 Design, Automation & Test in Europe Conference & Exhibition (DATE 2010), Dresden, Germany, 8–12 March 2010; pp. 673–678. [CrossRef]
4. Chua, L. Memristor-The missing circuit element. *IEEE Trans. Circuit Theory* **1971**, *18*, 507–519. [CrossRef]
5. Strukov, D.B.; Snider, G.S.; Stewart, D.R.; Williams, R.S. The missing memristor found. *Nature* **2008**, *453*, 80–83. [CrossRef]
6. Dong, Z.; Lai, C.S.; He, Y.; Qi, D.; Duan, S. Hybrid dual-complementary metal-oxide-semiconductor/memristor synapse-based neural network with its applications in image super-resolution. *IET Circuits Devices Syst.* **2019**, *13*, 1241–1248. [CrossRef]
7. Dong, Z.; Lai, C.S.; Qi, D.; Xu, Z.; Li, C.; Duan, S. A general memristor-based pulse coupled neural network with variable linking coefficient for multi-focus image fusion. *Neurocomputing* **2018**, *308*, 172–183. [CrossRef]
8. Dong, Z.; Qi, D.; He, Y.; Xu, Z.; Hu, X.; Duan, S. Easily Cascaded Memristor-CMOS Hybrid Circuit for High-Efficiency Boolean Logic Implementation. *Int. J. Bifurc. Chaos* **2018**, *28*, 1850149. [CrossRef]
9. Boyd, S.; Chua, L. Fading memory and the problem of approximating nonlinear operators with Volterra series. *IEEE Trans. Circuits Syst.* **1985**, *32*, 1150–1161. [CrossRef]
10. Ascoli, A.; Tetzlaff, R.; Chua, L.O.; Strachan, J.P.; Williams, R.S. History Erase Effect in a Non-Volatile Memristor. *IEEE Trans. Circuits Syst. I Regul. Pap.* **2016**, *63*, 389–400. [CrossRef]
11. Ascoli, A.; Tetzlaff, R.; Chua, L.O. The First Ever Real BistableMemristors—Part I: Theoretical Insights on Local Fading Memory. *IEEE Trans. Circuits Syst. II Express Briefs* **2016**, *63*, 1091–1095. [CrossRef]
12. Ascoli, A.; Tetzlaff, R.; Chua, L.O. The First Ever Real BistableMemristors—Part II: Design and Analysis of a Local Fading Memory System. *IEEE Trans. Circuits Syst. II Express Briefs* **2016**, *63*, 1096–1100. [CrossRef]
13. Menzel, S.; Waser, R.; Siemon, A.; La Torre, C.; Schulten, M.; Ascoli, A.; Tetzlaff, R. On the origin of the fading memory effect in ReRAMs. In Proceedings of the 2017 27th International Symposium on Power and Timing Modeling, Optimization and Simulation (PATMOS), Thessaloniki, Greece, 25–27 September 2017; pp. 1–5. [CrossRef]
14. Ascoli, A.; Ntinas, V.; Tetzlaff, R.; Sirakoulis, G.C. Closed-form analytical solution for on-switching dynamics in a TaOmemristor. *Electron. Lett.* **2017**, *53*, 1125–1126. [CrossRef]
15. Ascoli, A.; Tetzlaff, R.; Menzel, S. Exploring the Dynamics of Real-World Memristors on the Basis of Circuit Theoretic Model Predictions. *IEEE Circuits Syst. Mag.* **2018**, *18*, 48–76. [CrossRef]
16. Campbell, K.A. Self-directed channel memristor for high temperature operation. *Microelectron. J.* **2017**, *59*, 10–14. [CrossRef]
17. Drake, K.; Lu, T.; Majumdar, M.; Kamrul, H.; Campbell, K.A. Comparison of the Electrical Response of Cu and Ag Ion-Conducting SDC Memristors Over the Temperature Range 6 K to 300 K. *Micromachines* **2019**, *10*, 663. [CrossRef]
18. Gomez, J.; Vourkas, I.; Abusleme, A.; Sirakoulis, G.C.; Rubio, A. Voltage Divider for Self-Limited Analog State Programing of Memristors. *IEEE Trans. Circuits Syst. II Express Briefs* **2020**, *67*, 620–624. [CrossRef]
19. Gomez, J.; Vourkas, I.; Abusleme, A. Exploring Memristor Multi-Level Tuning Dependencies on the Applied Pulse Properties via a Low Cost Instrumentation Setup. *IEEE Access* **2019**, *7*, 59413–59421. [CrossRef]
20. Kvatinsky, S.; Ramadan, M.; Friedman, E.G.; Kolodny, A. VTEAM: A General Model for Voltage-Controlled Memristors. *IEEE Trans. Circuits Syst. II Express Briefs* **2015**, *62*, 786–790. [CrossRef]
21. Kvatinsky, S.; Friedman, E.G.; Kolodny, A.; Weiser, U.C. TEAM: ThrEshold Adaptive Memristor Model. *IEEE Trans. Circuits Syst. I Regul. Pap.* **2013**, *60*, 211–221. [CrossRef]

22. Williams, R.S. How We Found the Missing Memristor. *IEEE Spectr.* **2008**, *45*, 28–35. [CrossRef]
23. Strachan, P.; Torrezan, A.C.; Miao, F.; Pickett, M.D.; Yang, J.J.; Yi, W.; Medeiros-Ribeiro, G.; Williams, R.S. State dynamics and modelling of Tantalum oxide memristors. *IEEE Trans. Electron Devices* **2013**, *60*, 2194–2202. [CrossRef]
24. Pickett, M. The Materials Science of Titanium Dioxide Memristors. Ph.D. Thesis, University of California, Berkeley, CA, USA, 2010.

Article

Toward Reliable Compact Modeling of Multilevel 1T-1R RRAM Devices for Neuromorphic Systems

Emilio Pérez-Bosch Quesada [1,*], Rocío Romero-Zaliz [2], Eduardo Pérez [1],
Mamathamba Kalishettyhalli Mahadevaiah [1], John Reuben [3], Markus Andreas Schubert [1],
Francisco Jiménez-Molinos [4], Juan Bautista Roldán [4] and Christian Wenger [1,5]

[1] IHP-Leibniz-Institut für Innovative Mikroelektronik, 15230 Frankfurt, Germany;
     perez@ihp-microelectronics.com (E.P.); kalishettyhalli@ihp-microelectronics.com (M.K.M.);
     schuberta@ihp-microelectronics.com (M.A.S.); wenger@ihp-microelectronics.com (C.W.)
[2] Andalusian Research Institute on Data Science and Computational Intelligence, University of Granada,
     18071 Granada, Spain; rocio@decsai.ugr.es
[3] Computer Science 3—Computer Architecture, Friedrich-Alexander-Universität (FAU) Erlangen-Nürnberg,
     91058 Erlangen, Germany; johnreuben.prabahar@fau.de
[4] Department of Electronics and Computer Technology, University of Granada, 18071 Granada, Spain;
     jmolinos@ugr.es (F.J.-M.); jroldan@ugr.es (J.B.R.)
[5] BTU Cottbus-Senftenberg, 01968 Cottbus, Germany
*   Correspondence: quesada@ihp-microelectronics.com; Tel.: +49-335-5625-369

Abstract: In this work, three different RRAM compact models implemented in Verilog-A are analyzed and evaluated in order to reproduce the multilevel approach based on the switching capability of experimental devices. These models are integrated in 1T-1R cells to control their analog behavior by means of the compliance current imposed by the NMOS select transistor. Four different resistance levels are simulated and assessed with experimental verification to account for their multilevel capability. Further, an Artificial Neural Network study is carried out to evaluate in a real scenario the viability of the multilevel approach under study.

Keywords: RRAM; 1T-1R; multilevel; compact modeling; Verilog-A; artificial neural network

Citation: Pérez-Bosch Quesada, E.;
Romero-Zaliz, R.; Perez, E.;
Kalishettyhalli Mahadevaiah, M.;
Reuben, J.; Schubert, M.A.,
Jimenez-Molinos, F., Roldan, J.B.;
Wenger, C. Toward Reliable Compact
Modeling of Multilevel 1T-1R RRAM
Devices for Neuromorphic Systems.
Electronics **2021**, *10*, 645. https://
doi.org/10.3390/electronics10060645

Academic Editors: Chun Sing Lai,
Zhekang Dong and Donglian Qi

Received: 23 February 2021
Accepted: 7 March 2021
Published: 11 March 2021

1. Introduction

Considering the successful development of software-implemented Artificial Neural Networks (ANN) and their increasing integration in the commercial market, it is of special interest to consider the subsequent drawbacks that the performance of these kind of neuro-inspired networks entails. The gap between the off-chip memory units and the processing units is a clear example of one of the main shortcomings that the von Neuman architectures present. Owing to this fact, for instance, the downscaling of the electronic devices and the implementation of learning algorithms for mobile applications could be compromised. Among others, the reduction of the overall energy consumption and computation time within these networks are key aspects that the artificial intelligence research community has had in its scope for several years now [1]. As an alternative, the hardware-based neuromorphic networks have shown to be the precursor elements to step up onto a new stage in the integration of biological-inspired systems. Particularly, the RRAM technology has been taken into consideration, not only because of its integration as CMOS-compatible non-volatile memory (NVM) arrays but also because of its synaptic-like analog properties [2–7].

The implementation of RRAM cells as synaptic unions between artificial neurons allows the possibility to design and carry out ANNs featured by low-power consumption and low integration area [8]. This is possible due to the multilevel approach also known as Multi-Level Cell (MLC) behavior, consisting of modulating in multiple states the resistance/conductance of the dielectric layer in the RRAM cell. To obtain the mentioned MLC behavior, different programming techniques have been reported lately, such as gradual Reset process by consecutive identical pulses [9], applying positive and negative voltage sweeps with different stop values during the Set and Reset operations [10–12], modifying the compliance current imposed to the cell during the Set transition [13–15] and implementing multilevel incremental step pulses with verify algorithm (M-ISPVA) [16,17], among others.

Nevertheless, the existence of a gap between the device and circuit/system levels challenges the implementation of hardware-based ANN, thus the development of accurate and time-efficient RRAM models for circuit design simulations is an issue that must be tackled.

Several RRAM compact models have been developed and reviewed throughout the history of the memristor [18–23] also in the MLC approach [24]. Regarding the switching behavior of the RRAM devices, different phenomena and equations govern the implementation of each reported model and therefore, they account for different computational cost, yield and accuracy.

This work is focused on modeling the MLC by using the change in the compliance current in order to switch the RRAM device between various low resistance states (LRS) and a single high resistance state (HRS). This is accomplished by connecting a NMOS select transistor in series with the memristor, constituting the so called one-transistor-one-resistor (1T-1R) structure [17]. Furthermore, three different compact models implemented in Verilog-A are integrated in the mentioned 1T-1R structure to give insight into their viability in the multilevel approach. First of all, the Stanford-PKU model [25] including the modification provided by Reuben et al. for multilevel operation [26] is taken into account. Based on similar physical aspects, the Valence Change Memory model with Cylindrical shaped Filament (UGR-VCMCF) and Valence Change Memory model with Truncated-Cone shaped Filament (UGR-VCMTCF) developed by Gonzalez-Cordero et al. [27] are analyzed as well. For other modeling approaches, a revision paper was published recently [28], where the different memristor models were compared and described in depth. The experimental verification is driven by the RRAM devices fabricated using the 130 nm CMOS technology of IHP.

The abovementioned compact models are briefly detailed in Section 2, while the experimental samples characteristics are described in Section 3. The experimental verification and the subsequent modeling results are presented and discussed within Section 4. Later, an ANN study to assess the presented multilevel approach is described in Section 5. Finally, a set of conclusions are drawn in Section 6.

2. Compact Models Description

In this section, the implementation of three compact models extracted from [25–27] is presented. As it is indicated below, the behavior of the mentioned compact models is based on the increase and decrease of the gap distance between the tip of the conductive filament (CF) and the bottom electrode (BE). Although these compact models are based on similar physical phenomena, they consider different CF geometries, as it can be appreciated in Figure 1 and, thus, different results concerning the MLC behavior can be obtained.

2.1. Stanford-PKU Model Extended with Multilevel Capability

This model is based on the growth and disruption of a CF when an appropriate electric field is generated in the dielectric layer of the RRAM cell. This phenomenon is described by the increase or decrease of the gap distance between the CF tip and the bottom electrode within the dielectric layer (see Figure 1a) when a Reset or a Set operation take place, respectively. The gap evolution is modeled as follows:

$$\frac{dg}{dt} = -v_0 e^{\frac{E_a}{k_b T}} sinh\left(\frac{\gamma a_0 q V}{t_{ox} k_b T}\right), \tag{1}$$

where g stands for the gap distance between the CF and the bottom electrode, v_0 is a fitting parameter that accounts for the velocity dependent on the attempt-to-escape frequency, E_a is the effective activation energy for vacancy generation, k_b is the Boltzmann constant, a_0 is the atom spacing, q is the electron charge, V is the voltage applied to the device, t_{ox} is the dielectric thickness and T is the device temperature, which is computed as follows:

$$T = T_0 + V \cdot I \cdot R_{th},\qquad(2)$$

where T_0 is the room temperature, I is the current through the device and R_{th} accounts for its thermal resistance.

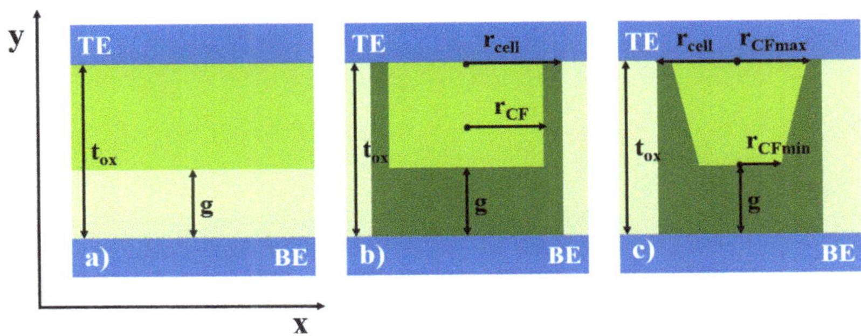

Figure 1. Three-dimensional geometrical representation of (**a**) the Stanford-PKU model, (**b**) the UGR-VCMCF model and (c) the UGR-VCMTCF.

The variable γ is the field local enhancement factor that accounts for the material polarizability [29], which depends on the current gap distance g. The latest is bounded between g_{min} and g_{max}, which are the minimum and maximum possible values that this variable can have.

Finally, the total current flowing through the cell is computed as

$$I = I_0 \cdot e^{\frac{-g}{g_0}} \cdot sinh\left(\frac{V}{V_0}\right),\qquad(3)$$

where I_0, g_0 and V_0 are fitting coefficients.

Considering the standard Stanford-PKU model [25], we observed that the MLC behavior can be accomplished by imposing different compliance currents through the cell during the Set operation, allowing the CF to grow accordingly to the electric field within the insulator layer of the MIM stack. Thus, the gap boundaries are constant during the whole simulation time. However, regarding the proposed modification by Reuben et al. [26], the minimum gap distance (g_{min}) turns into a variable. Owing to the fact that this model modification was implemented to be integrated within a 1T-1R structure, this variable depends on the voltage applied to the gate terminal of the transistor (V_{gate}). In this way, the NMOS transistor is assumed to be operating in the triode region and thus, it behaves as a linear resistor. Taking into account these statements, g_{min} is computed as follows:

$$g_{min} = K_{th} \cdot \frac{(W/L)}{V_{gate}} + C,\qquad(4)$$

where K_{th} and C are fitting constants for a particular 1T-1R cell, which are computed taking into account the experimental multilevel measurements. W/L accounts for the aspect ratio of the NMOS transistor. It has to be considered that an extra input terminal is added to this model modification to feed the gate voltage of the NMOS transistor into its algorithm, as it can be appreciated in (4).

Thereof, in this modeling approach the minimum gap distance variable determines the LRS level of the cell depending on V_{gate}. That is to say, the multilevel behavior of the RRAM model is fully controlled by limiting the CF growth and thus, the minimum gap distance between the filament and the bottom electrode.

2.2. Valence Change Memory Model with Cylindrical Shaped Filament (UGR-VCMCF)

This model accounts for the same physical phenomena presented in the standard Stanford-PKU model, in which the gap distance between the CF and the opposite electrode determines the current flow through the cell as indicated in (3). However, the CF is modeled following a cylindrical geometry (see Figure 1b) and thus, its ohmic and thermal properties are conditioned to the mentioned structure.

A more accurate thermal description is performed by using a unidimensional version of the heat equation. In particular, for a cylindrical CF, the maximum temperature along its y axis can be solved as indicated in [30]:

$$T = T_0 + \frac{\sigma_{eq} \cdot \zeta^2 \cdot r_{CF} \cdot (e^\alpha - 1)^2}{2 \cdot h \cdot (e^{2 \cdot \alpha} + 1)},\tag{5}$$

where σ_{eq} stands for the CF conductivity, computed considering the CF radius r_{CF} and the cell radius r_{cell} (see Figure 1b). ζ is the average electric field in the CF and h represents the heat transfer coefficient that accounts for the lateral heat dissipation from the CF to the dielectric. Owing to the fact that the shape of the CF is assumed to be cylindrical in this model, r_{CF} is constant. This solution is assumed as the CF temperature T used to compute the gap evolution in (1). During the Reset operation, E_a is substituted by the parameter E_m (migration energy). The parameter α is calculated as follows:

$$\alpha = \frac{t_{ox}}{2} \sqrt{\frac{2 \cdot h}{k_{th} \cdot r_{CF}}},\tag{6}$$

where k_{th} accounts for the thermal conductivity of the CF.

Concerning the multilevel approach, unlike the modification of the Stanford-PKU model proposed by Reuben et al. presented in Section 2.1, the variation of the minimum gap distance that allows the existence of multiple LRS levels is obtained following the standard Stanford-PKU behavior, i.e., without imposing variable limits to the filament growth. Thus, the increment and decrease of the gap distance between the CF and the bottom electrode for the different conductive levels is fully accomplished by means of the compliance current imposed by the transistor integrated in the 1T-1R cell.

2.3. Valence Change Memory Model with Truncated-Cone Shaped Filament (UGR-VCMTCF)

This model considers the CF as a truncated cone and therefore, its radius is not constant along its y axis (see Figure 1c). It is supposed that a truncated-cone shaped CF with constant conductivity is analytically analogous to a cylindrical CF with a variable conductivity along its main axis, whose equivalent radius is computed as follows:

$$r_{CFg} = \sqrt{r_{CFmax} \cdot r_{CFmin}},\tag{7}$$

where r_{CFmax} and r_{CFmin} stand for the maximum and minimum radii of the truncated-cone CF.

Taking into consideration the previous assumptions, the maximum temperature obtained when solving the heat equation follows the structure indicated in (5) and (6). However, the new axis-dependent conductivity σ_{eq} and the equivalent cylindrical radius r_{CFg} have to be considered, allowing the electric field through the CF to be dependent on the y axis.

In a similar way, (1) and (3) can be solved once the new geometry of the CF is considered and so its equivalent ohmic and thermal properties.

It also has to be considered that this model enables the possibility to configure two different values for each of the fitting parameters I_0, V_0 and g_0, considering the positive or negative sense of the current through the cell. This eases the possibility to determine the behavior of the model for positive and negative voltage values individually.

Analogously to the UGR-VCMCF model, the UGR-VCMTCF model achieves the MLC behavior in the same way as the standard Stanford-PKU model does, that is to say, controlling the gap distance between the tip of the CF and the bottom electrode by means of the compliance current imposed by the NMOS transistor of the 1T-1R structure.

3. Experimental Samples Characteristics

The electrical measurements were performed on single 1T-1R RRAM structures. These structures consists of a NMOS transistor connected in series to a Metal-Insulator-Metal (MIM) cell as shown in Figure 2. The NMOS transistor acts as a current limiter. The devices are fabricated using the 130 nm technology of IHP. The MIM stack consists of TiN/Al:HfO$_2$/Ti/TiN with TiN top and bottom electrodes of 150 nm and Ti layer of 7 nm deposited by sputtering and an Al doped (about 10%) HfO$_2$ layer of 6 nm deposited by Atomic Layer Deposition (ALD). The MIM cells are fabricated with an area of about 0.4 µm^2. Additionally, all the cells are encapsulated with a SiNO layer.

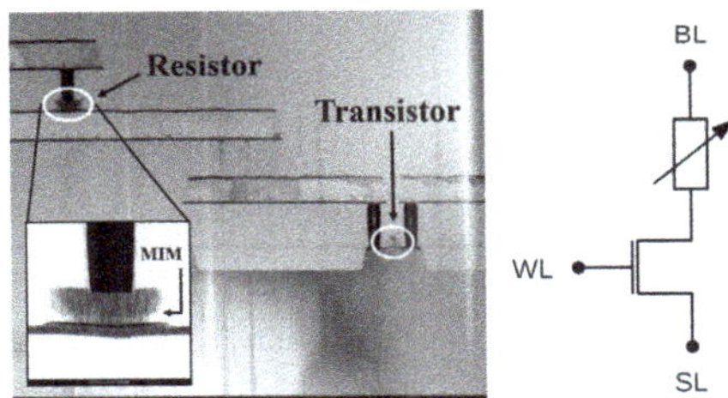

Figure 2. Cross-sectional TEM image of the 1T-1R structure and of the MIM stack in more detail (**left**). Schematic of the 1T-1R cell (**right**).

4. Modeling Results and Discussion

In order to validate the capability of the three described models to reproduce the MLC by changing the compliance current, we obtained a set of experimental measurements using a single 1T-1R structure detailed in Section 3. From the modeling point of view, in order to adapt the model parameters to the experimental measurements we are focused on the DC response of the devices with respect to the mentioned MLC behavior. This behavior is featured by the transition from a single HRS to three different LRSs and vice versa by modifying the compliance current imposed by the NMOS transistor during the Set operation.

The experiments were carried out by applying positive voltage sweeps between 0 V and 1.5 V with steps of 0.05 V to the top electrode (TE) of the 1T-1R cell to perform the Set operations and negative voltage sweeps between 0 V and −1.5 V with steps of −0.05 V to perform the Reset operations. The voltage sweep rate used was 0.6 V/s. The compliance current imposed to the cell was controlled by setting the gate voltage of the NMOS transistor to different values. Thereof, four conductive levels were accomplished: three LRSs corresponding to V_{gate} = 1 V, 1.2 V, 1.6 V, respectively, and one HRS in which V_{gate} = 2.7 V. We chose a gate voltage of 2.7 V to reduce as much as possible the resistance of the transistor during the Reset operation, easing the transition from the LRSs to the sole HRS. In total, 10 Set–Reset cycles were accomplished for each LRS level within the same sample.

The observed experimental results present typical RRAM behaviors such as abrupt Set operations at V_{Set} and a gradual Reset transition whose V_{Reset} value depends on the accomplished LRS level, i.e., the more conductive the LRS is, the higher the voltage amplitude required to Reset the device. Table 1 gathers the most remarkable electrical characteristics observed for each of the possible transitions between states.

Table 1. Electrical characteristics for the different resistance levels.

	V_{gate} (V)	V_{Set} (V)	V_{Reset} (V)	Resistance (kΩ) [a]
LRS1	1	0.65	0.6	16
LRS2	1.2	0.65	0.7	11
LRS3	1.6	0.75	0.9	8
HRS	2.7	-	-	170

[a] The resistance values were measured using a read voltage of 0.2 V.

Concerning the simulated results, the NMOS transistor model was provided by [31] and every RRAM compact model presented within this work is tuned in to fit the median DC characteristics of the experimental measurements. All the simulations were ran in Cadence Virtuoso Analog Design Environment (ADE) [32]. The aim of the fitting procedure is to provide with one set of parameters to each model to reproduce the MLC behavior indicated above. For this purpose, different steps were accomplished based on the methodology indicated in [25–27]:

First, one of the LRS levels is chosen to be the reference for the fitting procedure. In this case, the median values for LRS2 are taken as a reference to establish the initial set of parameters since it is the medium level of the LRSs under study. This first step can be accomplished following the methodology proposed in [25,26]. If the reference LRS is not the most conductive one, the minimum gap distance achieved during this first step cannot reach the atomic distance a_0 (0.25 nm). Otherwise, the multilevel approach cannot be accomplished for more conductive levels. Once the model is able to reproduce the reference LRS (in our case LRS2) and the single HRS, the rest of LRS levels are evaluated by modifying the gate voltage of the transistor during the Set operation and thus, the compliance current imposed to the cell. Further adjustments might be applied to the previous set of parameters in order to match the rest of the median curves for the different resistance levels. The fitting parameters I_0 and g_0 have special influence in the conductance difference between levels (see (3)). It is mandatory to achieve different minimum gap distances for every single LRS level to reproduce the MLC behavior due to the fact that the shorter the gap distance, the more conductive the resistance state is. Table 2 exposes the achieved gap distances for each conductive level for the three models.

Table 2. Gap distances for simulating the different resistance levels.

	Gap Distance (nm)		
	S-PKU	**UGR-VCMCF**	**UGR-VCMTCF**
LRS1	0.95	1	0.86
LRS2	0.85	0.86	0.65
LRS3	0.73	0.75	0.25
HRS	1.88	1.88	1.88

Figure 3 shows the comparison between the experimental and the simulated results extracted from the different models for each of the LRSs considered within this work. More precisely, Figure 3a–d make reference to the Stanford-PKU model extended with multilevel capability configured with the set of parameters exposed in Table 3. Figure 3e–h account for the UGR-VCMCF model fed with the parameters exposed in Table 4. Figure 3i–l make reference to the UGR-VCMTCF model configured as indicated in Table 5. All the simulations were carried out setting the HRS as the initial state ($gap_{ini} = gap_{max}$) thus, a Set operation is initially performed followed by a Reset of the cell.

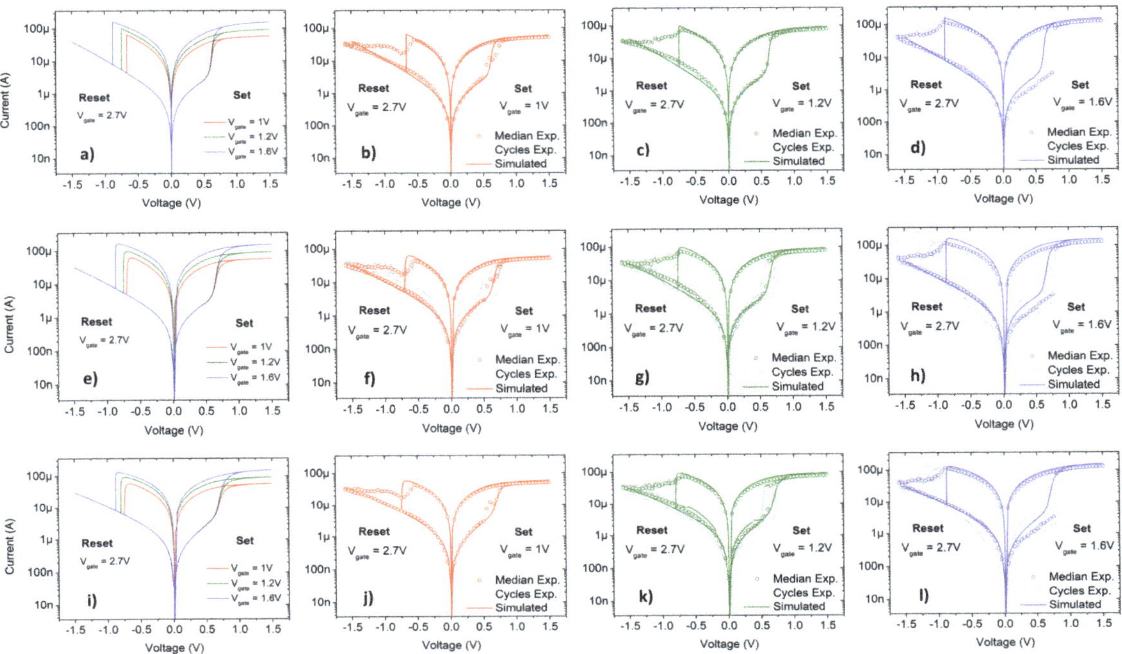

Figure 3. Comparison between the median values (doted line) of the experimental cycles (grey line) and the simulated results (straight line) concerning the Stanford-PKU model extended with multilevel capability (**a–d**), the UGR-VCMCF model (**e–h**) and the UGR-VCMTCF model (**i–l**) for the LRS1 (red), LRS2 (green) and LRS3 (blue).

Table 3. Fitting parameters for the Stanford-PKU modified model.

$g_0 = 0.28$ nm	$V_0 = 0.35$ V	$I_0 = 854$ μA
$v_0 = 0.4$ m/s	$\beta = 0.4$	$\alpha = 3$
$gap_{ini} = 1.8$ nm	$T_0 = 300$ K	$\gamma_0 = 20$
$gap_{max} = 1.8$ nm	$t_{ox} = 6$ nm	$K_{th} = 0.52$ nm·V
$E_a = 0.6$ eV	$R_{th} = 1500$ K/W	$C = 0.35$ nm

Table 4. Fitting parameters for the UGR-VCMCF model.

$g_0 = 0.275$ nm	$V_0 = 0.4$ V	$I_0 = 1.7$ mA
$v_0 = 0.8$ m/s	$\beta = 1$	$\alpha = 3$
$\gamma_0 = 18$	$gap_{ini} = 1.8$ nm	$gap_{min} = 0.25$ nm
$gap_{max} = 1.8$ nm	$t_{ox} = 6$ nm	$T_0 = 300$ K
$E_a = 0.65$ eV	$E_m = 0.65$ eV	$r_{CF} = 5$ nm
$h = 0.01\ \frac{K}{W \cdot \mu m^2}$	$k_{th} = 10\ \frac{K}{W \cdot m}$	$\sigma_{CF0} = 500$ kS/m

In general terms, despite the limited complexity of the models, the experimental results are reproduced in an acceptable way. The gap distances exposed in Table 2 match the required current levels for both, the LRSs and the sole HRS showing that this parameter is the key aspect to be considered in order to reproduce the MLC by changing the compliance current. Regarding the knee point of the Reset transition, it can be correctly simulated in all the conductive states except for LRS1 (see Figure 3b,f,j), where the maximum voltage difference for the simulated and experimental V_{Reset} is up to 0.15 V. Reducing I_0 helps to minimize the mentioned voltage difference, keeping the knee point of the Reset transition in LRS2 (reference state). However, due to the linear dependence of I with respect this parameter (see (3)), this would lead to a reduction of the V_{Reset} in LRS3 as well, enlarging the voltage difference between the simulated and the experimental results in this concrete point.

Table 5. Fitting parameters for the UGR-VCMTCF model.

$g_{0p} = 0.25$ nm	$g_{0n} = 0.28$ nm	$V_{0p} = 0.26$ V
$V_{0n} = 0.4$ V	$I_{0n} = 1.7$ mA	$I_{0p} = 1.7$ mA
$v_0 = 0.8$ m/s	$\beta = 1$	$\alpha = 3$
$\gamma_0 = 18$	$gap_{ini} = 1.8$ nm	$gap_{min} = 0.25$ nm
$gap_{max} = 1.8$ nm	$t_{ox} = 6$ nm	$T_0 = 300$ K
$E_a = 0.65$ eV	$E_m = 0.65$ eV	$r_{CFmax} = 5$ nm
$r_{CFmin} = 1$ nm	$\sigma_{CF0} = 500$ kS/m	$\sigma_{ox} = 1.65$ S/m

Due to the cycle-to-cycle variability of the experimental data measured in LRS3 (see the grey lines in Figure 3d,h,l), the HRS curve for positive voltage values presents lower median current levels as well as a higher V_{Set} with respect to LRS1 and LRS2. Thus, even though the simulated results fulfill the ideal behavior of LRS3, they differ from the experimental results in both mentioned aspects.

Lastly, since the memristor charge conduction is filamentary, device-to-device (D-D) variability, usually linked to technological differences in the fabrication process, is not studied. It is the cycle-to-cycle (C-C) variability during programming the factor that was considered in the experimental measurements. Modeling approaches linked to time series have been given in the literature [33,34] that can be implemented. Other modeling schemes linked to Gaussian distribution functions and Monte Carlo simulations within circuit simulators are also possible (in this case for D-D and C-C variability). The device-to-device variability that shows up during programming could be also modeled along the line described in [35], as well as the cycle-to-cycle variability. From the memory characterization point of view, retention time and endurance properties were previously studied both experimentally [16,36] and by simulation [37] from another perspective in our technology.

5. A Neural Network Study to Assess the Multilevel Approach

As it was mentioned in Section 1, the implementation of RRAM cells as synaptic unions within hardware-based ANNs is gaining momentum [2–6,8]. Regarding the MLC behavior taken into account in this work, it is convenient to study the performance of ANNs in this context. That implies the reduction of the precision of the synaptic weights from floating point numbers (used in conventional software-based ANNs) to a limited number of values, that is, the resistance levels defined in RRAM devices as shown in Section 4.

First, we used the SciKit-Learn [38] software tool to implement and train a Multilayer Perceptron (MLP) in order to recognize and classify handwritten digits. The selected dataset for the experiment is the well-known MNIST image dataset [39]. It is composed of 28×28 pixel images of 70,000 handwritten digits (labeled in the interval $[0, 9]$), divided into a training set of size 60,000 digits and a test set of size 10,000 digits. This classifier optimizes the log-loss function using stochastic gradient descent (with the back-propagation algorithm) and rectified linear unit functions (e.g., $f(x) = x^+ = max(0, x)$) [40]. The training process was performed during 500 epochs (cycles through the full training dataset) or until the loss (or score, a prediction error of the ANN) improves by a factor of 1×10^{-4}.

We have considered different quantization strategies making use of 2, 4 and 8 levels in order to explore the accuracy of the ANN inference. In this respect, we have employed a quantization approach in line with the multilevel implementation described above for our RRAM devices (4 levels), although we also included other multilevel possibilities for the sake of completeness (2 and 8 levels). The options considered here were: Uniform-ASYMM and Uniform-SYMM, as suggested in [41]. Both methods are linear, range-based and quantify a given input data x_f into x_q using n bits. For instance, for $n = 2$ they transform the original data into $2^n = 4$ levels, which is the case we have considered at the device resistance level in Section 4. Uniform-ASYMM (see (8)) is an asymmetric approach that maps the minimum and maximum of the float range to an integer range with a quantization bias term. Uniform-SYMM (see (9)) is a symmetric approach that maps the original data to the quantized range with the maximum absolute value of the minimum or maximum of the original data. In this latter case, there is no quantization bias term and it is symmetric around zero.

$$x_q = round\left((x_f - min_{x_f}) \frac{2^n - 1}{max_{x_f} - min_{x_f}} \right) \tag{8}$$

$$x_q = round\left(x_f \frac{2^{n-1} - 1}{max|x_f|} \right) \tag{9}$$

The architecture of the ANN adapted to the MNIST dataset is depicted in Figure 4. In order to simulate the usage of our ANN in a RRAM-based environment, all the synaptic weights were quantized by using the functions (8) and (9) and the number of levels (four) determined in our study, after the training process and before the final prediction on the performed tests.

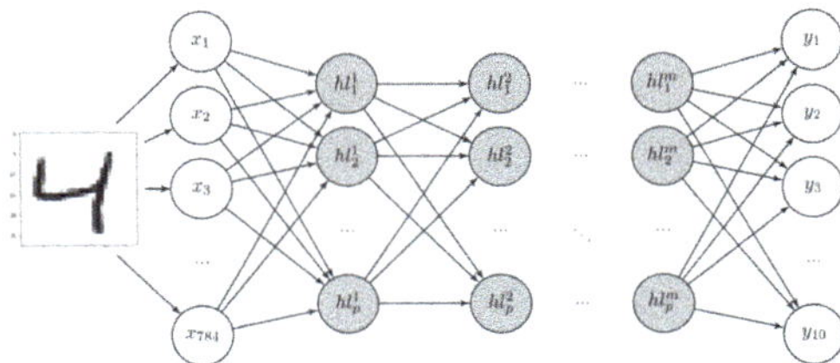

Figure 4. Artificial neural network architecture for the MNIST dataset classification. The input layer consists of 784 nodes (x_i), one for each of the 28×28 pixels of the input. The output layer consists of 10 nodes (y_i), one for each class label. There can be several m hidden layers (hl_i^j) with the same number of perceptrons units (p). The ANN is fully connected.

The resistance levels of the memristors are connected to the use of Equations (8) and (9) for the quantization of the ANN floating point weights obtained after training. In this respect, taking into consideration that weights can be positive and negative, the implementation could be performed using two memristive devices per synapsis as demonstrated in a previous work [42].

In summary, taking into account the mentioned quantization schemes, the inference was carried out with four synaptic levels (2 bits). In total, 1, 2 and up to 3 hidden layers and 32, 64, 128, 256 and 512 perceptrons (neurons) included within each hidden layer are considered as well.

The balanced accuracies (i.e., the number of true positives divided by the total number of elements that actually belong to a class digit in our case) obtained by using the described set of parameters are shown in Table 6. The Uniform SYMM works better than the Uniform ASYMM when using four levels of quantization by an average difference of 0.13, for all the number of hidden layers and perceptrons per layer. Nevertheless, the accuracy obtained by using the Uniform SYMM scheme with 4 levels is lower than a conventional MLP with no quantized synaptic weights by an average difference of 0.08.

Finally, in order to assess the impact of the number of conductive levels per device on the performance of the ANN, two additional number of levels, namely, 2 (1 bit) and 8 (3 bits), were tested. Considering the implementation of RRAM devices as synaptic unions between artificial neurons, it can be assumed that these numbers of levels correspond to the number of resistance levels achieved by the basic digital behavior of the RRAM devices (two levels) and by the MLC behavior with eight resistance levels. Table 6 shows the balanced accuracies obtained with the three mentioned number of levels.

Table 6. Balanced accuracies obtained on each class on the test set for the different parameters set.

Hidden Layers [a]	Levels	No Quantization	U-SYMM	U-ASYMM
1 (32)	2	0.86	0.13	0.21
	4		0.71	0.33
	8		0.88	0.88
2 (32)	2	0.80	0.09	0.33
	4		0.79	0.49
	8		0.91	0.87
3 (32)	2	0.94	0.15	0.21
	4		0.65	0.50
	8		0.84	0.69
1 (64)	2	0.93	0.17	0.45
	4		0.82	0.56
	8		0.93	0.93
2 (64)	2	0.92	0.15	0.48
	4		0.79	0.39
	8		0.95	0.89
3 (64)	2	0.94	0.23	0.38
	4		0.74	0.51
	8		0.92	0.87
1 (128)	2	0.94	0.08	0.25
	4		0.93	0.85
	8		0.97	0.96
2 (128)	2	0.94	0.18	0.26
	4		0.92	0.45
	8		0.96	0.95
3 (128)	2	0.95	0.11	0.51
	4		0.83	0.53
	8		0.97	0.86
1 (256)	2	0.95	0.08	0.19
	4		0.91	0.81
	8		0.97	0.96
2 (256)	2	0.95	0.11	0.84
	4		0.90	0.74
	8		0.98	0.96
3 (256)	2	0.96	0.17	0.75
	4		0.91	0.77
	8		0.97	0.95
1 (512)	2	0.96	0.11	0.41
	4		0.93	0.89
	8		0.97	0.97
2 (512)	2	0.97	0.10	0.77
	4		0.95	0.70
	8		0.98	0.97
3 (512)	2	0.97	0.23	0.66
	4		0.96	0.77
	8		0.98	0.97

[a] The number of perceptrons per hidden layer is indicated in parentheses.

The Uniform ASYMM scheme works better than the Uniform SYMM scheme only when the number of quantizied levels is 2 by an average difference of 0.28 in the balanced accuracy. However, it seems that the Uniform ASYMM improves slower than its counterpart with the increase of the number of levels.

Both quantization methods achieve almost the same results as the original ANN (without quantization) when the number of levels increases, as it is also observed in [41]. More precisely, when the number of levels is 8, the quantized version of the ANN may obtain better results than the conventional ANN. This is due to the fact that the ANN without quantization is overfitting the training data, a common consequence observed in ANN architectures which report small training error [43]. On the contrary, due to the downgrade of the weight precision in the quantized versions, the overfitting issue is mitigated during the training period and the balanced accuracies are slightly enhanced with respect to the conventional ANN. This highlights the advantages of increasing the number of levels from 4 to 8 in the MLC behavior of the RRAM cells.

An interesting observation is the fact that by using more perceptrons in each hidden layer the balanced accuracy tends to improve in all cases, while the number of hidden layers seems to mitigate the low number of perceptrons per layer (e.g., 32 perceptrons per layer).

While investigating the coefficient (weight) distribution in each level, we discovered that not all the levels were used for quantization in the Uniform SYMM quantization scheme (data not shown). This feature could be further studied to select a lower number of levels, not in the form of 2^n, which could be useful for tryouts.

Concerning device variability, further research in the ANN context indicates that it does not affect the outcome of ANNs in the same way as in other hardware systems [44]. Due to the particular features of neural networks, in some cases, variability in the synaptic weights produces better accuracy than the ideally quantized ANN without variability. This behavior is caused by the random nature of the variability introduced, which, in some cases may shift the values of the synaptic weights closer to the optimal ANN without quantization (data not shown). Furthermore, the variability introduced in the synaptic weights may cope with a small amount of overfitting. Finally, variability do not prevent the technology from being used in a multilevel scheme in different Artificial Intelligence hardware accelerators since some prototypes have been fabricated based on these type of devices [45,46].

6. Conclusions

In this paper, three physics-based compact models for RRAMs were studied and verified with experimental data to validate their capability to simulate the multilevel behavior of the 1T-1R cells. Each of the compact models were tuned following the proposed methodology to accomplish four resistance states (2 bits) making use of a single set of parameters per model. We found out that the three models, based on similar physical phenomena, still have margin for improvement concerning the multilevel behavior of RRAM cells. However, the UGR-VCMTCF and the Stanford-PKU model extended with multilevel capability present more accurate results in terms of multilevel performance. Both the UGR-VCMCF and the UGR-VCMTCF models reproduce the MLC behavior entirely by modifying the compliance current imposed by the transistor. On the other hand, the modification of the Stanford-PKU model makes use of an extra input terminal to limit the growth of the CF. Nevertheless, it requires less parameters to reproduce the RRAM behavior. Later, a brief ANN study was performed in order to assess the MLC behavior presented in this work, showing that even with just 4 levels of quantization, the performance in classifying the MNIST database is relatively good. Finally, we also demonstrated that the next step in the accomplishment of a higher number of conductance levels in the MLC behavior can be very beneficial for the implementation of reliable ANN based on RRAM cells.

Author Contributions: Conceptualization, C.W., J.B.R. and F.J.-M.; methodology, E.P.-B.Q., J.B.R., R.R.-Z.; physical analysis of the samples, M.A.S.; writing—original draft preparation, E.P.-B.Q. and R.R.-Z.; writing—review and editing, E.P.-B.Q., R.R.-Z., E.P., M.K.M., J.R., F.J.-M., J.B.R. and C.W.; supervision, C.W., E.P., F.J.-M. and J.B.R. All authors have read and agreed to the published version of the manuscript.

Funding: The authors would like to thank the financial support by Deutsche Forschungsgemeinschaft (German Research Foundation) with Project-ID SFB1461 and by the Federal Ministry of Education and Research of Germany under grant numbers 16ES1002, 16FMD01K, 16FMD02 and 16FMD03. The authors also gratefully acknowledge the support of the Spanish Ministry of Science, Innovation and Universities and the FEDER program through project TEC2017-84321-C4-3-R and project A.TIC.117.UGR18 funded by the government of Andalusia (Spain) and the FEDER program. The publication of this article was funded by the Open Access Fund of the Leibniz Association.

Data Availability Statement: The datasets generated during and/or analysed during the current study are available from the corresponding author on reasonable request.

Conflicts of Interest: The authors declare no conflict of interest regarding the publication of this paper.

References

1. Abiodun, O.I.; Jantan, A.; Omolara, A.E.; Dada, K.V.; Mohamed, N.A.; Arshad, H. State-of-the-art in artificial neural network applications: A survey. *Heliyon* **2018**, *4*, e00938. [CrossRef] [PubMed]
2. Valentian, A.; Rummens, F.; Vianello, E.; Mesquida, T.; de Boissac, C.L.; Bichler, O.; Reita, C. Fully Integrated Spiking Neural Network with Analog Neurons and RRAM Synapses. In Proceedings of the IEEE International Electron Devices Meeting (IEDM), San Francisco, CA, USA, 7–11 December 2019; pp. 14.3.1–14.3.4.
3. Yao, P.; Wu, H.; Gao, B.; Tang, J.; Zhang, Q.; Zhang, W.; Yang, J.J.; Qian, H. Fully hardware-implemented memristor convolutional neural network. *Nature* **2020**, *577*, 641–646. [CrossRef]
4. Prezioso, M.; Merrikh-Bayat, F.; Hoskins, B.; Adam, G.; Likharev, K.; Strukov, D. Training and Operation of an Integrated Neuromorphic Network Based on Metal-Oxide Memristors. *Nature* **2014**, *521*. [CrossRef]
5. Kang, J.F.; Gao, B.; Huang, P.; Liu, L.F.; Liu, X.Y.; Yu, H.Y.; Yu, S.; Wong, H.P. RRAM based synaptic devices for neuromorphic visual systems. In Proceedings of the IEEE International Conference on Digital Signal Processing (DSP), Singapore, 21–24 July 2015; pp. 1219–1222.
6. Zahari, F.; Hansen, M.; Mussenbrock, T.; Ziegler, M.; Kohlstedt, H. Pattern recognition with TiOx-based memristive devices. *AIMS Mater. Sci.* **2015**, *2*, 203–216. [CrossRef]
7. Ginnaram, S.; Qiu, J.T.; Maikap, S. Controlling Cu Migration on Resistive Switching, Artificial Synapse, and Glucose/Saliva Detection by Using an Optimized AlOx Interfacial Layer in a-COx-Based Conductive Bridge Random Access Memory. *ACS Omega* **2020**, *5*, 7032–7043. [CrossRef] [PubMed]
8. Ziegler, M.; Wenger, C.; Chicca, E.; Kohlstedt, H. Tutorial: Concepts for closely mimicking biological learning with memristive devices: Principles to emulate cellular forms of learning. *J. Appl. Phys.* **2018**, *124*, 152003. [CrossRef]
9. Huang, P.; Zhu, D.; Chen, S.; Zhou, Z.; Chen, Z.; Gao, B.; Liu, L.; Liu, X.; Kang, J. Compact Model of HfOx-Based Electronic Synaptic Devices for Neuromorphic Computing. *IEEE Trans. Electron Devices* **2017**, *64*, 614–621. [CrossRef]
10. Maestro-Izquierdo, M.; Gonzalez, M.; Campabadal, F. Mimicking the spike-timing dependent plasticity in HfO2-based memristors at multiple time scales. *Microelectron. Eng.* **2019**, *215*, 111014. [CrossRef]
11. Kim, W.; Menzel, S.; Wouters, D.J.; Waser, R.; Rana, V. 3-Bit Multilevel Switching by Deep Reset Phenomenon in Pt/W/TaOX/Pt-ReRAM Devices. *IEEE Electron Device Lett.* **2016**, *37*, 564–567. [CrossRef]
12. Larentis, S.; Nardi, F.; Balatti, S.; Gilmer, D.C.; Ielmini, D. Resistive Switching by Voltage-Driven Ion Migration in Bipolar RRAM—Part II: Modeling. *IEEE Trans. Electron Devices* **2012**, *59*, 2468–2475. [CrossRef]
13. Sedghi, N.; Li, H.; Brunell, I.; Dawson, K.; Potter, R.; Guo, Y.; Gibbon, J.; Dhanak, V.; Zhang, W.D.; Zhang, J.; et al. The role of nitrogen doping in ALD Ta2O5 and its influence on multilevel cell switching in RRAM. *Appl. Phys. Lett.* **2017**, *110*, 102902. [CrossRef]
14. Misha, S.H.; Tamanna, N.; Woo, J.; Lee, S.; Song, J.; Park, J.; Lim, S.; Park, J.; Hwang, H. Effect of nitrogen doping on variability of TaOx-RRAM for low-power 3-Bit MLC applications. *ECS Solid State Lett.* **2015**, *4*, 25–28. [CrossRef]
15. Prakash, A.; Deleruyelle, D.; Song, J.; Bocquet, M.; Hwang, H. Resistance controllability and variability improvement in a TaOx-based resistive memory for multilevel storage application. *Appl. Phys. Lett.* **2015**, *106*, 233104. [CrossRef]
16. Pérez, E.; Zambelli, C.; Mahadevaiah, M.K.; Olivo, P.; Wenger, C. Toward Reliable Multi-Level Operation in RRAM Arrays: Improving Post-Algorithm Stability and Assessing Endurance/Data Retention. *IEEE J. Electron Devices Soc.* **2019**, *7*, 740–747. [CrossRef]
17. Milo, V.; Zambelli, C.; Olivo, P.; Pérez, E.; Mahadevaiah, M.K.; Ossorio, O.G.; Wenger, C.; Ielmini, D. Multilevel HfO2-based RRAM devices for low-power neuromorphic networks. *APL Mater.* **2019**, *7*, 081120. [CrossRef]
18. Hajri, B.; Aziza, H.; Mansour, M.M.; Chehab, A. RRAM Device Models: A Comparative Analysis With Experimental Validation. *IEEE Access* **2019**, *7*, 168963–168980. [CrossRef]
19. Kuzum, D.; Yu, S.; Wong, H.P. Synaptic electronics: Materials, devices and applications. *Nanotechnology* **2013**, *24*, 382001. [CrossRef] [PubMed]
20. Lekshmi Jagath, A.; Hock Leong, C.; Kumar, T.N.; Almurib, H.F. Insight into physics-based RRAM models—Review. *J. Eng.* **2019**, *2019*, 4644–4652. [CrossRef]
21. Ielmini, D.; Milo, V. Physics-based modeling approaches of resistive switching devices for memory and in-memory computing applications. *J. Comput. Electron.* **2017**, *16*, 1121–1143. [CrossRef]
22. Linn, E.; Siemon, A.; Waser, R.; Menzel, S. Applicability of Well-Established Memristive Models for Simulations of Resistive Switching Devices. *IEEE Trans. Circuits Syst. Regul. Pap.* **2014**, *61*, 2402–2410. [CrossRef]
23. Menzel, S.; Siemon, A.; Ascoli, A.; Tetzlaff, R. Requirements and Challenges for Modelling Redox-based Memristive Devices. In Proceedings of the IEEE International Symposium on Circuits and Systems (ISCAS), Florence, Italy, 27–30 May 2018; pp. 1–5.
24. Li, H.; Jiang, Z.; Huang, P.; Wu, Y.; Chen, H.; Gao, B.; Liu, X.Y.; Kang, J.F.; Wong, H.P. Variation-aware, reliability-emphasized design and optimization of RRAM using SPICE model. In Proceedings of the Design, Automation Test in Europe Conference Exhibition (DATE), Grenoble, France, 9–13 March 2015; pp. 1425–1430. [CrossRef]

25. Jiang, Z.; Wu, Y.; Yu, S.; Yang, L.; Song, K.; Karim, Z.; Wong, H.P. A Compact Model for Metal–Oxide Resistive Random Access Memory With Experiment Verification. *IEEE Trans. Electron Devices* **2016**, *63*, 1884–1892. [CrossRef]
26. Reuben, J.; Fey, D.; Wenger, C. A Modeling Methodology for Resistive RAM Based on Stanford-PKU Model With Extended Multilevel Capability. *IEEE Trans. Nanotechnol.* **2019**, *18*, 647–656. [CrossRef]
27. González-Cordero, G.; Roldán, J.B.; Jiménez-Molinos, F. Simulation of RRAM memory circuits, a Verilog-A compact modeling approach. In Proceedings of the Conference on Design of Circuits and Integrated Systems (DCIS), Granada, Spain, 23–25 November 2016; pp. 1–6. [CrossRef]
28. Panda, D.; Sahu, P.P.; Tseng, T.Y. A collective study on modeling and simulation of resistive random access memory. *Nanoscale Res. Lett.* **2018**, *13*, 1–48. [CrossRef] [PubMed]
29. McPherson, J.; Kim, J.; Shanware, A.; Mogul, H. Thermochemical description of dielectric breakdown in high dielectric constant materials. *Appl. Phys. Lett.* **2003**, *82*, 2121–2123. [CrossRef]
30. González-Cordero, G.; González, M.; García, H.; Campabadal, F.; Dueñas, S.; Castán, H.; Jiménez-Molinos, F.; Roldán, J. A physically based model for resistive memories including a detailed temperature and variability description. *Microelectron. Eng.* **2017**, *178*, 26–29. [CrossRef]
31. AdMOS: Advanced Modeling Solutions. Available online: https://admos.de/en/home-en/ (accessed on 21 September 2020).
32. Virtuoso Analog Design Environment. Available online: https://www.cadence.com/ko_KR/home.html (accessed on 21 September 2020).
33. Roldán, J.B.; Alonso, F.J.; Aguilera, A.M.; Maldonado, D.; Lanza, M. Time series statistical analysis: A powerful tool to evaluate the variability of resistive switching memories. *J. Appl. Phys.* **2019**, *125*, 174504. [CrossRef]
34. Miranda, E.; Mehonic, A.; Ng, W.H.; Kenyon, A.J. Simulation of Cycle-to-Cycle Instabilities in SiO_x-Based ReRAM Devices Using a Self-Correlated Process With Long-Term Variation. *IEEE Electron Device Lett.* **2019**, *40*, 28–31. [CrossRef]
35. Pérez, E.; Maldonado, D.; Acal, C.; Ruiz-Castro, J.; Alonso, F.; Aguilera, A.; Jiménez-Molinos, F.; Wenger, C.; Roldán, J. Analysis of the statistics of device-to-device and cycle-to-cycle variability in TiN/Ti/Al:HfO2/TiN RRAMs. *Microelectron. Eng.* **2019**, *214*, 104–109. [CrossRef]
36. Pérez, E.; Kalishettyhalli Mahadevaiah, M.; Zambelli, C.; Olivo, P.; Wenger, C. Data retention investigation in Al:HfO2-based resistive random access memory arrays by using high-Temperature accelerated tests. *J. Vac. Sci. Technol. B* **2019**, *37*, 012202. [CrossRef]
37. Aldana, S.; Pérez, E.; Jiménez-Molinos, F.; Wenger, C.; Roldán, J.B. Kinetic Monte Carlo analysis of data retention in Al:HfO2-based resistive random access memories. *Semicond. Sci. Technol.* **2020**, *35*, 115012. [CrossRef]
38. Pedregosa, F.; Varoquaux, G.; Gramfort, A.; Michel, V.; Thirion, B.; Grisel, O.; Blondel, M.; Prettenhofer, P.; Weiss, R.; Dubourg, V.; et al. Scikit-learn: Machine Learning in Python. *J. Mach. Learn. Res.* **2011**, *12*, 2825–2830.
39. LeCun, Y.; Cortes, C.; Burges, C. MNIST Handwritten Digit Database. Available online: http://yann.lecun.com/exdb/mnist (accessed on 21 September 2020).
40. Popescu, M.C.; Balas, V.E.; Perescu-Popescu, L.; Mastorakis, N. Multilayer Perceptron and Neural Networks. *WSEAS Trans. Circ. Syst.* **2009**, *8*, 579–588.
41. Nayak, P.; Zhang, D.; Chai, S. Bit efficient quantization for deep neural networks. *arXiv* **2019**, arXiv:1910.04877.
42. Pérez-Ávila, A.J.; González-Cordero, G.; Pérez, E.; Pérez-Bosch, E.; Kalishettyhalli Mahadevaiah, M.; Wenger, C.; Roldán, J.B.; Jiménez-Molinos, F. Behavioral modeling of multilevel HfO2-based memristors for neuromorphic circuit simulation. In Proceedings of the XXXV Conference on Design of Circuits and Integrated Systems (DCIS), Segovia, Spain, 18–20 November 2020; pp. 1–6. [CrossRef]
43. Bilbao, I.; Bilbao, J. Overfitting problem and the over-training in the era of data: Particularly for Artificial Neural Networks. In Proceedings of the 8th International Conference on Intelligent Computing and Information Systems (ICICIS), Cairo, Egypt, 5–7 December 2017; pp. 173–177.
44. Covi, E.; Brivio, S.; Serb, A.; Prodromakis, T.; Fanciulli, M.; Spiga, S. Analog Memristive Synapse in Spiking Networks Implementing Unsupervised Learning. *Front. Neurosci.* **2016**, *10*, 482. [CrossRef] [PubMed]
45. Jeong, H.; Shi, L. Memristor devices for neural networks. *J. Phys. D Appl. Phys.* **2018**, *52*, 023003. [CrossRef]
46. Tang, J.; Yuan, F.; Shen, X.; Wang, Z.; Rao, M.; He, Y.; Sun, Y.; Li, X.; Zhang, W.; Li, Y.; et al. Bridging Biological and Artificial Neural Networks with Emerging Neuromorphic Devices: Fundamentals, Progress, and Challenges. *Adv. Mater.* **2019**, *31*, 1902761. [CrossRef] [PubMed]

MDPI

St. Alban-Anlage 66

4052 Basel

Switzerland

Tel. +41 61 683 77 34

Fax +41 61 302 89 18

www.mdpi.com

Electronics Editorial Office

E-mail: electronics@mdpi.com

www.mdpi.com/journal/electronics